ORDERING KNOWLEDGE IN THE ROMAN EMPIRE

The Romans commanded the largest and most complex empire the world had ever seen, or would see until modern times. The challenges, however, were not just political, economic and military: Rome was also the hub of a vast information network, drawing in worldwide expertise and refashioning it for its own purposes. This groundbreaking collection of essays considers the dialogue between technical literature and imperial society, drawing on, developing and critiquing a range of modern cultural theories (including those of Michel Foucault and Edward Said). How was knowledge shaped into textual forms, and how did those forms encode relationships between emperor and subjects, theory and practice, Roman and Greek, centre and periphery? *Ordering Knowledge in the Roman Empire* will be required reading for those concerned with the intellectual and cultural history of the Roman Empire, and its lasting legacy in the medieval world and beyond.

JASON KÖNIG is Senior Lecturer in Greek and Classical Studies at the University of St Andrews. He is author of *Athletics and Literature in the Roman Empire* (2005), and of a wide range of articles on the Greek literature and culture of the Roman world.

TIM WHITMARSH is E. P. Warren Praelector in Classics at Corpus Christi College and Lecturer in Greek Language and Literature at the University of Oxford. His publications include *Greek Literature and the Roman Empire* (2001), *Ancient Greek Literature* (2004) and *The Second Sophistic* (2005).

ORDERING KNOWLEDGE IN THE ROMAN EMPIRE

EDITED BY

JASON KÖNIG AND TIM WHITMARSH

CAMBRIDGE
UNIVERSITY PRESS

CAMBRIDGE UNIVERSITY PRESS
Cambridge, New York, Melbourne, Madrid, Cape Town,
Singapore, São Paulo, Delhi, Tokyo, Mexico City

Cambridge University Press
The Edinburgh Building, Cambridge CB2 8RU, UK

Published in the United States of America by Cambridge University Press, New York

www.cambridge.org
Information on this title: www.cambridge.org/9780521296939

First published 2007
First paperback edition 2011

A catalogue record for this publication is available from the British Library

ISBN 978-0-521-85969-1 Hardback
ISBN 978-0-521-29693-9 Paperback

Contents

Preface

We are grateful to the Master and Fellow of St John's College, Cambridge, for funding of the December 2001 conference on which this volume is based; and to all who participated in that event. We would also like to thank Michael Sharp and Sarah Parker at Cambridge University Press, and the anonymous readers for the volume; and also colleagues at Cambridge, Exeter and St Andrews, many of them working on related projects, for ideas and support (within the Exeter Centre for Hellenistic and Romano-Greek Studies and the St Andrews Logos Centre for study of ancient systems of knowledge). We are grateful especially to Simon Goldhill for comments on Chapter 1.

Contributors

Serafina Cuomo is Reader at Imperial College London, and a historian of ancient Greek and Roman science and technology. She works in particular on the political, social and economic significance of ancient forms of knowledge, and has written on science in late antiquity, on ancient mathematics, on military technology and on Roman land-surveying. Her third book, *Technology and Culture in Greek and Roman Antiquity*, is forthcoming with Cambridge University Press.

Rebecca Flemming is Lecturer in Ancient History at the University of Cambridge. She is the author of *Medicine and the Making of Roman Women: Gender, Nature, and Authority from Celsus to Galen* (Oxford University Press, 2000), and a range of essays and articles on women and medicine in the ancient world, both jointly and separately. She is currently writing a book on medicine and empire in the Roman world.

Thomas Habinek is Professor of Classics at the University of Southern California. He has published extensively on Latin literature and Roman cultural history. His most recent book is *The World of Roman Song: From Ritualized Speech to Social Order* (Johns Hopkins University Press, 2005). He is currently at work on an interdisciplinary project linking humanistic and natural scientific approaches to the human capacity for imitation and its role in cultural change.

John Henderson is Professor of Classics at the University of Cambridge and a Fellow of King's College. His books include monographs on Plautus, Phaedrus, Seneca, Statius, Pliny and Juvenal, besides general studies of epic, comedy, satire, history, art, culture and the history of classics, and *The Medieval World of Isidore of Seville: Creating Truth through Words* (Cambridge University Press, 2007).

Alice König is Lecturer in Latin at the University of St Andrews. Her recent research has focused on Latin 'technical' literature, particularly the works of Frontinus and Vitruvius. She is currently revising her PhD thesis, on Frontinus' three surviving treatises, for publication.

JASON KÖNIG is Senior Lecturer in Greek and Classical Studies at the University of St Andrews. He is author of *Athletics and Literature in the Roman Empire* (Cambridge University Press, 2006), and of a wide range of articles on the Greek literature and culture of the Roman world. He is currently engaged, with Greg Woolf, in a project funded by the Leverhulme Trust on Science and Empire in the Ancient World.

ANDREW M. RIGGSBY is Associate Professor of Classics and of Art and Art History at the University of Texas at Austin. He is the author of *Crime and Community in Ciceronian Rome* (University of Texas Press, 1999) and *Caesar in Gaul and Rome: War in Words* (University of Texas Press, 2006). He works on the cultural history of Republican Roman political institutions and on the cognitive history of the Roman world.

VICTORIA RIMELL teaches Latin literature at the University of Rome, La Sapienza. She is author of *Petronius and the Anatomy of Fiction* (Cambridge University Press, 2002) and *Ovid's Lovers: Desire, Difference and the Poetic Imagination* (Cambridge University Press, 2006), and editor of *Orality and Representation in the Ancient Novel (Ancient Narrative* Supplement, 2007).

JAMES WARREN is Senior Lecturer in Classics at the University of Cambridge and Fellow and Director of Studies in Philosophy at Corpus Christi College. He is the author of *Epicurus and Democritean Ethics: an Archaeology of Ataraxia* (Cambridge University Press, 2002) and *Facing Death: Epicurus and his Critics* (Oxford University Press, 2004).

TIM WHITMARSH is E. P. Warren Praelector in Classics at Corpus Christi College and Lecturer in Greek Language and Literature at the University of Oxford. A specialist in the Greek literature and culture of the imperial period, he is the author of *Greek Literature and the Roman Empire* (Oxford University Press, 2001), *Ancient Greek Literature* (Polity Press, 2004), *The Second Sophistic* (Cambridge University Press, 2005), and *Reading the Self in the Ancient Greek Novel* (Cambridge University Press, forthcoming). He is currently working on interactions between Greek and Semitic narrative.

JOHN WILKINS is Professor of Greek Culture at the University of Exeter. His books include *Euripides*: Heraclidae (Oxford University Press, 1993),

The Boastful Chef: The Discourse of Food in Ancient Greek Comedy (Oxford University Press, 2000) and (with Shaun Hill) *Food in the Ancient World* (Blackwell, 2006). He edited (with David Braund) *Athenaeus and his World* (University of Exeter Press, 2000) and is currently preparing editions of Galen, *De alimentorum facultatibus* for the Budé series and the *Corpus Medicorum Graecorum.*

Abbreviations

Abbreviations for Greek and Latin authors, and for scholarly resources, follow those used in S. Hornblower and A. Spawforth (eds.) (1996) *The Oxford Classical Dictionary* (3rd edn) Oxford; where author abbreviations are not found in *OCD*, usual conventions are followed. Exception is made in the case of Galen's works, for which a full list of abbreviations used in this volume is given below. Journal abbreviations follow *Année Philologique*, with occasional anglicisations. All other abbreviations are listed below.

CIL = *Corpus Inscriptionum Latinarum* (1863–)
CISem. = *Corpus Inscriptionum Semiticarum* (1881–)
CMG = *Corpus Medicorum Graecorum* (1908–)
CML = *Corpus Medicorum Latinorum* (1915–)
FGrH = *Die Fragmente der griechischen Historiker* (1923–) eds. F. Jacoby *et al.*
IG = *Inscriptiones Graecae*, 2nd edn (1924–)
ILS = *Inscriptiones Latinae Selectae*, ed. H. Dessau (1892–1916)
K = *Claudii Galeni Opera Omnia*, 20 vols., ed. C. G. Kühn (1821–33)
LSJ = *A Greek–English Lexicon*, 9th edn, H. G. Liddell, R. Scott *et al.* (1996)
Migne *PL* = *Patrologiae Cursus, Series Latina*, ed. J.-P. Migne (1863–)
OCD = *Oxford Classical Dictionary*
OCT = *Oxford Classical Text*
OLD = *Oxford Latin Dictionary*, ed. P. G. W. Glare, corrected edn (1996)
PVindob. = *Papyrus Vindobonensis*
PHerc. = *Papyrus Herculanensis*
POxy. = *Oxyrhynchus Papyrus*
SM = *Claudii Galeni Pergameni Scripta Minora*, 3 vols. (1884–93) eds. J. Marquardt, I. Müller and G. Helmreich, Leipzig.

GALEN

Abbreviations and editions used for the works of Galen and other medical writers.

AA	*De anatomicis administrationibus* ('On anatomical procedures'), books 1–8 from II K, books 9–15 (extant only in Arabic) from Simon (1906)
Alim. fac.	*De alimentorum facultatibus* ('On the properties of foodstuffs'), *CMG* 5.4.2
Ant.	*De antidotis* ('On antidotes'), XIV K
Ars med.	*Ars medica* ('The medical art'), Boudon (2000)
Comp. med. gen.	*De compositione medicamentorum secundum genera* ('On the compounding of drugs according to kinds'), XIII K
Comp. med. loc.	*De compositione medicamentorum secundum locos* ('On the compounding of drugs according to places'), XII–XIII K
Cris.	*De crisibus* ('On crises'), Alexanderson (1967)
Foet. form.	*De foetuum formatione* ('On the formation of the foetus'), *CMG* 5.3.3
Food	see under *Alim. fac.*
Lib. prop.	*De libris propriis* ('On my own books'), *SM* II
Loc. aff.	*De locis affectis* ('On the affected parts'), VIII K
MM	*De methodo medendi* ('On the therapeutic method'), X K
Nerv. diss.	*De nervorum dissectione* ('On the dissection of the nerves'), II K
Ord. lib. prop.	*De ordine librorum propriorum* ('On the order of my own books'), *SM* II
Part. art. med.	*De partibus artis medicativae* ('On the parts of the art of medicine'), *CMG supp. or.* II
PHP	*De placitis Hippocratis et Platonis* ('On the doctrines of Hippocrates and Plato'), *CMG* V 4.1.1–3
Praen.	*De praenotione ad Epigenem* ('On prognosis'), *CMG* V.8.1
Prop. plac.	*De propriis placitis* ('On my own opinions'), *CMG* V.3.2
Simples	see under *SMT*

SMT	*De simplicium medicamentorum temperamentis ac facultatibus* ('On the mixtures and properties of simple drugs'), XI–XII K
Ther.	*De theriaca ad Pisonem* ('On theriac to Piso'), XIV K
Thras.	*Thrasybulus*, *SM* III
UP	*De usu partium* ('On the usefulness of parts'), Helmreich (1907–9)

EDITIONS

Alexanderson, B. (1967) *Peri Kriseôn Galenos*. Stockholm

Boudon, V. (2000) *Galien II: Exhortation à la Médecine. Art Médical.* Paris

Daremberg, C. and Ruelle, E. (1879) *Oeuvres de Rufus d'Ephèse*. Paris

Garofalo, I. (ed.) (1997) *Anonymi Medici de Morbis acutis et chroniis.* Leiden

Helmreich, G. (1907–9) *Galeni De Usu Partium libri XVII*, 2 vols. Leipzig

Muhaqqiq, M. (ed.) (1993) *Kitab al-Shukuk 'ala Jalinus li-Muhammad ibn Zakariyya al-Razi*. Tehran

Sconocchia, S. (1983) *Scribonii Largi Compositiones*. Leipzig

Simon, M, (1906) *Sieben Bücher Anatomie des Galen*, 2 vols. Leipzig

Wellmann, M. (1906–14) *Pedanius Dioscorides. De Materia Medica*, 3 vols. Berlin

PART I

Introduction

CHAPTER I

Ordering knowledge

Jason König and Tim Whitmarsh

IMPERIAL KNOWLEDGE

This volume seeks to explore the ways in which particular conceptions of knowledge and particular ways of textualising knowledge were entwined with social and political practices and ideals within the Roman Imperial period. In the process, we explore the possibility that the Roman Empire brought with it distinctive forms of knowledge, and, in particular, distinctive ways of ordering knowledge in textual form.

The chapters following this one contain a series of case studies, examining the politics and poetics of knowledge-ordering within a wide range of texts, testing out each of them carefully for signs of their engagement with other works of similar type, and with the world around them. Our principal interest is in texts that follow a broadly 'compilatory' aesthetic, accumulating information in often enormous bulk, in ways that may look unwieldy or purely functional to modern eyes, but which in the ancient world clearly had a much higher prestige than modern criticism has allowed them. The prevalence of this mode of composition in the Roman world is astonishing, as will become clear in the course of this discussion. It is sometimes hard to avoid the impression that accumulation of knowledge is the driving force for all of Imperial prose literature. That obsession also makes its mark on verse, for example within the scrolls of didactic epic or in the anthologisation of epigrams. In this volume, we range across miscellanistic, encyclopedic, biographical, novelistic, philosophical, scientific, technical, didactic and historical works (insofar as these generic distinctions can be maintained), in Greek and Latin.[1] Inevitably we cover only a tiny fraction of the texts such a project might engage with, picking especially works

[1] Many of these areas have been largely neglected in recent scholarship, especially by scholars working in the area of cultural history, although in some cases that has begun to change. To take just one example, the field of ancient technical writing has seen a recent expansion of interest; relevant works not discussed further below include the following: Fögen (ed.) (2005), Horster and Reitz (eds.) (2003), Santini, Mastrorosa and Zumbo (eds.) (2002), Formisano (2001), Long (2001), Meissner

which seem to us to have paradigmatic status for habits of compilation in this period – although we have tried to convey something of the enormous (if inevitably unquantifiable) scale of this compilatory industry in our footnoted lists of known authors and works within a range of genres.

The essays in Part 2, following this introduction, are focused especially on the way in which authors order their own texts and the writings of others. All of these chapters start by teasing out some of the ordering, structuring principles and patterns of the texts they examine, and move from there to discuss the cultural and political resonances of those patterns, and the ways in which they contribute to authorial self-positioning. The essays in Part 3 in addition address more head-on the question of how compilatory texts impose order on the extra-textual world. These chapters are generally more interested, in other words, in the way in which texts deal with practical challenges, and the way in which they take on images and ideals from the world around them – especially the world of empire – reshaping them and using them as structuring reference-points for their own projects. Needless to say, there can be no firm dividing line between those two approaches.

However, the broad question of the 'Imperialness' or otherwise of these knowledge-ordering strategies – which is a central preoccupation of many (though not all) of the chapters which follow – cannot simply be left to emerge from these individual readings. This introduction attempts a preliminary answer to that question.

The idea of an interrelation between knowledge and empire in the modern world is not new.[2] Edward Said has shown how imperial ideologies shaped and were shaped by the rhetoric of modern European ethnography, and how they seeped into many other areas of discourse.[3] There are countless studies, many of them drawing on Said's work, which show how European scientific knowledge, and the knowledge of colonised cultures within European empires, developed step by step with the institutions and assumptions of empire.[4] Those enquiries have illuminated, amongst other things, the role of science as a tool of empire; the influence of European science on conquered populations; the ways in which local knowledge

(1999), Nicolet (ed.) (1995). All of those volumes share the aim of comparing and juxtaposing a range of different technical authors; many of them bring out vividly the way in which these at-first-sight purely functional texts manipulate shared tropes of structuration and authorial self-representation, often with a high degree of ingenuity (e.g., see Formisano (2001), esp. 27–31, on recurrent use of the rhetoric of *utilitas, sollertia, diligentia* and *dissimulatio* in late-antique technical writing).

[2] See Flemming (2003) for an attempt to relate work on modern empires to Hellenistic knowledge.

[3] Said (1978) and (1993).

[4] See, amongst many others, Stafford (1989), Macleod (1993), Bayly (1996), Miller and Reill (eds.) (1996), Washbrook (1999), Drayton (2000).

influenced metropolitan scientific practice; the ways in which increased knowledge of the globe opened up new areas for scientific study; and the ways in which ideals of scientific progress and ambition were intertwined with metropolitan justifications of imperial domination.

Moreover, modern practices of scientific writing have been significantly shaped by ancient models of objective and exhaustive compilation of knowledge within textual form – although this volume for the most part leaves to one side the question of the reception of ancient knowledge-ordering in the post-classical world.[5] The structures of post-classical knowledge-ordering – in the Arabic, medieval and Renaissance worlds and beyond – are indebted to ancient models.[6] Modern encyclopedism follows the encyclopedic projects of Pliny and others, despite the great differences between modern and ancient conceptions of what an 'encyclopedia' comprises.[7]

One might therefore expect to see similar links between knowledge-ordering texts and imperial ambitions in both the ancient and modern worlds. And yet when we read the knowledge-bearing texts of the Roman Empire, it is often difficult – more difficult than for much of the scientific writing of the British Empire, for example – to ground their relation with the imperial project in detailed analysis. Some ancient authors shun the impression of being implicated in the realities of imperial power. Many avoid the appearance of radical innovation, advertising instead their close relationship with the accumulated knowledge of the past. That difficulty can be partly explained by the tendency for imperialist rhetoric to conceal itself beneath the mask of objectivity or aesthetic elevation (as Said and others have shown). This point is crucial for ancient and modern empires alike. But that explanation is not on its own enough. We also need to acknowledge that the Roman Empire poses its own very particular problems of analysis – that the mutually parasitic relationship between ancient empire and knowledge arose from rhetorical traditions and institutional structures very different from anything familiar in the experience of modern European empires. Most obviously, the cultural impositions and interventionist strategies of administration that have characterised many

[5] Equally we leave to one side any attempt at comparative approaches of the kind Geoffrey Lloyd has pioneered in juxtaposing Chinese science, and its context of empire, with Greek science and society: see esp. Lloyd (1996).

[6] See, e.g., Koerner (1999) on the influence of ancient knowledge-ordering texts on Linnaeus.

[7] See Collison (1964); McArthur (1986), esp. 38–56, who traces the development of compilatory writing from Aristotle and Pliny, through Christian compilers like Cassiodorus and Isidore of Seville, to the scholasticism of Thomas Aquinas and beyond; Arnar (1990); Yeo (2001) 5–12 on the descent of modern encyclopedism from ancient precedents, and *passim* on development of conceptions of encyclopedism in eighteenth-century Europe; also Murphy (2004) 11–12.

of those empires find almost their inverse in the relatively light touch, in cultural terms, of Roman rule. What we need, then, is a set of questions sensitive to that specificity. That is the task of this introduction.

KNOWLEDGE AND POWER

The links between knowledge and power more generally – putting aside for now the specific context of empire – have of course been much theorised. For Michel Foucault, most influentially, power is not simply a commodity, possessed by governments and influential individuals and exercised by them from above. Rather it is a complex network of relationships constantly being acted out and reshaped within even the smallest encounters of everyday life. Moreover, knowledge and its 'will to truth' are central to Foucauldian power. Epistemology cannot be divorced from particular social relations and situations. It is not some abstract activity, practised from a position of detachment; rather it is enacted within all institutions of social encounter. Each society, Foucault argues, has its own conditions for truth:

> that is, the type of discourse it harbours and causes to function as true; the mechanisms and instances which enable one to distinguish true from false statements, the way in which each is sanctioned; the techniques and procedures which are valorised for obtaining the truth; the status of those who are charged with saying what counts as true.[8]

Those who have access to the knowledge that holds a social and political system together necessarily control the distribution of power within that system. And yet truth is never stable and monolithic. Rather it is something open to debate and renegotiation, shaped and enacted through and within the workings of power. The systems of thought identifying individuals with certain roles do so not bluntly and coercively, but rather with the collusion of those individuals – through the creation of desire for particular subject positions. Negotiation of truth and power are thus ingrained in the textures of everyday life. When people act out particular roles, as parents and children, teachers and students, doctors and patients, they are constantly negotiating 'questions of power, authority, and the control of definitions of reality'.[9] Knowledge-bearing institutions and bodies of thought – medicine, hospitals, prisons, asylums – are embedded in and founded upon these relationships of power; and knowledge-bearing texts, often the texts that

[8] Quotation from an interview with Foucault published in Gordon (1980) 131.

[9] Dirks, Eley and Ortner (1994) 4, part of a good brief discussion, setting Foucault's work in the context of wider developments in anthropology, history and the social sciences; see also McNay (1994) 48–132.

provide theoretical backing for those institutions, are profoundly marked by them, able to reveal beneath their dispassionate surfaces something of what it is possible to say or to think within the societies and disciplines from which they arise.

The broad relevance of those points will be clear. The world of knowledge – comprising both the institutions defining it and the texts embodying it – is never neutral, detached, objective. The assumption that the textual compilation of knowledge is a practice distinct from political power will not stand. All of the texts examined in this volume are embedded both within the overarching hierarchies and patterns of thought of Roman-empire society and within the power relations and power struggles of specific intellectual disciplines (more on that below)[10] – although here again we should acknowledge how far our own experiences differ from those of the ancient world, where official institutionalisation of knowledge production was in general more localised and circumscribed. Similar conclusions – both inspired by Foucault's work and developed in parallel to it – have increasingly preoccupied a whole range of modern academic disciplines. Feminist scholarship has revealed the gendered assumptions deeply rooted within centuries of male-produced and male-centred discourse.[11] Anthropology has shown how the structuring hierarchies and thought patterns of a society may be ingrained even – or perhaps especially – within its most frivolous and abstract habits of cultural activity.[12]

Foucault's challenging work is not without its difficulties, of course – in fact Foucault himself constantly struggled to revise and update his models during the course of his career.[13] Most importantly for this volume, Foucault's model of the functioning of power and knowledge on some readings leaves little or no room for the agency of individuals. Foucault's insistence that resistance to power is always bound up in and reproductive of the systems it challenges has been thought to have pessimistic implications for the possibility of resistance to social injustice.[14] Many of the essays in this volume address that problem, particularly through questioning the degree to which encyclopedic styles of composition allow and provoke varied reader response to the patterns of thought they showcase. How far, in other words, does knowledge imply subjection to historically determined forces? How do individuals carve out their own spaces within the overarching structures

[10] Pp. 24–7; cf. Barton (1994b) on the scientific writing of the Roman Empire.

[11] See, e.g., Dirks, Eley and Ortner (1994) 32–6.

[12] E.g., see Geertz (1973).

[13] See McNay (1994) 66–9 on Foucault's attempts in *The Archaeology of Knowledge* (1972) to nuance his rather monolithic concept of the 'episteme' in *The Order of Things* (1970).

[14] See McNay (1994) 100–102.

which they are formed by? And what role does textual presentation of knowledge play within those processes?

Examination of the Roman Empire as a specific context for knowledge production also has relevance for Foucault's conceptions of chronological change. Does Foucault's model of 'epistemic shifts' between different periods with different systems of logic[15] offer insight into the post-Augustan world, where the rhetoric of a 'new start' was paraded so widely? Or does that model play into the hands of a naïve historicism, resting on simplistic modern periodisations of the ancient world? Should we be looking instead for a model that accounts for change in conceptions of knowledge as a gradual and painstaking evolution impelled by the pressures and innovations of competitive elite self-assertion?

HELLENISTIC/REPUBLICAN KNOWLEDGE

One way of assessing the cultural and historical specificity of knowledge-systems of the Roman Empire is to view its relation with what had come before it. Certainly, they did not emerge *e nihilo*. Aristotle's project of systematising knowledge across an enormous range of different subjects lies behind all of the texts we discuss in later chapters. Equally influential was the culture of Hellenistic Alexandria, which both inherited and developed Aristotelian scholarly practice. Here we see uniquely concrete links between the projects of political organisation and cultural systematisation. The Alexandrian library (later imitated in Pergamum and elsewhere) brought the whole world into a single city, broadcasting the glory of the Ptolemaic rule that had provided the conditions for its possibility. And a whole range of scholars imitated and influenced that totalising gesture in their individual works, covering a range of subjects inconceivable within the hyper-specialised world of modern academic writing: Zenodotus, for example, Homeric editor and lexicographer and first head of the Library; Callimachus, whose poetry flaunts its own dazzling generic flexibility, in combination with designedly abstruse bibliographical and historical knowledge; and most prodigiously of all, Eratosthenes, whose work covers mathematical, chronographical, geographical, philosophical and literary scholarship.[16] Others outside Alexandria followed similar paths: Theophrastus, the successor of Aristotle in the Athenian Lyceum; Aratus, the poet-scholar based

[15] See McNay (1994) 64–6.

[16] See Pfeiffer (1968), and now Erskine (ed.) (2003) (especially the chapters by Hunter (2003) and Flemming (2003)); for Eratosthenes, see the rich account of Geus (2002).

in Pergamum; and Posidonius, the extraordinary polymath of the second to first centuries BCE, who prospered in Rome. Many Imperial Greek writers depended heavily on their Hellenistic predecessors for both form and content. Similarly, their Latin counterparts often drew heavily from Hellenistic Greek work, while also following the agendas laid out by great Republican systematisers like Cicero and especially Varro, whose work covered history, grammar, geography, agriculture, law, philosophy, medicine and other fields.[17]

On that evidence, modern scholars of ancient science have sometimes concluded that Imperial compilers of knowledge were merely derivative.[18] That approach, however, drastically underestimates the potential for innovativeness in compilatory styles of composition, as well as failing to examine the key questions of synchronic cultural analysis which this volume addresses.

For one thing, it mistakes the rhetoric of conservatism often paraded by ancient scientific discourse for the real thing. The importance of rhetorical self-promotion within ancient science and medicine encouraged a degree of originality; but also paradoxically suppressed excessive inventiveness, as speakers and writers went out of their way to avoid the impression of showy innovation.[19] It also ignores the opportunities for inventive reshaping embedded within the techniques of editing and compiling – inventiveness which several of the following chapters explore. And it fails to consider the ways in which even texts following broadly Hellenistic or Republican structures or styles of composition so often bring out the tension between older and newer configurations of knowledge. That is clear, for example, in works where the concept of geographical scope is an important structuring principle.[20] Strabo's geographical history,[21] for instance, or Pausanias' *Periegesis*,[22] work with fundamentally Hellenistic conceptions of space, but are also acutely aware of the way in which Roman rule has reconfigured the geography of the Greek east. Pliny's *Natural history* draws into itself the accumulated erudition of the Greek and Roman past, but in doing so it

[17] On the late-Republican intellectual scene see esp. Rawson (1985).

[18] On modern scholarship's deprecation of Imperial literature on the grounds of derivativeness, see Whitmarsh (2001) 41–5.

[19] See Lloyd (1996) 74–92 (esp. 90–92) on medical writers. On the ambiguities of innovation in rhetorical theory, see Whitmarsh (2005a) 54–6.

[20] For the general point, see Momigliano (1974) 27–49.

[21] See Clarke (1999), esp. 193–244.

[22] See Cohen (2001) for the argument that Pausanias' worldview is more 'Hellenistic' than, for example, Strabo's, less comfortably integrated with Roman imperial geography; see, however, Elsner (1992) and (1994), and (from a different perspective) Arafat (1996) for Pausanias' engagement with the realities of the Roman present.

repeatedly invites us to compare this accumulation with patterns of Roman topographical dominance.[23]

A number of scholars have also suggested causal links between the political and cultural conditions which framed the transition from Republic to Empire and the emergence of distinctive knowledge-ordering genres. Claudia Moatti has argued that the drive to assemble disparate strands of knowledge was a response to the fragmentation of late-Republican society and political culture.[24] Andrew Wallace-Hadrill has linked the move towards specialised knowledge under Augustus with shifting ideas of political authority.[25] Trevor Murphy has pointed out that the 'encyclopedia' is a Roman invention, but also a product of the Roman encounter with Greek ideals of all-embracing education (*enkyklios paideia*) – the alienness of this concept for Romans drove them to attempt a fixed, textual version of it, as opposed to the more fluid version which was enshrined within centuries of Greek educational tradition – and dependent on the territorial and intellectual ambitions of a unified empire. In the process he shows how Pliny's encyclopedic project in particular is adapted for the context of the Roman Empire, drawing, for example, on the rhetoric of imperial conquest and the emperor's authority (more on that below).[26]

LOCAL KNOWLEDGE

One of the most distinctive features of Roman Imperial conceptions of geographical space was its insistence on the co-existence of overarching identities with local ones, in line with both the inclusive ideology of Roman rule and Panhellenic visions of the world, where civic individuality is compatible with, even necessary for, the perpetuation of shared Greek identity. How far can we see those tensions reflected in Imperial textualisations of knowledge? And how far should we distinguish between different contexts for local knowledge within the melting-pot of Roman culture?

There are some signs of regional clusters of specialisms. For example, Athens, Alexandria, Tarsus, Aegae and Pergamum were all thriving centres of rhetorical and philosophical education.[27] And yet those concentrations

[23] See French (1994) 207–18; Murphy (2003) and (2004); Carey (2003), esp. 32–40.

[24] Moatti (1988), (1991) and (1997).

[25] See Wallace-Hadrill (1997), discussed further below, p. 21.

[26] See pp. 20–2, and Murphy (2004), esp. 13–14 and 194–6 on the origins of Roman encyclopedism; cf. McEwen (2003) on the way in which Vitruvius' project links itself with its political context by appropriating the metaphor of the empire as a unified body in order to apply that to the discipline of architecture.

[27] See Natali (2000) 210.

leave very few textual traces. Roman Empire writing tends to emphasise (variably conceived) intellectual cosmopolitanism ahead of provincial specificity[28]. There are some exceptions, where an insistence on cosmopolitanism leads paradoxically to a strong sense of place, focused on iconic cultural centres like Rome and Athens. Galen, who is reticent about his medical training in Pergamum but conjures up a vivid portrait of the medical and philosophical scene in Rome,[29] is a case in point – although even he is often vague about the precise setting of the medical debates he describes, conjuring up an imagined, utopian landscape of shared intellectual endeavour, which also stretches back over the centuries, allowing him to enter into dialogue with his medical predecessors. Aulus Gellius, similarly, implies a cosmopolitan but specifically Athenian setting for the miscellaneous collection of conversations and reminiscences in his *Attic nights*. Plutarch does much the same with Delphi in his *Delphic dialogues*. But Athens and Delphi and Rome were unusual cases.

Where Imperial writers do grant specific forms of local knowledge to provincial contexts, it is usually to tease them for their failure to match normal Panhellenic standards, as in Dio Chrysostom's comical portrait of the cultural backwaters of Borysthenes and Euboea (although for the Cynic moralist, such places also offer positive lessons for his Prusan audience).[30] There is evidence for the continuing importance of local history, but with the near-total loss of this genre, and few signs of its lateral impact on other literature, it is hard to press any strong claims on its behalf.[31]

That relative invisibility of local context does at least have some resonance with the increasing emphasis within anthropology and modern history on the importance of seeing 'local knowledges' not as self-contained and inward-looking ways of seeing the world, but rather as bodies of thought which engage with and contribute to universal knowledge.[32] But it may well make us uncomfortable even so, trained as we are to insist on the potentially disruptive power of local, marginal voices within the homogenising textures

[28] See pp. 18–20 below. [29] See Nutton (1972). [30] Trapp (1995).

[31] See Bowie (1974) 184–8. Others local historians dated by Jacoby to the Imperial period might be added to Bowie's list: e.g., Lyceas of Argos, ὁ τῶν ἐπιχωρίων ἐξηγητής (Paus. 1.13.8 = *FGrH* 312); Posidonius of Olbia, author of *Attic histories* (*FGrH* 335 = 279 T1); Glaucippus, author of a tract on the religion of Athens (*FGrH* 363); Telephanes, author of *On the city* (*FGrH* 371); Menelaus of Aegae, author of a work on Boeotia (*FGrH* 384); Callippus of Corinth, author of a history of Orchomenoi (*FGrH* 384); Timagenes or Timogenes of Miletus, author of *On Heracleia in Pontus* (*FGrH* 435); Theseus, author of *Corinthian matters* (*FGrH* 453); Crito, author of *Sicilian matters*, *Foundations of Syracuse* and a *Tour of Syracuse* (*FGrH* 277 T1); Phlegon of Tralles, author of a description of Sicily (*FGrH* 257 T1).

[32] See, e.g., Moore (1996).

of global and imperial culture. Are all the textual traces of knowledge found across the Empire invariably in collusion with the globalising ideals of the centre? Were there other bodies of local knowledge separate from the Empire's literate, intellectual culture, which are simply too faint for us to bring back to life?

Even here, of course, there are exceptions, texts that take on the cosmopolitan tropes of elite culture and twist them in order to speak from resistant corners of the Mediterranean world. Lucian's satirical insight into Greco-Roman elite culture is founded on his pose of being a Syrian outsider to the cultural centres of the Empire[33]. But such cases, far from representing indigenous tradition countering imperial superimposition, clearly demonstrate that they are *always already imperialised.* The concept of the local only becomes operative when globalisation is already at work.

There is also a different order of issue, focused on the relationship between textual and ritual knowledge. Jack Goody has influentially emphasised the role of listing as a literate technique, a technology that produces habits of thought connected specifically with literature cultures.[34] And yet in the ancient world listing and cataloguing have strong links with orality (from Homer's catalogue of ships onwards) and with ritual (for example, with the kinds of enumeration which guided and memorialised processional activity). We should perhaps give more weight to the ritual overtones of listing even within the apparently functional pages of the Roman Empire's scientific and miscellanistic writing. For example, Plutarch's *Sympotic questions* (as Jason König argues further in his chapter) records philosophical conversations set in specific Greek cities, often at festival banquets. Plutarch thus aligns his own compilatory work with the rhythms of festival life, casting it as a performance of cultural memory to match the habits of cultural memorialisation which were ingrained in local life. Thomas Habinek's chapter shows how Manilius uses the image of sacrificial ritual both for his astrological knowledge and for his own activity as a *vates*, a poet-prophet figure. Ovid's poetic exposition of the Roman calendar in the *Fasti*, again, is a subversive meditation on the ritual and theological culture of Rome, built around the defamiliarising juxtaposition of Roman ritual patterns with a Greek framework of astrological and mythographical knowledge.[35] In these works, at least, the stark details of scientific and biographical compilation engage with the distinctive and familiar contours of local life and ritual experience in more sustained ways than is initially obvious.

[33] See further below, pp. 13–14.

[34] See Goody (1977) 74–111; for criticisms, see Miyoshi (1994).

[35] See Feeney (1998) 123–33.

KNOWING PHILOSOPHY

Questions of knowledge will inevitably end up confronting philosophy. Our concern is not here with philosophical epistemology as such;[36] it is rather with the cultural valency of philosophical knowledge, its quasi-institutionalised status within society. Ancient philosophical theory, almost by definition, often aimed at totalisation: adherents of one view held it to the reasoned exclusion of other alternatives, personally committing to the idea of its superiority. (This goes even for the Pyrrhonists, sceptical anti-dogmatists who disdained all philosophical positions.) Yet by virtue of its exclusions, spoken or unspoken, philosophy necessarily acknowledged the co-existence (albeit not the equal value) of alternative perspectives; ongoing border disputes implied that the process of totalisation was never complete. In many cases these border disputes were all the more urgent for the fact that mutual influence between different schools was so strong. By the time of the empire, the consolidation of philosophical schools, each with its own tenets and dogma, had created a market-place in knowledge.[37] No philosopher was *just* a philosopher: s/he was a Stoic, Epicurean, Cynic, Sceptic, Academician . . .

Viewed from outside, however, the conflicts between the schools could be considered evidence for the impossibility of totalisation: if philosophers cannot agree what they know, can there be anything at all to know? This position is dramatised perhaps most eloquently by the satirist Lucian. His *Sale of lives*, for example, presents a slave-market where a potential buyer surveys a series of potential philosophers desperate to whore their trade.[38] His *Symposium*, meanwhile, works playfully against Plato's and Xenophon's texts of the same name: in contrast with their paradigms of social and intellectual order, Lucian represents his philosophers – all from different schools – warring drunkenly and bitterly.[39] These powerfully vivid narrative metaphors (the slave auction, convivial disharmony) for the philosophical market-place do more than simply debunk the authority of philosophers;

36 Schofield, Burnyeat and Barnes (eds.) (1980); Everson (1990); Striker (1996).

37 On the development of philosophical schools, cf. Boys-Stones (2001); also Hahn (1989).

38 See further Whitmarsh (2001) 258–60.

39 See Branham (1989); similarly Alciphron *Letters* 3.19 – in a closely related narrative probably written in imitation of Lucian – portrays philosophers brawling at a dinner party, each of them misbehaving in ways appropriate to his own philosophical school. Lucian and Alciphron are here parodying ideals of sympotic co-operation between different specialists, where playful rivalry between a range of professional viewpoints contributes to an atmosphere of convivial harmony: e.g., see Jacob (2001) xxii–xxvi on Athenaeus, and Hardie (1992) 4754–6 (discussed further in Jason König's chapter, below) on Plutarch's *Sympotic questions*.

they also seek to comprehend and express the bewildering variety of claims to philosophical knowledge under the Empire.

In doing that, Lucian is staking his own claim to knowledge – although knowledge of a different kind, to be sure. To figure, to allegorise, to encapsulate in narrative, is to master. In *Icaromenippus*, Menippus (the primary narrator) describes flying to the moon, while looking down upon the world and the philosophers' parochial discussions. Lucian's satire offers a different – and by implication, 'loftier' – epistemological order to philosophy. Here, as elsewhere in his work,[40] Lucianic knowingness *looks down upon* philosophical knowledge. The *Hermotimus*, to take another example, stages a dialectic struggle between Lycinus (the Lucianic figure in the text) and the eponymous Stoic; inevitably Lycinus wins out, converting Hermotimus to his view. This is not, however, a simple conversion from one allegiance to another, but a radical refocusing: 'do not think I am set against the Stoa . . . my discourse is equally hostile to all' (κοινὸς ἐπὶ πάντας ὁ λόγος, 85).

Lucian's negative epistemology, however, is itself parasitical upon philosophy. In reducing Hermotimus to tears within a dialectic framework, Lycinus is replaying the role of Socrates in the early, 'aporetic' dialogues of Plato (and, indeed, there are also Stoic and Sceptical elements to his arguments throughout).[41] The famous apophthegm in the *True stories*, 'the one true thing I will say is that I am telling lies' (1.4), is a calculated echo of Socrates' claim to wisdom on the grounds that 'what I do not know I do not think I know' (Plato, *Apology* 21d).[42] Elsewhere, this self-consciously fickle author assumes the guise of an Epicurean (in the *Alexander*), a Sceptic or a Cynic,[43] ransacking the closet of philosophical masks in the service of anti-philosophical satire. Some works (*Demonax, Nigrinus*) even portray certain philosophers with approbation – philosophers who are, of course, critical of (among other social institutions) other philosophers.

Lucian construes satire, we might say, as metaphilosophy. It is centrally preoccupied with philosophical questions of truth and knowledge; but at the same time, it exists above and beyond the mundane, interdogmatic squabbles of the philosophical sects. This is the point of that imagery of lunar travel in the *Icaromenippus*: satire is (or presents itself as) a cosmic, universalising vision of humanity in its most pared-down form, unencumbered by issues of cultural, social, political or sectarian difference. In that sense, the Lucianic worldview is as universalising, even totalising, as the philosophical systems upon which it feeds so voraciously.

40 Cf. *Nigr.* 18; *Astr.* 13–19; *Somn.* 13; more generally, Georgiadou and Larmour (1998) 15–16.

41 Möllendorff (2000) 197–210.

42 Rütten (1997) 30–1; Georgiadou and Larmour (1998) 57–8.

43 For Lucian's personae, see Whitmarsh (2001) 247–94, Goldhill (2002) 63–7.

HELLENISED KNOWLEDGE

Satire was not the only form of metaphilosophy. The project of comprehending philosophy within an overarching framework, of unifying wisdom into a meaningful whole that transcended the sum of its parts, can be seen in a range of synthetic works of what we have come to call (following Hermann Diels) 'doxography'. The genre developed in the Hellenistic period, but most of the key figures of the early phase – Antigonus of Carystus, Hermippus, Sotion, Sosicrates of Rhodes, Diocles of Rhodes – are known to us only sciagraphically from fragments preserved in later doxographers.[44] Doxography flourished in the Imperial period, notable cases being Alcinous' *Handbook of Platonism*, Arius Didymus' *On the philosophical sects*,[45] Aetius' *Collection of doctrines*,[46] and, most of all, Diogenes Laertius' *Lives and opinions of the philosophers* (see James Warren's chapter).[47] Much doxographical material can also be found in miscellanies such as Favorinus' *Memorabilia*,[48] and Aulus Gellius' *Attic nights*.[49] These works have appealed to scholars particularly as sources for earlier thought.

Certainly they had their roots in the traditions of Aristotle and the Hellenistic scholars,[50] but there was also a strong contemporary cultural imperative lying behind them, particularly relating to issues of Greek cultural identity. In his prologue (as James Warren emphasises) Diogenes argues, against those who see its origins as lying in the East, that philosophy is definitively and constitutively Greek; indeed, the human race itself began in Greece (1.3). This metaphilosophical project may be universalising in one sense, in that it synthesises philosophy across time and place, but it is also closely integrated with Diogenes' ideological programme for the present. Diogenes constructs a symbolic empire of knowledge that emblematises the aspirations of contemporary Hellenism.

Diogenes' attempt to 'purify' philosophy represents an extreme case of the 'invention of tradition'.[51] At the other pole, we find a radical emphasis upon philosophical hybridity: Greek wisdom is variously said to be rooted in, or no better than, Egyptian, Babylonian, Jewish, Indian and others. Lucian

[44] Philodemus of Gadara, some of whose works are partially preserved among the Herculaneum papyri, was also concerned with the synthesis of philosophy; for an overview see Obbink (1996) 81–3.

[45] Della Corte (1991). [46] Mansfeld and Runia (1997); Bremmer (1998).

[47] Other doxographical works include, e.g., Albinus or Alcinous of Smyrna (2nd cent. CE), *Digest of Plato's philosophy* and *Introduction to Plato's dialogues*; Atticus' (2nd cent. CE) work (title lost) on the categories of philosophy; Pseudo-Galen (2nd cent. CE?), *On the history of philosophy*; Hierocles (2nd cent. CE) *Exposition of ethics* and other works; Pseudo-Plutarch (2nd cent. CE?); Sextus Empiricus (2nd–3rd cent. CE), *Outlines of Pyrrhonism*.

[48] Holford-Strevens (1997), (2003) 98–130.

[49] Holford-Strevens (2003); Holford-Strevens and Vardi (eds.) (2004). [50] See pp. 8–10 above.

[51] Hobsbawm and Ranger (eds.) (1983).

claims that philosophy originated in India (*Runaways* 6–7); Numenius of Apamea, the Pythagorean or Platonist of the second or third century CE, famously asks 'what is Plato but an Atticising Moses?' (fr. 13 Guthrie). An alternative strategy is to include exoticising elements within Greek philosophy. Thus, for example, Dio Chrysostom includes what he calls a Zoroastrian myth (though it looks, to the trained eye, rather Stoic)[52] in his Borysthenic oration (36.40–54); and Plutarch's *On Isis and Osiris* develops a reading of Egyptian theology that embodies nicely his own Platonist philosophy. What we see in both positions, the purist and the exoticist, is a metaphilosophical concern to narrativise philosophy. Philosophising is not just an abstract intellectual practice; it also, necessarily, invites the individual to position him- or herself within a wider web of debates. Where does philosophy come from? How definitively Greek is it? What does our response to these questions tell us about ourselves?

It is tempting to evaluate 'purists' like Diogenes and 'hybridists' like Numenius in cultural-political terms. Thus Festugière, in his influential and (in many ways) brilliant *La Révélation d'Hermès Trismégiste*, presents the turn to alien wisdom as a failure of Hellenism, symptomatic of a general collapse of Greek values.[53] Modern readers, by way of contrast, may think of Stuart Hall's distinction between conservative narratives of tradition and pluralist narratives of cultural 'translation' and hybridity.[54] Certainly, this knot of concerns over the cultural value of knowledge needs to be located against the backdrop of the enormous, varied empire, with its slick lines of communication and trade routes: the experience of 'globalisation' induces both a heightened awareness of what is shared between cultures and an increased desire to insist on singularity. Yet it would be naïve to see the matter in simple terms of a battle between conservatives (or cultural fundamentalists) and multiculturalists. Both strategies are, ultimately, attempts to encapsulate the global-imperial status of philosophy, and both are centrally concerned to explore the role of Greekness in the modern world. Even Numenius' apparent degradation of Plato is also an implicit argument for the capaciousness and adaptability of Greek philosophy; it is, after all, the form that he has chosen to express himself in.

Was the Latin language capable of accommodating philosophy? The intellectual relationship of Rome to Greece was, broadly, that of scribal culture to reference culture: Greece was conceived of as the originator of

52 de Jong (2003).

53 Festugière (1944–54); the book evinces a tangible sense of anxiety about cultural loss, perhaps not surprisingly given its publication date.

54 Hall (1992).

ideas, Rome as the translator and interpreter. Though marked as a relative latecomer, Roman culture could nevertheless adopt a range of strategies in relation to Greek philosophy. Lucretius, who in the first century BCE rendered Epicurean philosophy into Latin hexameters, famously protests the 'poverty' of his native tongue in relation to the task in hand (*On the nature of things* 1.136–9) – but still, of course, produces one of the most linguistically adventurous and adept poems of antiquity. Lucretius' contemporary Cicero, on the other hand, argues that Rome can now overtake Greece in this field (*Tusculan disputations* 1.5–6). Both writers confront the paradox later to be articulated so famously – in elegant, Hellenising hexameters – by Virgil's Anchises that Romans should leave the arts to others, and focus on *imperium* (*Aeneid* 6.851–2).[55] Latin philosophy, however, could and did exist in mature and confident forms, particularly in the post-Ciceronian tradition: in the first century CE, we find, among others, the younger Seneca writing Stoic works *On anger*, *On clemency* and on numerous other topics; in the calmer second century, Apuleius could ruminate philosophically in Latin in north Africa, and Aulus Gellius in Athens. Even so, each of these writers, in different ways, manifests an anxiety, or at least a negotiation, of Greek influence. The most extreme case is perhaps that of Apuleius, whose philosophical works vary from creative studies of Greek thinkers (principally Socrates and Plato) through to translations (a lost version of Plato's *Phaedo* and – if it is genuinely Apuleian – an extant rendering of the pseudo-Aristotelian *On the cosmos*).

It is remarkable, however, how little canonical authority Latin philosophy achieved in the first three centuries CE (in marked contrast to later, Christian antiquity, when Latin philosophers and theologians were the intellectual colossi of the day). Roman philosophers did not have significant acolytes, nor were their books translated or commented on. The reasons for this should be sought in the ancient practice of the division of intellectual labour into cultural 'zones'. For elite Romans, philosophy was a sophisticated, but on occasion dangerously unRoman, pastime that could safely be practised only in the circumscribed leisure zone of a rural retreat or a day off from the business of empire. Through the early empire, this strict separation between (Greek) philosophy and the proper (Roman) activity of imperial management became all the more pronounced. The one exception is a telling one: in the field of law alone – that most pragmatic and politicised of intellectual disciplines – Greeks ceded conceptual mastery to Rome.[56]

[55] See further Petrochilos (1974) 58–62. [56] See Millar (1999), esp. 105.

The best-known philosophers at Rome in the first century CE were thought of, in general, as either opponents of or advisors to the emperor; either way, the allocation of roles emphasised the differential between philosophy and power.[57] Despite sporadic appearances of Latin philosophy, Greek was almost always (in the pagan era) considered the appropriate language for philosophy. The first-century Romano-Etruscan knight Musonius Rufus turned to Greek to express his Stoic thoughts.[58] The second-century Romano-Gallic knight Favorinus also chose Greek (though he could discourse with equal competence in Latin).[59] The most prodigious example is that of the emperor himself, Marcus Aurelius, who chose to write the work of Stoic philosophy we call the *Meditations* in Greek – a work composed, so he claims, for his own benefit alone. Marcus effectively united the roles of Roman emperor and Greek adviser within a single persona.[60]

COSMIC KNOWLEDGE

This tension between philosophy as a specifically Greek cultural signifier and its ambitions to a cosmic, supraparochial knowledge is exemplified most powerfully in the aspiration to what is conventionally called 'cosmopolitanism', a philosophy that has its roots in Hellenistic Cynicism and Stoicism but takes root in a number of Imperial genres.[61] In particular, the exilic discourses of Musonius Rufus, Dio, Plutarch and Favorinus point to a new intensification of concerns.[62] For these writers, the fact of penal exile stimulates reflection upon the limitations placed upon knowledge and understanding by family, community, city and state. State persecution thus becomes a rite of passage, an opening to a new, more intense knowledge of the structure of nature, the world, the cosmos. As Musonius puts it,

> Why should anyone who is not devoid of understanding be grieved by exile? It does not deprive us of water, earth, air, or the sun and the other planets, or indeed, even of the society of men, for everywhere and in every way there is opportunity for association with them. (fr.9 = p. 41 Hense)

Empires of philosophical knowledge are evidently constructed to vie with, even outdo, the scale of the quasi-global Roman empire: to philosophise is to control, at least epistemologically, a territorial space that

[57] Opposition: Macmullen (1992); Rudich (1993); advice: Rawson (1989). [58] Geytenbeek (1963).
[59] See most recently Holford-Strevens (2003) 98–144, with 118–29 on his use of Latin.
[60] See Whitmarsh (2001) 216–25, with further references.
[61] Baldry (1965); Stanton (1968); Schofield (1991) 57–92; Moles (1993). [62] Whitmarsh (2001) 133–80.

exceeds the boundaries of mere political space.[63] This theme is played out to brilliant effect in Philostratus' *In honour of Apollonius Tyana*, where the philosopher's travels pointedly take him beyond the outer limits of Roman imperial control, and indeed Macedonian conquest.[64] In the eastern voyage of the early books (a voyage he conceptualised in terms of 'border-crossing', 1.18), Apollonius passes through a succession of boundaries symbolically marking the journey into the unknown. At the 'borders' of Babylonia – in both Apollonius' and Philostratus' time, the nerve centre of the defiant Parthian empire – he meets a frontier control (1.21). Indeed, the narrative itself is organised around the theme of border-crossing. At the end of the first book, Apollonius resolves to leave Babylon; at the beginning of the second book, Philostratus refers to the Caucasus as the 'beginning' of the Taurus (2.1–2). At the end of the second book, Apollonius reaches a column inscribed 'Alexander got this far' which Philostratus supposes to have been erected either by Alexander to mark the 'limit' of his empire, or by the Indians out of pride that he 'got no further' (2.43). The words 'got no further' close book 2, so that Alexander's column also marks the end of a book. For Greek readers, the next book travels into the radical unknown; and it is significant that book 3 begins with a description of the wonders of India: the river Hyphasis, and extraordinary trees, fish, worms and wild asses, pepper trees and dragons (3.1–9). Philostratus offers a compendium of knowledge that takes its readers beyond the confines of Greco-Roman political, military and epistemological control.

Apollonius' knowledge is predicated on his grasp of 'the world' in all its polymorphous variety.[65] The Brahmans, the sages of India who teach him, realise that true wisdom lies in understanding not 'the parts of the cosmos' but 'the intelligence that lies in it' (ὁ ἐν αὐτῶι νοῦς 3.34). Greek philosophy – not only Apollonius', but also of course the reader's inherited paradigm – is said by the Indians to be insufficient (3.18, 3.27; cf. 2.29). Yet as ever, the tyranny of Hellenocentrism is not so much overthrown as subtly reconfigured. Indian wisdom does turn out to be suspiciously familiar: in terms of kingship theory (2.26–9), the Brahmans are broadly Platonist; in terms of cosmology, broadly Stoic (3.34–5);[66] in terms of communist utopianism, broadly Cynic.[67] Meanwhile, we find the Indian king

[63] Nicolet (1991). [64] See further Elsner (1997).

[65] For the proclaimed unity of the world, see 1.15, 1.21, 6.2 (ἡ γῆ πᾶσα ἐδόκει μία), 8.5, 8.7(iv).

[66] For στοιχεῖα and νοῦς in Stoicism see Long and Sedley (1987) 46B, 47 (though the Indians have an extra element (στοιχεῖον)).

[67] Cf. the Spartanising ὡς ἐν ξυσσιτίωι (3.27), with Dawson (1992) 28; cf. Onesicritus at *FGrH* 134 F20. For Indian gymnosophists (as the Brahmans are usually called, though Philostratus locates *his* gymnosophists in Egypt) as Cynics, see Muckensturm (1993).

Phraotes reading Euripides (2.32), Greek-speaking locals (3.12), statues of Greek gods (3.13), and a whole series of assimilative comparisons between Greek and Indian buildings and city structures (2.20, 2.23, 2.27, 3.13). Greek wisdom does not evaporate in this text, rather it expands to colonise all other knowledge systems: 'to the wise man', Apollonius famously comments, 'everything is Greek' (1.35). Apollonius' cosmic knowledge extends effortlessly beyond the realms of mortal empire.

PRINCELY KNOWLEDGE

The empire of knowledge did not have to vie directly with the political empire of Rome; it could also serve it, as we saw a moment ago in the role of philosophical advisor. The emperor and his servants employed battalions of experts and scholars. For these individuals, the figure of the emperor loomed large: partly because emperors were often patrons of scientific and literary endeavour, and because the adminstrators and technicians of imperial power were always ultimately answerable to the emperor himself; but also for the less immediately tangible reason that ideas of writerly authority and ambition were often explored through and against the image of imperial control.

How much did the emperors of Rome themselves know, or need to know? In some cases it seems that a particular emperor's choice of which kinds of knowledge to patronise could be used – either by the emperor himself or by those who wrote about him – as an index of his most distinctive characteristics.[68] In Suetonius' biography of Claudius, for example, the emperor chooses historical knowledge, leaving behind him an enormous history of Rome in many volumes, begun in his youth with the encouragement of the historian Livy.[69] The most distinctive feature of this history, however, is its incompleteness, the result of continual badgering by his mother and grandmother, who persuade him to leave out sensitive topics from recent history. In Suetonius' account, Claudius' failure to match the exhaustive historical ambitions of his mentor, Livy, is used as a sign of his lack of independence, which is a dominant theme of the biography as a whole.[70] Nero, by contrast, chooses performance expertise, in his increasingly obsessive interest in Greek musical competition, as Suetonius again makes clear.[71] Marcus

[68] Cf. Woolf (1994) 135 on selective appropriations of Greek culture by successive emperors.

[69] Suet. *Claud.* 41. See also Wallace-Hadrill (1983) 72–96 on the fact that Suetonius' own career as a 'scholar at court' was itself in part a result of imperial patronage of scholarship on the part of Trajan and Hadrian, a good example of imperial patronage of a particular type of knowledge.

[70] See esp. Suet. *Claud.* 29.

[71] See esp. Suet. *Ner.* 20–25 and 41–3, with Edwards (1993) 135–6.

Aurelius, as his own writings testify, chooses philosophical self-knowledge above everything. Elsewhere, by contrast, we find the assumption that the emperor should know everything. Suetonius' Augustus, for example, is characterised by his ability to focus on many different areas of government simultaneously.[72] Nero, by contrast, at least on Suetonius' account, falls short of that ideal through the narrowness of his concentration on musical skill, which leads him to neglect the military crises which are brewing all around him.[73] To some extent that image of imperial omniscience is grounded in administrative reality, given what we know of the involvement of emperors in hearing judicial appeals and diplomatic embassies from across the Mediterranean world – although this ideal of the emperor as ubiquitous was itself a carefully orchestrated one, not least through the omnipresence of imperial statues and inscriptions.[74] It was also an ideal that must have relied in practice on a massive exercise of delegation, a sharing of expertise between many different specialists.

Texts dedicated to emperors often reflect that process of jostling for position between rival specialisms keen to gain imperial favour. The idea of interrelation between author's knowledge and emperor's needs is manipulated in a range of different ways. Often the precise specialism of the author is represented as the thing the emperor most needs to know. That trope casts the author as subordinate, but also allows him (it is always 'him') to claim a kind of patriotic usefulness for his own writing, and sometimes also to equate his own compilatory ambitions with the emperor's territorial and administrative grasp.[75]

Some authors even take on the imagery of imperial omniscience, applying it to their own wide-ranging erudition. In this volume, for example, Alice König examines the imperial dedications of Frontinus and Vitruvius. For Vitruvius, she suggests, architecture lies at the heart of empire; it is also equivalent to the constructive political skills of the emperor Augustus. Frontinus uses the same words – 'diligent' and 'loving' – to describe both himself, in his care of the aqueduct system of Rome, and the Emperor Nerva. Again, Maecianus' metrological treatise, addressed to Marcus Aurelius, explores (as Serafina Cuomo's chapter shows) connections between standardisation of measures and the establishment of political order. That

[72] For just one example, see Suet. *Aug.* 33 on Augustus' painstaking personal involvement in the administration of justice; and cf. Wallace-Hadrill (1983) 119–25 on Suetonius' awareness of the wide range of areas covered by imperial administration and his tendency to value personal participation of emperors in them.

[73] Cf. König (2005) 229–33.

[74] Ando (2000), esp. 206–73, on images of emperor and of empire.

[75] See esp. Murphy (2004), esp. 203–9, on Pliny's manipulation of the image of Titus' imperial authority in the *HN*.

strategy can in some cases take on subversive overtones. Petronius' *Satyrica*, as Victoria Rimell argues in her chapter in this volume, has no sign of any specific reference to the emperor, but it too is daringly parasitic upon the tropes of imperial self-presentation, grounding its grotesque vision of excessive knowledge in the image of Neronian over-consumption.

KNOWLEDGE, SOCIAL STATUS AND CULTURAL AFFILIATION

What you know says a great deal about who you are. Knowledge is intimately tied up with social self-positioning. In the east of the empire, for example, mastery of abstruse rhetorical and literary knowledge was widely associated with social distinction.[76] Socially empowering rhetorical expertise – publicised in its most extravagant form in the display speeches of sophists in front of huge audiences – was not only about intellectual agility, although there are certainly numerous handbooks on matters of style, grammar and rhetoric surviving from the period;[77] it was also embodied, displayed through posture and gesture and style of voice as much as through words, absorbed and learned and constantly reperformed within the encounters of everyday interaction and repetitive training.[78] In that sense, sophistic skill was simply a more intense form of the skills of social self-presentation which all elite men (women too – though there is less sense in ancient sources of female identities being forged within the rhythms of public display)[79] had to learn.

[76] See esp. Schmitz (1997). A huge body of technical rhetorical writing existed in the imperial period, encompassing, e.g., works by (in Greek) Valerius Apsines of Gadara (3rd cent. CE), Aelius Aristides (2nd cent. CE), Dionysius of Halicarnassus (1st cent. BCE–1st cent. CE), Hermogenes of Tarsus (2nd–3rd cent. CE), Aelius Herodian of Alexandria (2nd cent. CE), Lesbonax (2nd cent. CE), Dionysius Cassius Longinus of Athens or Palmyra (3rd cent. CE), Menander 'rhetor' of Laodicea (3rd–4th cent. CE), Minucianus the younger of Athens (3rd cent. CE), Aelius Theon of Alexandria (1st–2nd cent. CE), Tiberius (3rd CE?), Potamon of Mytilene (1st cent. BCE–1st cent. CE), as well as the 'Anonymous Seguerianus' (3rd CE); and in Latin, by Rutilius Lupus (1st cent. CE), Suetonius and Tacitus.

[77] See Kaster (1988); Swain (1996), 43–64. For rhetorical works, see previous n.; for lexicographical works, see below, n. 92. Grammatical works in Greek are transmitted or attested by, e.g., Ammonius (1st–2nd cent. CE), Apollonius 'Dyscolus' of Alexandria (2nd cent. CE), Aristonicus of Alexandria (1st cent. BCE–1st CE), Hephaestion of Alexandria (2nd cent. CE), Heraclides of Miletus (1st–2nd cent. CE), Herennius (or Eranius) Philo of Byblis (1st–2nd cent. CE), Aelius Herodian of Alexandria (2nd cent. CE), Lesbonax (2nd cent. CE), Nicanor of Alexandria (2nd cent. CE), Polybius of Sardis (2nd cent. CE?), Telephus of Pergamum (2nd cent. CE), Theon of Alexandria (1st cent. BCE–1st cent. CE), and Tyrannion the younger or Diocles (1st cent. BCE–1st cent. CE); in Latin, by Flavius Caper (2nd cent. CE), Censorinus (3rd cent. CE), Verrius Flaccus (1st cent. CE), Fronto and Velius Longus (2nd cent. CE).

[78] See Gleason (1995); Whitmarsh (2005a) 23–34.

[79] One female rhetorician, Aufria, is recorded at Delphi in an inscription of the second century CE: see Puech (2002) 156–7. In general on female education see Hemelrijk (2004).

Nor is that concept of knowledge as bodily practice confined to the Greek part of the empire; Thomas Habinek, for example, shows in this volume how a vision of knowledge as embodied experience structures Manilius' *Astronomica*. Those patterns exemplify strikingly Bourdieu's notion of social knowledge as something formed through the repetitions of everyday life, experienced bodily as well as intellectually.[80] Many of the encyclopedic and scientific texts we examine here at first sight seem far removed from the bodily experience of knowledge, in their dry and disordered surfaces. And yet if we surrender ourselves to the repetitive patterns of these texts we can perhaps begin to see how they mirror the processes by which social knowledge is acquired, offering their readers a cumulative experience of knowledge, gradually imprinting the grooves of knowledge on to the reader's mind through their relentlessly recurring yet endlessly varying rhythms.[81]

We find similar social pressures within late-Republican and early-Imperial Rome. Here – as within Greek tradition – manual work and specialised, technical knowledge were generally represented as unsuitable for men and women of high status; while certain other forms of expertise – military, rhetorical, agricultural – were more highly prized, though never unequivocally so.[82] Moreover, the social acceptability or otherwise of a particular body of knowledge was often linked with its perceived cultural affiliations. Social status and gestures of cultural affiliation were closely intertwined with each other, although the precise nature of these links was constantly open to restatement and reperformance. Greek knowledge in some forms was treated with suspicion – philosophy or astrology, for example, whose reputation for subversive potential, leading to sporadic banishments of philosophers and astrologers by successive emperors, was partly linked with its Hellenic associations.[83] But in other forms it held social cachet. Elite Romans had to tread a delicate balance between excessive devotion to Greek knowledge and ignorance of it (as we have seen

80 See esp. Bourdieu (1977); cf. Crick (1982) 300 on embodied knowledge.

81 E.g., see Jacob (2001) on the painstaking, repeated practices of quoting, filtering and juxtaposing, by which Athenaeus' *Sophists at dinner* draw meaning from the Hellenic literary heritage.

82 See Rawson (1985) for exhaustive discussion of the status map of Roman Republican disciplines.

83 On the ambiguous relationship between astrology and imperial power, see Barton (1994a) 32–63 and (1994b) 27–94. Astrological and astronomical texts from the period are numerous: e.g. (in Greek) Achilles Tatius (2nd cent. CE), *On the sphere*; Apollonius of Tyana (1st cent. CE), *Celestial influence*; Cleomedes (2nd cent. CE), *On the circular motion of the heavenly bodies*; Dorotheus of Sidon (1st cent. BCE-1st cent. CE), various astrological works; Manetho (3rd CE), *Celestial influence*; Maximus (2nd cent. CE?), *On forecasts*; Claudius Ptolemy (2nd cent. CE), *Celestial influence*; Teucer of Babylon or Egypt (1st cent. CE) *On the zodiac* and *On the seven stars*; Thrasyllus of Alexandria (1st cent. CE), a work on astrology; Vettius Valens (2nd cent. CE), *Anthology*; in Latin, Apuleius (2nd cent. CE), *Astronomy*; Germanicus Caesar (1st cent. CE), *Prognostications*, *Celestial phenomena*; Hyginus (2nd cent. CE), *Astronomy*; Manilius (1st cent. CE), *Astronomy*.

already for Greek philosophy).[84] These considerations continued to exercise a powerful influence well into the Imperial period, and many texts continued to advertise the distinctive Romanness of their own reconfigurations of the Greek systematising project.

The recurrent attraction of agriculture as a subject for Roman knowledge-orderers is a good example. Columella, for example, writing in the first century CE, knows and discusses at length Greek traditions of writing on agriculture,[85] but he also goes out of his way to mark the Romanness of his text throughout his preface, for example by representing his own project as an attempt to reinvigorate the productiveness of Italian farmland, and by looking back to the hardy stock of Romulus who tilled the fields in the beginnings of Roman history. For Columella, advertising one's relation with the myths of early Roman frugality and self-sufficiency, and with the rigours of specifically Roman erudition, is an essential authorising gesture for wealthy, landed, elite status in the present (a gesture which is parodied in Petronius' portrayal of Trimalchio's self-sufficiency (*Sat.* 37–8) – not the self-sufficiency of the stereotypically modest Roman market-gardener, but rather of the man who produces everything he could desire on his own massive estates).[86]

DISCIPLINARY KNOWLEDGE

Unwritten social rules, however, are notoriously unstable. The hierarchy of disciplines was constantly being restated and refashioned. Changes in political climate and in administrative conventions led to changes in the prestige of specific groups of practitioners. For example, the increasing popularity – and, at times, official disapproval – of astrology in Rome from the late first century BCE onwards has conventionally been explained by the fact that it thrived on making predictions related to powerful political figures, and to Augustus' exploitation of that focus (although Thomas Habinek's chapter in this volume nuances that explanation).[87] Under the Flavians the backlash against empowerment of freedmen employed by the Julio-Claudians made it increasingly common for senators to be given administrative posts, and prompted reformulation of conventional senatorial antipathies towards applied knowledge.[88]

84 See pp. 16–18, above and, e.g., Gruen (1993).

85 See esp. Columella *Rust.* 1.1.7–11, and Henderson (2004). Other agricultural texts of the period include, e.g., Apuleius (2nd cent. CE), *On rustic matters*; Columella, *On trees*; Siculus Flaccus (2nd cent. CE), *On the status of fields*; Julius Graecinus (2nd cent. CE), *On vines*.

86 See Garnsey (1999) 23–4.

87 See Barton (1994a), esp. 38–49.

88 See Talbert (1984), with 15–16 and 134 on Vespasian's adlection of new senators; and 372–407 for senatorial duties.

Other changes were less clearly anchored in specific institutional adjustments, but not for that reason any less firmly grounded in political reality. It is clear, for example, that some bodies of Greek knowledge (e.g., medicine, siegecraft, geography) were widely appropriated and adapted within Rome, while others (e.g., philology, literary criticism) were relatively neglected. Those variations in treatment of Greek intellectual material were partly a result of struggles for political and cultural authority within the Roman elite, where choices about which resources to exploit and which not were always determined in part by strategic aims.[89] Similar considerations seem to have shaped shifting attitudes towards traditionally Roman forms of expertise and cultural authority. On one argument, for example, the first century BCE saw an erosion of the link between historical/religious knowledge and political authority within Rome, at a time when appeals to the past history of individual families were losing their moral authority, in the turmoil of civil war. That shift was then manipulated by Augustus and his successors by bringing antiquarian experts – who no longer tended to come from the politically active elite – under direct imperial control, thus creating a new vision of universalising knowledge, where political authority was based on delegation of knowledge rather than possession of it.[90] Whatever the virtues of that model in its precise details, it is important for its attempts to ground changing conceptions and textual manifestations of Roman knowledge materially in the continually evolving struggles for elite prestige and authority.

Much of this shifting landscape of disciplinary self-presentation was also due to the fact that the Roman Empire – unlike the modern world – had few explicit professional qualifications, institutional structures for controlling and guaranteeing expertise. That situation led to heavy reliance on rhetorical means of self-legitimation, developed within a system where experts had to compete for adherents and clients.[91] For that reason we often see knowledge-ordering writers jostling for position against their rivals. Galen, for example, argues in his *Protrepticus* for a separation between good arts and bad arts; and then subdivides the former category, listing medicine as the best art of all (*Protrepticus* 14).[92] Galen's equation of medicine with philosophy – both there and elsewhere (most obviously in *That the best doctor is also a philosopher*) – allows him to separate his own expertise not only from other disciplines, but also from the activity of those he represents as more disreputable and incompetent claimants to medical knowledge, whose expertise is not worthy of that label. Philostratus responds to that

[89] See Wallace-Hadrill (1988). [90] See Wallace-Hadrill (1997).
[91] See Lloyd (1979) 86–98. [92] See König (2005) 291–300.

scheme in his *Gymnasticus* by downgrading medicine, praising instead the – socially marginal – art of athletic training which had been one of Galen's main targets.[93]

That is not to say that we should equate these kinds of disciplinary self-definition with the sharp disciplinary divides with which we are familiar in the modern world. It was standard for learned people to write treatises on a broad range of topics. The common habit of addressing works to private recipients, and that of drawing attention to one's own reluctance to publish, together contribute to an impression of the absence of rigid disciplinary boundaries. There were also pressures towards totalising knowledge for ancient writers, pressures to integrate different bodies of thought into a single system, 'metaphilosophical' or otherwise. Strabo, for example, represents his combination of geographical and historical knowledge as a kind of philosophy, grounded in Stoic assumptions that the cosmos is a unified body whose disparate parts are held together by a single binding force.[94] Galen – as Rebecca Flemming shows in her chapter – takes a more Platonic view, seeing the workings of divine order within the order of the human body, and drawing on many different areas of expertise – natural-historical, philosophical, philological – to convey that vision throughout his voluminous oeuvre.[95] And yet despite all of these attempts to harmonise different bodies of knowledge, it is also clear – not least in the example just quoted from Galen's *Protrepticus* – that the gesture of equating one's own expertise with the overarching label of philosophical expertise could often serve partisan aims.

This distinction between different disciplines is important partly just because it reminds us of the dangers of characterising knowledge systems of the Roman Empire as monolithic entities, even though that is a characterisation these texts themselves often construct rhetorically (as we have already seen above[96] for the all-embracing nature of philosophical self-definition). Moreover, we need to recognise not only that different bodies of knowledge were formed very differently through battles for political and social authority, but also that they in turn fed diverse models and impulses of social and political interaction back into the cultures that produced them. Not all disciplines created the same kinds of social positions for the human subjects and objects of their practice. Medicine, as it was both practised and theorised, offered distinctive visions of the human body and of

[93] See König (2005) 315–25.

[94] See French (1994) 123–30; Clarke (1999), esp. 216; cf. 185–90 on similar conceptions in the geographical work of Posidonius.

[95] See Hankinson (1988). [96] Pp. 13–20.

human subjectivity, both male and female, distinctive roles for doctors and patients alike to inhabit;[97] the increasingly systematised body of Roman law did the same;[98] and the repetitive techniques of Roman declamation, as they are revealed to us in surviving treatises of rhetorical exercises, had encoded within them particular visions of the social and gender hierarchies of Roman society.[99] All of these disciplines were very different from each other in the experiences of knowledge they offered (as well as being differently experienced in different contexts), though all of them also reflected and contributed in interlocking ways to the overarching power relations of Greco-Roman society.

THE WORLD IN THE TEXT

The discussion so far prompts wider synchronic reflections on the nature of textuality and authorship in the period. We have advanced strong claims for the interrelationship between varieties of the ordering of knowledge and of the ordering of empire. But what of the presentation of material within texts, and within those social institutions that allow for the reception and circulation of texts? As Foucault argues in a well-known essay, the 'author function' is historically variable, to the effect that different forces (ideological, political, cultural, economic) shape the way that textual significance and textual ownership are imagined in different periods.[100] Should we expect to see the empire of knowledge mirrored in the mini-empires of textual ordering?

[97] See Flemming (2000) on medical visions of the female body, and more broadly Barton (1994b). Medical authors in the period, in addition to Galen, are numerous: works transmitted or known include, e.g., in Greek, Aglais of Byzantium (1st cent. CE?), *On cataracts*; Antyllus (2nd cent. CE), *On enemas*; Archigenes of Apamea (1st–2nd cent. CE), various; Aretaeus (2nd cent. CE), various; Cassius (2nd–3rd cent. CE), *Medical questions* and other works; Titus Statilius Crito of Heracleia (1st–2nd cent. CE), *On cosmetics* and *On the composition of drugs*; Dioscorides 'Pedianus' (1st CE), various; Marcellinus (2nd cent. CE), *On pulses*; Philo of Tarsus (1st–2nd cent. CE), a medical poem; Plutarch of Chaeroneia (1st–2nd cent. CE), *Precepts for good health*; Rufus of Ephesus (1st–2nd cent. CE), various; Severus (1st cent. CE), *On cauterisation*; Soranus of Ephesus (1st–2nd cent. CE), various; in Latin, Celsus (1st cent. CE), *On medicine*; Scribonius Largus (1st cent. CE), *Prescriptions*.

[98] A massive and tangled corpus of Roman jurisprudential material survives from the period. The most important works are: Gaius (2nd cent. CE), *Institutes* and other works; Lucius Volusius Maecianus (2nd cent. CE), various legal and other works (see further Cuomo in this volume); Aelius Marcianus (3rd cent. CE), *Institutes* and other works; Masurius Sabinus (1st cent. CE), several jurisprudential works; Domitius Ulpian (3rd CE), *Institutes* and other works. Harries (1999) offers many suggestive insights into the ways in which late-antique legal convention influenced and was influenced by conceptualisations of authority in other areas of public and private life.

[99] See especially the elder Seneca's *Utterances, categories and techniques of the orators*, with Bloomer (1997); Dupont (1997); Gunderson (2003). For other rhetorical works, see above, n. 73.

[100] Foucault (1986). See further Whitmarsh (2004) 8–9.

Can we see, for example, a shift of emphasis from 'work' to 'text'? A 'work', on this interpretation (mimicking Roland Barthes),[101] would be an apparatus of signs viewed principally as the manifestation of the author's privileged intelligence; a 'text', on the other hand, would be a resource made available for the reader. Clearly, any such historical shift would be supported by the (self-diagnosed) domination of prose over verse in this period:[102] there is no expectation of 'inspiration' attached to technical prose. Texts like Frontinus' *On aqueducts*, Artemidorus' *The interpretation of dreams*, Apollodorus' *Library*, Polemo's *Physiognomics*, Aelian's *On the nature of animals*, Ampelius' *Book of memory*, Pollux's *Onomasticon* or Ulpian's *Digest* are self-consciously utilitarian, anticipating a reader who consults judiciously rather than capitulating to the linear textual narrative.

Of course, at another level, each of these texts is a virtuoso authorial performance of mastery in the spheres of research, synthesis and exposition. It is not that the author recedes in such texts, more that the role of the author is reconceived: new virtues are located in the arts of editing and the organisation of pre-existing units of knowledge. We can identify this phenomenon across a range of cases, from the Hellenistic period onwards, but with particular intensity in the Roman period. Nor is this process confined to the enormous range of massive, synthetic texts alluded to in the previous paragraph. Philosophers' words were edited by their disciples, playing Plato/Xenophon to their teachers' Socrates (the most prominent cases are those of Lucius/Musonius Rufus and Arrian/Epictetus). Philostratus' *In honour of Apollonius of Tyana*, a text that we have already considered, represents itself as the result of the 'rewriting' (μεταγράψαι) for imperial consumption of the memoirs of Damis, synthesised with the works of Moeragenes and Maximus of Aegeae (1.2–3). Labour could profitably be invested in ordering the books of intellectual avatars: Porphyry, for example, edited and arranged Plotinus' books (the full title of his biography is *On the life of Porphyry and the order of his books*).[103] The self-regarding Galen, indeed, even wrote *On the order of my own books*. In the sphere of poetry, 'garlands' of short poems by established authors were collected by such figures as Meleager (in the first century BCE) and (in the first century CE) Philip. It is as though the world of knowledge was now fully bounded and all that remained was to debate its *arrangement*.

[101] Barthes (1986). [102] Whitmarsh (2005b).

[103] Porphyry's catalogue of Plotinian book titles is discussed at Nachmanson (1941) 26–7.

This is a period that also sees a profusion of works of commentary and (particularly in Latin) of translation, epitomisation and summary.[104] What is extraordinary in these cases is that we almost always know the identity of the commentator, translator, epitomiser or summariser in question: thus, for example, Julius Florus is known as the epitomiser of Livy, Justinus as the author of the *Prologues* of Trogus' *Philippic histories*, Sulpicius Apollinaris as the author of verse summaries of Virgil's *Aeneid* and Terence's plays.[105] The reconfiguration of pre-existing texts is viewed not simply as second-order intellectual parasitism, but as a major intellectual project in its own right. This, of course, is not *per se* new: after all, the traditional founder of Latin literature (Livius Andronicus) was a translator from the Greek. The crucial point, however, is to note the rich, thick context in which this imperial literary production was appearing. This was not a secondary culture held in thrall to its originating predecessors (as older scholars, preoccupied with the romantic ideal of creative originality, sometimes assumed): it was rather an imperial power mapping and colonising the enormous expanse of pre-existing knowledge. From the time of Trajan onwards, the boundaries of the Roman Empire were not significantly expanded; likewise, the textual world was conceived of as sufficient and fully formed.

Inevitably, counterexamples can be advanced.[106] In the field of geography, exotic new parts were being discovered at the margins of the world: this sense of excitement at the discovery of new spaces feeds both the burgeoning genre of *periplous* ('circumnavigation') literature[107] and other travel-narrative forms, such as the novel.[108] Yet the exceptions prove the rule: that new discoveries were consigned to the exotic margins of the world is an indication of the epistemological exhaustion of the Empire. The fantastical

[104] See appendix for full details. For bilingualism in the period, see Adams, Janse and Swain (eds.) (2002), and Adams (2003).

[105] Cf., e.g., Apuleius' (2nd cent. CE) translations (noted above) of Plato's *Phaedo* and (though authorship is debated) Aristotle's *On the universe*, and also a verse translation from Menander; Quintus Asconius Pedianus' (1st cent. CE) commentaries on Cicero's orations; Baebius Italicus' (1st cent. CE) epitome of Homer's *Iliad* in Latin; Germanicus Caesar's (1st cent. CE) translation of Aratus' *Celestial phenomena*; Marcus Cetius Faventinus' (3rd cent. CE) epitome of Vitruvius' *On architecture*; Sextus Pompeius Festus' (2nd cent. CE) epitome of Verrius Flaccus' *On the meaning of words*; 'Septimius'' (3rd CE?) translation of Dictys of Crete's *Journal of the Trojan war*. The 'Arguments' (2nd cent. CE) for Plautus' plays, on the other hand, are anonymous.

[106] Arrian's book on hunting with dogs, for example, stakes its claim to surpass Xenophon precisely on the superior knowledge of dog breeds available to the later author: full discussion at Stadter (1980) 53–4.

[107] Cf. the anonymous *Circumnavigation of the Red Sea* (2nd cent. CE) and *Circumnavigation of the Great Sea* (3nd cent. CE), Arrian's *Circumnavigation of the Black Sea*, Dionysius of Byzantium's (2nd cent. CE) *Navigation up the Bosporus*.

[108] Especially Lucian's *True stories* and Antonius Diogenes' *Wonders beyond Thule*: see Romm (1991) 172–214.

invention of new places to visit plays the same imaginary role as a dream that one discovers a new room in one's house: the thrill of the new derives in part from the familiar structures of the established.

What we are proposing is that the period of the first three centuries of the Roman Empire saw a large-scale shift in the perception of intellectual labour: the fact of empire crucially changed the way in which knowledge was used, abused, presented, represented. In his classic discussion *The Archaeology of Knowledge*, Michel Foucault proposes the idea of 'the archive' as a historical phenomenon (with roots, he argues, in the eighteenth century) that enables a certain way of understanding language and its relationship to the world: not simply the physical institutions that store and disseminate knowledge, but an entire 'system of discursivity':

> The archive is first the law of what can be said, the system that governs the appearance of statements as unique events. But the archive is also that which determines that all these things said do not accumulate endlessly in an amorphous mass, nor are they inscribed in an unbroken linearity, nor do they disappear at the mercy of chance external accidents; but they are grouped together in distinct figures, composed together in accordance with multiple relations, maintained or blurred in accordance with specific regularities.[109]

For Foucault – here in characteristically oracular mode – 'the archive' is a 'historical *a priori*', a dynamic force that drives all intellectual production within a period. This is clearly both overstated and undernuanced: societies are more fractured and embattled than Foucault allows (and indeed later writings such as *La volonté du savoir* reflect his increasing awareness of precisely this point). We might also question the validity of construing the archive as an '*a priori*' cause that has (apparently) only limited, localised negotiability:[110] surely there is no single system that 'governs' or 'determines' (to use Foucault's phraseology) thought, without itself becoming subject to immediate and radical revision.

Still, the archive – as a habit of thought, an intellectual genre, an interrelated set of culturally operative, but also embattled, propositions as to the necessary properties and social roles of language – provides a useful model for conceptualising the order of knowledge. Though Foucault seeks to explicate a much later period, archival thinking can be detected behind a range of textual practices in the Roman Empire, where (as we have seen) the desire to itemise and order knowledge reaches a new peak of intensity.

109 Foucault (1972) 129. 110 *Ibid.* 127.

ORDER IN DIVERSITY

How can diverse objects of knowledge be synthesised into a single textual form? Common terms like 'miscellany', used particularly of texts like Pamphila's *Collection of historical reminiscences*, Favorinus' *Miscellaneous history*, Aulus Gellius' *Attic nights* and Aelian's *Miscellaneous history*, give the impression of a random aggregration of unconnected phenomena.[111] And indeed the texts themselves often foster that impression. According to Photius (*Library* 119b), Pamphila claimed to have compiled her work 'at random, as each thing came to her' (εἰκῆι καὶ ὡς ἕκαστον ἐπῆλθεν). But this kind of claim should not be taken as evidence for incompetence: Photius goes on to report her assertion that it would have been easy to structure her work by topic (κατ' εἶδος), but more pleasant (ἐπιτερπέστερον . . . καὶ χαριέστερον) to present a polymorphous variety. Miscellanism was a conscious, deliberate and motivated choice for Pamphila. This example should warn us against taking the semblance (and indeed the protestation) of randomness too literally. There was a knowing, controlling intelligence behind Pamphila's 'miscellany' of knowledge, even if too little of the text survives today to be able to judge it.[112]

The other thing which unifies miscellanistic knowledge, of course, is its openness to being manipulated within specific moments of performance. It is striking that many of the miscellanistic works of the Roman Empire offer themselves as resources to be used – collections of anecdotes or exempla to be reactivated and reshaped by the reader in other contexts. Elsewhere, we see that kind of performance of knowledge in action, for example in representations of sympotic conversation of the kind we find in Plutarch's *Sympotic questions* or Macrobius' *Saturnalia*, or in many of the dialogues of Aulus Gellius' *Attic nights*.

But how do we read such diversity when we are confronted with it? What kind of sense can we make from accumulations and lists as we read (as opposed to making sense of them through our own re-use of the anecdotes and facts they offer)? Let us turn, by way of example, to one of

111 Other contemporary miscellanies include, e.g., Lucius Ampelius (3rd CE?), *Book of memory*; Apuleius, *Florida*; Censorinus (3rd cent. CE), *On the birthday*; Clement of Alexandria (2nd–3rd cent. CE), *Miscellanies*; and the pseudepigraphical 'Fragmentum Censorini'. Earlier works entitled 'mixtures' (σύμμικτα) are known of by Aristoxenus (frr.122–7 Wehrli), Istrus (*FGrH* 334 F57) and Callistratus 'the Aristophanean' (*FGrH* 348 F2–3).

112 For further discussion of the often disingenuous nature of claims about random composition in miscellanistic texts, see Jason König's chapter in this volume; for Pamphila's knowingness, note her claim (again reported by Photius) to have learned her subject matter from her husband, from visitors and from books: we take this a playful allusion to Ar. *Lys.* 1125–7.

the paradigm cases of archival thinking under the empire, Julius Pollux's *Onomasticon* (addressed, as we shall shortly discuss, to the Emperor Commodus). The history of ancient lexicography is still largely unwritten – indeed, as John Henderson observes in his chapter, scholarship has largely connived to repress its visibility, while simultaneously exploiting it as a resource – but it seems as though Pollux's work innovated radically, in that it orders its words rigorously by theme, grouping the themes together into books.[113] To that extent, it seems to borrow substantially from the conventions of 'encyclopedic' knowledge-orderers like Pliny, who deal with a wide range of different types of expertise in successive sections. Phrynichus' contemporary *Atticist* (from which we have an extant *Eclogue* or 'selection'), by way of contrast, proceeds by picking its way apparently at random through the many phrases it amasses (though there may be traces of an original alphabetical order).

Pollux's technique is to enter lists of words under broad rubrics. Though the *Onomasticon* is not the kind of text that many readers will choose to read sequentially, the sequence is in fact crucial: the mapping of language is also an exercise in mapping the hierarchies of the world.[114] Let us take an example, the entry under names for girls and women (which, predictably, comes after that for men):

In the case of females, the first names, up to *paidarion*, are the same as for men (for this is common to both, to females and males). From there on: *paidiskē, korion* (which is found in Eupolis' *Goats*), *korē, koriskion*. *Korasion* is used but it is a mean word, as is *koridion*. The concrete noun is for the state of a virgin is *korikon*; but I do not admit this. Phrynichus the comic calls young girls *aphēlikas*: 'there were also *aphêlikes* women there'. Pherecrates calls an older woman *aphēlikesteran*, just as Cratinus calls an old man *aphēlika*. You will also say 'virgin in the season for marriage'. Aristophanes also says that girls who are of age 'are ripe as beans' (*kuamizein*):

[113] The text as we have it probably derives from an early epitome by Bishop Arethas in the tenth century, but it is still plenty ample enough to allow for general comments on structure. Lexicography was a vibrant intellectual industry, particularly in the Greek-speaking world: cf., e.g., Apion of Alexandria (1st cent. CE), *On Homeric language*; Apollonius (1st–2nd cent. CE), *Lexicon to Homer*; Aelius Dionysius of Halicarnassus (2nd cent. CE – not to be confused with his first-century homonym), *Attic words*, Erotianus (1st CE) *Collection of Hippocratic words*; Eudemus of Argos (2nd cent. CE?), *On rhetorical language*; Galen, *Explanation of Hippocrates' language*; Pseudo-Galen (2nd cent. CE?), *Names of plants*; Harpocration of Alexandria (1st–2nd cent. CE), *Lexicon of the ten orators*; Herodian of Alexandria (2nd cent. CE), *Homeric schematisms*; Moeris (2nd–3rd cent. CE), *Lexicon of Atticism*; Pausanias (2nd cent. CE), *Collection of Attic words*; Phrynichus 'the Arab' (2nd cent. CE), *Preparation for sophistry* and *Selection*; Ptolemy of Ascalon (2nd cent. CE?), *On the difference between words*. There is no comparable field in Latin, although some shared ground is covered by grammatical works.

[114] 'Simple classification is hard to achieve without the imposition of special value judgements like "higher" or "lower"' (McArthur (1986) 35).

'Others of them are ripe as beans, they are now nearly taking wing towards their husbands'. Lass, lassie, unmarried girl, nubile, newly-wed, woman/wife (*gunē*), woman hitched to a man (*ēndrōmenē*), woman fused with a man (*andri memigmenē*), young girl, girl of age, young woman, girl who has reached her age (*aphēbēkuia*), girl who has gone past her age (*parhēbēkuia*), one who is inclining towards old age, old woman, and (as in Isaeus) older woman, oldie, and (as in Theopompus the comic) aged woman lover of wine, drunkard, lusting after wine, perineum. The rest are the same as for men, such as on the edge of old age, weighed down with age and so forth. (2.17–18)

This entry is structured (notwithstanding an interlude in the middle) as a progression through a woman's life, from birth to death. The terms for very young and very old are common to women and men alike. Gender differentiation is presented as a feature of culture, not biology: it only begins when the child begins to be socialised, and ceases to apply in very old age. Maturation (*hēbē*) is closely linked to marriage, which is constructed as the natural goal of female existence. This connection is underlined in the quotation from Aristophanes, which is not necessary for lexical purposes: what it does, for Pollux, is to underscore the leap from the biological (ripening like beans) to the socially programmatic (taking wing towards husbands). But even the lexical point does cultural work: the reference to beans (*kuamoi*) may well demand comparison with other fruit and vegetable words used by the comedians to describe the firm body of a young girl.[115] In other words, the young girl is being presented in salacious, titillating terms, as a definitively embodied being, and available to the touch and control of the male subject.

Particularly interesting is the citation from Theopompus that describes the old woman: the quotation is designed, *prima facie*, to introduce the word *presbutis* (which I have translated 'aged woman'), but it brings with it a host of abusive adjectives activating the comic stereotype of the old drunk. These words are lexically superfluous; nor do they complete a metrical line (the quotation is ametric). The concluding word, κοχώνη, is a medical word for the perineum, also found in the comedians meaning 'arse' and (according to Jeffrey Henderson) 'almost always refers to anal intercourse'.[116] These debasing words serve, in Theopompus and Pollux alike, the function of abjecting the old woman, again in strikingly corporeal terms; and such abjection is a crucial social technology because old women are often free of husbands and thus uncontrolled (another comic motif).

Pollux's quotations are primarily introduced to anchor his discourse in the literature of the prestigious past: they are cited to exemplify lexical

[115] Henderson (1991) 149. [116] Henderson (1991) 200.

points, not for their content. But content cannot be so quickly deactivated: citation introduces intertextual pluralism, opens up possibilities for multiple hermeneutic alleys. It is notable that Pollux's woman is an almost exclusively comic creation (Eupolis, Phrynichus, Pherecrates, Cratinus, Aristophanes, Theopompus; only Isaeus strikes a different note), and this steers his reader towards a certain culturally enshrined vision of the pleasures and threats of females. Intertextuality is often presented as a form of interpretative pluralism and liberation (particularly from the tyranny of 'allusion and influence'), but in this case – and there may be a more general underlying point about all conceptions of 'freedom' – it is a carefully controlled, and indeed carefully controlling, mode of operation.

Pollux' work, then, is not simply a collection of miscellaneous synonyms: it provides an idealised map of society, a vision of *les mots et les choses* that performs and manipulates the paradigmatic relationships at the heart of Romano-Greek society. This lexicon is thus an archive in action: here you learn through words about the world, its deep structures and unspoken orders, its hierarchies, equivalences, symbolic parataxeis, and – not least – its subtle equivocations.

TEXTUAL REVOLUTIONS

During this period there also arose a series of revolutions in textual technology. Perhaps the most important was the appearance of the codex (or book), which gradually replaced the unwieldy papyrus scroll.[117] Scrolls were designed for information storage in libraries; they slotted neatly into horizontal slots, with an identification tag visible at the end. The codex, however, allowed for quick scanning back and forth across several pages – an early form of what we now call 'hypertextuality'. It is surely no coincidence that the earliest codices contained Christian and technical material, two genres of discourse that privilege, indeed insist upon, cross-referencing and non-linear reading.[118] The Christian Bible, in particular, was a text that many exegetes wanted to read hypertextually, as they grappled with its multiple authors and often conflicting demands. The Bible was, indeed, perhaps the first book conceived of as a textual embodiment of the cosmos. In the beginning was the word: God's language was the sacred transcription of the mysteries of the universe, of human society and mortality, of bodily suffering and spiritual redemption.

[117] Roberts and Skeat (1983).

[118] See Habinek (1998) 117–21 for a different argument, that the codex was closely associated with sub-elite identity, in contrast with the high-status connotations of the papyrus roll.

Another innovation – albeit less dramatic, perhaps – was the table of contents, which we find first in extant literature in the works of Scribonius Largus, the Elder Pliny, Aulus Gellius and Columella.[119] At first sight, we might want to take these TOCs as consolidation of a new movement under the empire towards what we have called 'hypertextual' reading. As Andrew Riggsby's contribution to this volume shows, however, Pliny's TOC is not pragmatically useful in finding one's way around this massive work; rather, it serves the rhetorical end of displaying his mastery. Yet although it would be hopelessly crude to see these devices as unilateral reflexes in response to a single, unified urge underlying the multifarious textual practices of the empire, it is clear that paratextual phenomena of this kind do begin to gain a hold in this period, and that they are intimately related to the intensification of activity associated with archival thinking.

ITEMISING KNOWLEDGE

Archival thinking encourages a specific approach to knowledge, as manipulable, discrete fragments. Like Propp's structuralism, Lévi-Strauss's mythography or Barthes's cultural semiology, the texts analysed in this volume characteristically conceive of their primary operation as the analysis of raw material (whether 'reality' or pre-existing text) by a process of *itemisation.* 'Knowledge' is to be conceived of as an aggregate of discrete particles that are to be subjected to a process of analytical ordering.

How can we conceptualise the relationship of this process of itemisation of knowledge to the imperial project? One metaphor that recurs in this book is that of 'mapping' knowledge.[120] This is not an innocent image: the map is a central image of Roman imperial rhetoric, a taxonomy of the subject states of the global empire. Not all scholars are equally convinced that Agrippa's famous dedication (Pliny, *Natural history* 3.17) was a map in our sense,[121] but it certainly itemised the nations of the world. Similarly, the preface to *The achievements of the divine Augustus* (*Res gestae*) – an inscription copies of which were set up throughout the empire – speaks of the emperor's subjection of 'the whole world' (*orbem terrarum*) to Roman *imperium.* The extraordinary fact of the subsumption of the multifarious

119 Pliny does, however, claim a precedent, for Latin literature at any rate (*in litteris nostris*), in the *Epoptides* of the late-Republican author Valerius Soranus, on whom see Holford-Strevens (2003) 30–1.

120 Cf. Murphy (2004) 131–7 on the metaphor of mapping in Pliny's *Natural history*.

121 See most recently Brodersen (1995) 275–8 (arguing against the map), and more generally Nicolet (1991).

nations of the earth under a single political framework was a repeated motif of imperial ideology, whether in the artistic representations carried in triumphs,[122] or more formally in the development of world-maps. The relationship in a map between discrete parts and architectonic whole is simultaneously 'imperial' and 'archival': 'unity in diversity' is the rhetoric common to both modes of thinking.

The manner of imperialism, though, was not the sole prerogative of the emperor: through the complex 'chandelier' structure of imperial bureaucracy, patronage and influence, it percolated down all the way to provincial aristocrats. In the atria of Roman noblemen across the empire, looted artworks signalled at once their exoticism and their submission to the necessity of the new economic, political and military order. Art galleries, such as we encounter in Petronius' *Satyricon* and Philostratus' *Portraits*, encourage the viewer to respond both to the individual artwork and to the act of assemblage that has created (and resourced) the viewing space. From the late Republic, meanwhile, we find the Roman nobles collecting books and hoarding them in purpose-built libraries in country villas: trophies of conquest, now largely tamed of their political and ethical urgency and consigned to the leisure space of the dominant elite.

The archive controls time as well as space. The rows of statues in the forum of Augustus, for example, present a historico-temporal 'map' of ancestors, simultaneously individuated and unified in the service of their support for Augustus. Processions performed a similar rhetoric of universalism, whether the imperialist processions of the Hellenistic kings (most memorably that of Ptolemy Philadelphus, as described by Callixenus) or, on a smaller and more local scale, Roman aristocratic funeral processions, with their displays of ancestral masks.[123] It is this capacity to control the representation of space and time, to figure its complex diversity in a single, appropriative space, that hallmarks imperial power.

It is this same imperialist impulse that underlies much of the intellectual habit of knowledge collection under the empire. Large-scale compendia of knowledge (e.g., Seneca's, Pliny's and Apuleius' *Natural histories*, Pamphila's *Collection of historical reminiscences*, Favorinus' *Miscellaneous history*, Aulus Gellius' *Attic nights*, Aelian's *Miscellaneous history*) perform, in their encyclopedic accumulation of diverse phenomena, the aggregative rhetoric of empire. Athenaeus' *Sophists at dinner*, most notably, is dramatised (in emulation of Plato's *Symposium*) at the dinner table of a wealthy Roman

[122] E.g., see Murphy (2004) 154–60.

[123] See Rice (1983) for Callixenus; Flower (1995) on Roman funeral masks.

patron, Larensis, 'who outdid all of those who have inspired awe in their collection of books' (3a).[124] Larensis' personal library, along with its textual transcription (the *Sophists at dinner* itself), emblazons not only his cultured refinement but also his power within the imperial hierarchy.[125]

Edward Said writes as follows of nineteenth-century French imperialism:

> the power even in casual conversation to represent what is beyond metropolitan borders derives from the power of an imperial society, and that power takes the discursive form of a reshaping or reordering of 'raw' or primitive data into the local conventions of European narrative and formal utterance, or, in the case of France, the systematics of disciplinary order.[126]

The brilliance of Said's various analyses of orientalist discourse lies in their ability to show exactly how politically active the concept of knowledge is, and particularly how the organisation of knowledge relates to the organisation of empire. Knowledge of the East, for Said, is driven by what Foucault would call the 'historical *a priori*' of empire. It is easy to see how parallels with Romano-Greek ethnography might be drawn: figures like Strabo, Arrian, Ptolemy and the *periplous* writers could be absorbed into the post-Hartog school of analyses of 'the other' (and, indeed, this process is already well underway).[127] The 'conquest of the world' (Roman imperialism never sacrificed the rhetoric of world-empire to the truth, which was less flattering and certainly messier) evidently relates closely to both the literal mapping of the world and its symbolic mapping through the discourse of ethnography.[128]

But this volume stakes the case for something else, something both deeper and thicker. It is not, ultimately, the *ideality* of knowledge that interests us here, so much as its *embodiment*: the modes of selection, the processes of aggregation, the formal techniques for its presentation, the cultural meaning of the work that lends it its flesh. The two are, of course, mutually reciprocal: there is no matter without form, as Aristotle would say, just as there is no form without matter. The aim of this volume, however, is to rectify an imbalance. Classical scholarship has traditionally privileged

124 For Larensis, see Braund (2000a).

125 According to Athenaeus, Marcus Aurelius had set Larensis in charge of 'temples and sacrifices' (ἱερῶν καὶ θυσιῶν, 3c); Braund (2000a) 6 argues that Larensis was in reality only a *pontifex minor*, the status of which was 'rather less than the glorious supervision of Roman religion suggested in the eulogy'. The extent to which Athenaeus' characters map onto real individuals, however, is radically uncertain.

126 Said (1993) 119.

127 On 'Greeks and barbarians' in the Roman period, see esp. Schmidt (1999); Hartog (2001).

128 Momigliano (1974); Nicolet (1991).

idealised knowledge, because it has found the myth of disembodied knowledge so congenial to its own aspirations. The essays in this volume seek to re-embody knowledge within the rich cultural context of the early Roman Empire. Our central proposition is that there is a 'discursive form' (to borrow Said's phrase) of knowledge that is characteristically imperial, which is to say the typical modes of operation of the archive: it rests upon itemisation, analysis, ordering, hierarchisation, synthesis, synopsis.

'Imperial', of course, does not necessarily mean 'pro-imperial': the opposition between consolidation and challenging of society is too crude. An author like Lucian offers a number of archival, 'synoptic' figures that mimic the rhetoric of empire, such as the auction of philosophers in the *Sale of lives* and the aerial view in *Icaromenippus*. But it would be hopelessly simplistic to decide on this basis that he was either 'supporting' or 'criticising': as we have seen above, his knowing satire is subtle and dangerously broad-ranging. Other modes of knowledge, particularly philosophical and theological, simultaneously borrow the symbolic force of imperial-colonialist rhetoric and seek to transcend it: Philostratus' *Life of Apollonius*, also discussed above, is an excellent example. Still other authors express the frail inability of language to match the diversity of the world: Pollux' word book, for example, does not include 'every word' (πάντα τὰ ὀνόματα), for it is not easy to 'gather together' (συλλαβεῖν) 'everything' (πάντα) into a single book (1.2). Rhetorical *recusatio*, for sure; but in an address to the emperor (as his dedicatory epistle is), Pollux is conspicuously rendering unto Caesar what is Caesar's. Moreover, the fact of cultural and linguistic differences between Romans, Greeks and other groups means that this volume is best viewed as a complex tessellation of interrelated responses and reactions to the imperial order. The ordering of knowledge is an 'imperial' phenomenon in that it is mired in the real world of the Roman Empire – but that observation does not mean that we should confine ourselves to crudely characterising texts variously as 'consolidatory' or 'subversive'.

The process that we are describing as the 'ordering' of knowledge, thus, operates on two levels. First, knowledge is commissioned, impelled, commanded, by or in competition with the authoritarian edicts of empire. Secondly, however, and no less importantly, it is also given its own distinctive matrix of intratextual 'orders': items of knowledge are isolated, lemmatised, structured, ranked. It is the interaction between these two levels, the political and the textual, that forms the substance of this book.

As we have stressed, the Roman Empire was not the first example of a culture where the organisation of knowledge reflected (in the complex ways we have described) the political order: while many of its practices were distinctive (or gained new significance from their new context), many others were inherited from earlier Rome or adapted from Hellenistic Greece. Nor was it the last. As John Henderson's chapter shows, mediaeval Europe took over Rome's rhetoric of intertwined knowledge and empire and put it to new uses in the service of the Christian God. Rome's imperial legacy to the mediaeval world (and indeed beyond) was not confined to the spheres of politics, economics and warfare: she also showed that an empire must be an empire of knowledge. What the essays in this volume show is just how powerful, intense, multifarious, durable and – above all – intellectually captivating was this society's engagement in the ordering of knowledge.

PART II

Knowledge and textual order

CHAPTER 2

Fragmentation and coherence in Plutarch's Sympotic Questions

Jason König

READING MISCELLANISM

This volume attempts to draw out some of the ordering principles which lie beneath the surface of the Roman Empire's compilatory writing. The difficulty of identifying any such principles is particularly acute for works which have a strongly miscellanistic quality. I should say at the outset that it is hard to isolate any clearly bounded ancient genre of the 'miscellany'. It seems more fruitful instead to recognise the recurring presence of a range of miscellanistic characteristics across many different kinds of writing. Miscellanistic works – in the sense in which I understand that term here – are marked primarily by the disparateness of the material they accumulate. In some cases that quality of disparateness is supplemented by other markers: for example, many miscellanistic texts claim that their primary aim is to give pleasure to their readers', rather than to instruct or to be comprehensive; many make claims about the randomness of their own structures. Sometimes, for sure, all of these characteristics are combined with each other. Moreover, in some cases we find authors situating their own texts in relation to other miscellanistic writing. For example, Aulus Gellius, *Attic nights* pr. 4–10, not only chooses a title which evokes the idea of variety (the many different nights the author has spent in reading and compiling), but also compares his title with the titles other miscellanistic writers have chosen, in a way which suggests a high degree of self-consciousness about his work's place among a series of other similar texts.[1] At other times, however, these miscellanistic characteristics find their way in a diluted form into works

[1] Vardi (2004) usefully discusses the difficulty of defining any genre of 'miscellanism', while also at the same time mapping out some of the recurring tropes of miscellanistic writing in Gellius' preface and elsewhere. It is worth noting, however, that even Gellius, who is one of the ancient writers who comes closest to identifying a genre of miscellanism and identifying his own work as part of it, insists on undermining that identification even as he gestures towards it, since one of his main aims in this preface is actually to distinguish his own work from the others he lists, which he criticises for their excessive bulk (e.g., Gell. *NA* pr. 11–12).

which fit (similarly fluid) categories like encyclopedic or technical writing. In that sense I hope the problems this chapter raises will have resonances for a wide range of different kinds of compilatory writing, not only for those who make it into Gellius' list of rival miscellanists.

How can we make sense of writing which is apparently marked by lack of system and lack of order? There are many possible approaches: one might look, for example, for underlying ideological coherence – a sense that disparate material is unified through being imbued with distinctive ways of viewing the world; such analysis might reveal the unseen effects of particular ethical priorities or particular assumptions and anxieties about hierarchies of social status, gender or cultural superiority (as argued for Pollux's lexicographical compilation in the introduction to this volume). One might also look for recurring images and thematic patterns lying beneath the apparently chaotic surfaces of these texts – despite the fact that they so often claim not to have any such patterning. We should perhaps be cautious of that approach: the gesture of rehabilitating texts on the grounds of their thematic coherence is in some ways a relic of old-fashioned literary criticism,[2] and there is an obvious danger of anachronistically mapping our own critical preoccupation with making sense of ancient literature on to ancient readers. I argue here, however, that the idea of thematic order does nonetheless have some applicability for the miscellanistic writing of the Roman Empire. Many ancient miscellanists, I suggest, gesture towards thematic order, drawing us into a search for patterns while also at the same time disrupting and frustrating that search. On that argument, the claim many miscellanists make, that they are composing at random, turns out, at least in some cases, to be a matter of convention, a miscellanistic pose which can hide careful structuring beneath it.[3] Perhaps most importantly, one might think about the way in which disparate material may be unified by a consistent methodology of reading. In particular, the image of the active reader, who must use his or her reading as a resource, a starting-point for his or her own coherent philosophical development, is a common one

[2] E.g., see Eagleton (1996) 40–4 for a convenient account of the importance of coherence for the New Criticism of the mid-twentieth century.

[3] For claims about random composition, see, for example, Gell. *NA* pr. 2–3, discussed by Holford-Strevens (2003) 34, who cites a number of parallels, including Pamphile (attested by Phot. *Bibl.* 17: 119^b 27–32), Clem. Al. *Strom.* 6.2.1, Plin. *Ep.* 1.1.1; Pliny's claim in particular has been shown to be dubious: see Sherwin-White (1966) 21–3 and 42–51; cf. Vardi (2004) 169–79 who draws a contrast between the genuinely random structure of Gellius' miscellany, and other miscellanistic works where we find much clearer signs of thematic grouping (with brief mention (169–70) of Plutarch's *Quaest. conv.*, along with works by Athenaeus, Macrobius, Clement and Solinus). Cf., p. 62, below, for discussion of the disingenuous nature of Plutarch's claims about the randomness of his own composition in *Quaest. conv.*

in ancient philosophical literature. Here one of the most obvious Imperial examples – albeit not a miscellanistic example – is in the work of Galen, who often represents his medical writing as provisional, stressing the fact that each reader must reach a full understanding of each individual subject, and of the medical art as a whole, for him- or herself, via proper application of logical method.[4]

This chapter takes Plutarch's *Sympotic questions* (*Quaest. conv.*) – an enormous accumulation of dinner-party conversations on scientific, literary and sympotic topics, recorded accurately, so Plutarch claims, from several decades of symposium-going – as a test-case for those approaches. I want to suggest that this work exemplifies all of the different kinds of order outlined in the previous paragraph. I also want to suggest, however, that Plutarch is in some ways highly untypical, especially in the degree to which he is self-conscious about his own project of conjuring order from diversity.[5] More specifically, I argue that the *Sympotic questions* does offer us, contrary to first impressions, a carefully orchestrated vision of how we can draw coherence out of its own fragmented aggregation of material, if only we read with proper philosophical attention. In order to achieve that effect, it draws on models of how to read which are carefully theorised elsewhere in Plutarch's oeuvre (more on that in the next section). The *Sympotic questions* prompts us to read actively – in other words to respond creatively and philosophically for ourselves to the many different questions under discussion, and to stay alert to the recurring themes and patterns of the

[4] E.g., see Gal., *Thras.* 3–4 for one good example of that.

[5] The *Quaest. conv.* had demonstrable influence over later miscellanism, but none of its imitators quite matches Plutarch's fascination with the tension between order and disorder: see Gell. *NA* 3.6 and 17.11 for essays which take their material from the *Quaest. conv.*; and cf. n. 3, above, for Vardi's argument that Gellius on the whole resists the underlying coherence of the *Quaest. conv.*; however, see also Morgan (2004) on the underlying ethical coherence of Gellius' work; also Gell. *NA* pr. 16–18, where Gellius emphasises, like Plutarch, his hope that the reader will be inspired to personal reflection and improvement by his reading of the work, a passage which shows some traces of Plutarchan requirements for the reader to create his or her own coherence. Macrobius draws on the *Quaest. conv.* heavily in *Saturnalia* book 7, but he is much less interested than Plutarch in showing his guests indulging in inventive speculation (e.g., the Greek guests in the *Saturnalia* are repeatedly criticised by other speakers for their ingenuity and inventive styles of argumentation (e.g., 7.5.1, 7.9.9, 7.16.1)). At first sight, he seems to fall far short of Plutarch's ideals of active reading (i.e., the idea that each individual – both the symposium guests and the reader of the *Quaest. conv.* – should value the *process* of thinking creatively more than getting the right answer); on closer inspection, however, it becomes clear that Macrobius is committed to the principle that verbatim quotation of the literature of the past is quite compatible with creative, original, personally distinctive expression: 'language, for Macrobius, was what the present user made of it, even though the thoughts and expressions of the present were inseparable from what had been thought and written earlier by others' (MacCormack (1998) 82). In that sense, as for Gellius, we may be seeing the traces of a Plutarchan insistence on the way in which the interpretations of the individual reader or sympotic speaker brings a kind of order to diverse material.

texts. Plutarch also shows us his fellow dinner-guests learning that style of active response for themselves, using the topics they discuss as springboards for personal response, as stepping-stones in their philosophical lives. The work demonstrates, in other words, how processes of universally relevant philosophical enquiry can start from frivolous snatches of conversation. In that sense, it follows the principle stated in Xenophon, *Symposium* 1.1, that the true philosopher can do philosophy anywhere.[6]

In addition, I also argue that Plutarch hints at parallels between those patterns of philosophical learning, and the organising patterns of social and political life in Roman Greece. Plutarch sets all of these discussions on specific occasions, many of them in specific cities, contexts which are briefly but vividly sketched in their opening lines. In doing so, as we shall see, he not only foregrounds the links between fragmented conversational subject matter and all-empowering philosophy, but also, in a way which is closely parallel with that, insists on the power of fragmented local identities within the all-embracing political and philosophical culture of the Roman Empire. It is a vision of overarching Greek culture as something which depends on and encompasses local specificity, and which is in tune with the prominence Plutarch gives elsewhere to the intertwining of local identity with philosophical cosmopolitanism within his own life.[7] And that vision, as we shall see in the final section of this chapter, frames and enhances his insistence on engagement with detail in the quest for overarching philosophical knowledge.

What implications does that parallel have for our understanding of Plutarch's view of the cultural and political hierarchies of his own contemporary Greco-Roman world? We have suggested in our introduction that the archival patterns of thought which map unity through diversity may be fundamentally 'imperial' patterns, developed in the service of empire. We have also suggested that they are available to be reshaped in ways which subvert or redirect the rhetoric of imperial dominance. Plutarch's use of the themes of unity and diversity is one such reshaping, based on the conviction that the final unified framework within which the fragmented diversity of the world can most powerfully be contained will be a philosophical one. And that philosophical framework, he suggests, finds not only its most

[6] The question of whether it is right or possible to combine philosophical speech with the playful atmosphere of the symposium is the subject of both the preface and first dialogue of book 1; in the preface (612d) Plutarch justifies that combination with reference to the philosophical symposia of Plato and Xenophon and others.

[7] Cf. Plutarch's *Greek questions*, where his exploration of Greek tradition takes the form of inquiry into obscure local customs and local terminology.

fertile ground but also its most powerful guiding metaphor in the Panhellenic interweaving of local commitment with overarching Greek identity. That is not to say that Plutarch's philosophical project in the *Sympotic questions* is insulated from the realities of Roman power. On the contrary, he is obsessed with its capacity to encompass and explain Roman culture as well as Greek,[8] and with the many things it has in common with non-Greek thought, Roman, Egyptian and otherwise.[9] But it is nevertheless strongly marked as a Greek project, dependent upon patterns of thought whose basic technique of seeking unity in diversity resembles the unity-in-diversity of Greek Panhellenic experience.

PLUTARCH ON READING

Plutarch is repeatedly interested in giving us guidelines for proper philosophical response to texts and speeches. That insistence on personal response as a central part of philosophy, is one of the things which unites his many writings – whether historical, scientific, ethical – as part of a broader philosophical project. The text which lays out those principles in most detail is Plutarch's *On listening*. In the traditional order of the *Moralia* the work comes close to the beginning of the collection, preceded only by *On the education of children*, and *On how the young man should listen to poetry*. Whoever arranged these treatises seems to have seen these three works as programmatic and interconnected, moving as they do from the techniques of education and interpretation suitable for the very youngest children, through to the approaches which are appropriate for young men, and indeed all men, once they graduate to proper study of philosophy. That assumption of coherence is in some ways unconvincing, not least because the first work, *On the education of children*, is generally believed to be by someone other than Plutarch.[10] But there are clearly signalled overlaps between the second and third works in the series, on poetry and listening respectively. The work on poetry suggests strategies of reading suitable for the young, who listen to poetry before they graduate to philosophical subject matter, and who should accustom themselves to reading creatively, imposing ethically edifying interpretations even on passages which at first sight seem unsuited to such interpretation, in order that they will be more prepared for philosophical ideas once they are exposed to

[8] E.g., see Preston (2001), Boulogne (1992) and (1987) on those themes in the *Greek and Roman questions*.

[9] E.g., in his work *On Isis and Osiris*.

[10] See Whitmarsh (2001) 98–100 for brief discussion.

them.[11] *On listening* deals with the next step on that path, as the very opening sentence of the work suggests in offering advice to a young man named Nicander, who has just reached adulthood, with the freedom to manage his own education which that implies. Plutarch emphasises first (*On listening* 1 (37c–e)) the need for Nicander to take reason as his controlling guide, rather than revelling in the sense of freedom from guidance which adulthood might be thought to bring with it. He then suggests (2 (37e–38a)) that Nicander will be familiar with philosophical reasoning already because of the way in which his early training has been saturated with it. Clearly the addressee is envisaged as someone who has been brought up according to the precepts of *On how the young man should listen to poetry*; the techniques recommended in *On listening* are part of a lifelong project of philosophical education.

After this prefatory address to Nicander, Plutarch then stresses both the benefits and the dangers the sense of hearing can bring with it, arguing for a style of listening that is obedient and attentive, but also at the same time selective and sceptical.[12] The whole of the rest of the dialogue is dedicated to illustrating those principles, and above all to demonstrating the way in which listening should be an active process, which involves responding for oneself to the arguments one has heard. It is a technique which may not come easily to the young, he explains, but which can be developed with perseverance (17 (47b–d)): 'For the mind is not like a vessel in need of filling, but rather, like wood, needs only a spark to kindle it, to produce an impulse towards inventiveness, and a desire for the truth' (18 (48c)).[13] Passive, unreflective listening, by that standard, can never be adequate for anyone who aspires to philosophical progress.

One of the work's many striking features – which hints at the relevance of these principles to the *Sympotic questions* – is the recurring presence of the symposium as a point of reference. For Plutarch, in this text at least, the symposium is both an imagined context for the styles of listening and response he recommends, and at the same time an important metaphor for those styles. In 6, for example, he suggests that one should listen affably, 'as though one is a guest at a dinner or a festival banquet', in other words not in a spirit of rivalry, but also not in a way which buries one's capacity for criticism:

[11] See Whitmarsh (2001) 49–54; cf. Zadorojnyi (2002) for the argument that this stress on ethical response in *On how a young man should listen to poetry* is Platonic in character.

[12] See Goldhill (1999) 106–7 for brief discussion.

[13] οὐ γὰρ ὡς ἀγγεῖον ὁ νοῦς ἀποπληρώσεως ἀλλ' ὑπεκκαύματος μόνον ὥσπερ ὕλη δεῖται, ὁρμὴν ἐμποιοῦντος εὑρετικὴν καὶ ὄρεξιν ἐπὶ τὴν ἀλήθειαν.

> When speakers are successful, we should assume that they are successful not by chance or by accident, but rather through their care and hard work and study, and we should imitate these qualities, feeling admiration for them and envy. When speakers make mistakes, on the other hand, we must turn our minds to the question of what the reason for the error was and where it came from. (6 (40b))[14]

Similarly in 10 we hear that we should be willing to listen respectfully, but also be ready to contribute problems for discussion when that is appropriate, just as an ideal symposium guest would do. And in 14 Plutarch explains that we should avoid the temptation of passive listening, like those who sit back and enjoy themselves at a dinner party while others do the work. Rather we must work together with the speaker, criticising our own arguments as much as his. Mutual respect and co-operation between listener and speaker are the hallmarks of Plutarchan listening, as they are of all sympotic conversation. And these skills of responsive, self-reflexive interpretation are precisely the things which allow us to draw together the varied impressions our experience of the world confronts us with, just as they allow us to draw together in a morally coherent way the varied material of Plutarch's oeuvre, and the ostentatiously varied miscellanism of the *Sympotic questions*.

What relevance do these principles of responsive reading have, in practice, for Plutarch's massive enterprise of knowledge aggregation? For one thing they hint at ethical significance lying behind Plutarch's agglomerations of detail, which have the potential to spark self-reflection and morally admirable lifestyle in the responsive reader. That is most obvious in his collections of historical material, both in the *Lives* and elsewhere, with their focus on the deeds and sayings of individuals, which offer both positive and negative examples for the reader to decipher and assess. A number of these historical compilations actually underline the disjointed nature of the excerpted material they present us with, and yet at the same time prompt us to see an underlying potential for unity. In the prefatory letter of his *Sayings of kings and commanders*,[15] for example, Plutarch draws attention to the way in which the emperor Trajan will be able to read these snippets briefly and yet also profitably: 'taking away from these brief words (ἐν βραχέσι) the opportunity for reflection (ἀναθεώρησιν) on many men who have been worthy of memory' (172e).[16] Those closing phrases of the work's

[14] τοῖς μὲν οὖν κατορθουμένοις ἐπιλογιστέον ὡς οὐκ ἀπὸ τύχης οὐδ᾽ αὐτομάτως ἀλλ᾽ ἐπιμελείᾳ καὶ πόνῳ καὶ μαθήσει κατορθοῦται, καὶ μιμητέον ταῦτα θαυμάζοντάς γε δὴ καὶ ζηλοῦντας· τοῖς δ᾽ ἁμαρτανομένοις ἐφιστάναι χρὴ τὴν διάνοιαν, ὑφ᾽ ὧν αἰτιῶν καὶ ὅθεν ἡ παρατροπὴ γέγονεν.

[15] See Beck (2002) for arguments in favour of viewing the prefatory letter as Plutarch's own work.

[16] ἐν βραχέσι πολλῶν ἀναθεώρησιν ἀνδρῶν ἀξίων μνήμης γενομένων λαμβάνοντι.

prologue draw attention to the paradoxical combination of brevity with lasting value, whose attainment will be dependent on the reader's capacity to respond through active 'reflection' (ἀναθεώρησις), a word Plutarch uses similarly elsewhere to describe the most desirable kind of reflective response to reading.[17] In the prologue to *Bravery of women*, similarly, Plutarch proclaims both the disjointed nature of his narrative, and at the same time the need to look for a defining essence of bravery which underlies the superficial differences between the many examples he is presenting us with, and which is the same for women as for men, though it may not at first sight seem so.[18] Coherence comes, then, in part from the capacity of disparate material to be interpreted within a consistent moral framework.

Secondly, and perhaps less obviously, it has increasingly been recognised that Plutarch embeds the requirement for personal response in the very form of his writing, forcing us to take up the provocative challenges of interpretation precisely through his arrangement of material. In the *Lives*, for example, the final passages of *synkrisis* – where the pairs of biographical subjects are compared with each other at length, after they have been individually biographised – not only prompt reflection on similarities and differences between the men in question, but also force us to reassess each of them individually, through their frequent inconsistency with the material we have already encountered.[19] In that sense the ordering of the work's details is very far from being neutral and artless, but rather makes a central contribution in provoking response.

LEARNING TO READ IN THE *SYMPOTIC QUESTIONS*

The *Sympotic questions*, I will argue here, is the among most intricate and self-conscious of all Plutarch's actualisations of those principles. And yet, despite that, the text has frequently had a bad press.[20] The negative attention it has received is typical of common criticisms of encyclopedic and miscellanistic writing. Plutarch's arguments, for example, are branded ineffective, even frivolous. François Fuhrmann, not untypically, laments as follows:

[17] E.g., the same word is used in *Quomod. adul.* 19e to describe the process of creative reading, which goes beyond face value in its search for meaning in a text.

[18] Plut. *De mul. vir.* 243b–d. For the general point, see McInerney (2003) on the way in which Plutarch's new understanding of conjugal relations emerges (but only partly) from beneath this apparently disparate collection of conventional moralizing material.

[19] See Duff (1999), esp. 243–86.

[20] For an important exception to that, see Romeri (2002), esp. 109–89, who analyses at length the way in which Plutarch privileges speech ahead of consumption, drawing on Platonic precedents, in this work and others.

Au lieu de chercher les causes véritables des phénonèmes, Plutarque se contente en général de la vraisemblance, en citant plusieurs théories qui s'y rapportent, ou en rappelant ce que divers auteurs en ont dit. Les différentes opinions se succèdent ainsi sans aucune analyse et le plus souvent sans solution, comme si ceux qui sont chargés de les défendre s'amusaient avec elles.'[21]

In this case, the fault is attributed not so much to Plutarch himself as to the generic assumptions he is working with[22] and to the 'affaiblissement général de l'esprit scientifique',[23] an assumption which exemplifies a common failure to understand the rhetorical idiom of so much ancient scientific writing.[24] And second, closely related to that criticism, is the suggestion that Plutarch's main interest is in the indiscriminate amassing of information. Michel Jeanneret, for example, categorises Plutarch with Athenaeus and Macrobius, as writers who aim for quantity and variety of material, in an 'orgy' of erudition, rather than seeking narrative realism or convincing argumentation.[25] Both of those criticisms, I suggest, underestimate more than anything the importance of Plutarch's self-conscious exploration of the activities of reading, listening and interpreting within this work. And both of them are criticisms to which the *Sympotic questions* has powerful in-built replies.

The first point to make is that the *Sympotic questions* shows us how comprehensive erudition can be adapted for specific social situations, through the symposiasts' capacity for creative manipulation of their wide reading. Knowledge in the Plutarchan symposium is always a performance.[26] In that sense, the *Sympotic questions* resists commonly stated modern assumptions that the project of compiling knowledge in textual form, and the practice of exhaustive reading, are faceless exercises of indiscriminate absorption and accumulation.[27] In addition, Plutarch draws on the traditional status of the symposium as a space for elite initiation in representing these conversations as occasions for himself and his fellow symposiasts to learn the distinctive

[21] Fuhrmann (ed.) (1972) xxiv, quoted approvingly by Teixeira (1992) 221; and by Flacelière and Irigoin (eds.) (1987) lxxxiii. Cf. similar criticisms elsewhere, e.g., Barrow (1967) 21.

[22] See Fuhrmann (ed.) (1972) xxiii. [23] *Ibid.*: 'cette "triviality" était, hélas, le lot du genre'.

[24] On the rhetorical character of Imperial scientific writing, see esp. Barton (1994b).

[25] E.g., see Jeanneret (1991) 166–7; for criticism of Jeanneret's assumption, see Relihan (1992) 218.

[26] Martin (1998) discusses the way in which performance, often within a sympotic context, is a central part of the wisdom of the seven sages of Greek tradition; Plutarch's engagement with that tradition is clear from his work *Symposium of the seven wise men*, which depicts the seven sages drinking and talking together, and which has many similarities with the guiding principles of sympotic, philosophical discussion in the *Quaest. conv.* – especially in those passages where the sages take it in turns to offer opinions on ethical and political problems – as Romeri (2002) 109–89 shows.

[27] Cf. n. 29, below, for Jeanneret's claims about the facelessness of the *Quaest. conv.* and other sympotic compilations.

styles of ingenious analysis with which the work is saturated, not only by listening but also by responding in a spirit both of imitation and of friendly rivalry to the conversations they hear. For example, he repeatedly returns to the scene of young men learning appropriate styles of speech from their older companions, or of Roman readers working hard to learn and participate in Greek styles of speech.[28] In doing so, he draws attention to his own involvement,[29] and the involvement of his Roman addressee, Sosius Senecio, in many of the conversations he describes. Often, for example, as we shall see further in the next section, Plutarch is himself the figure who speaks last and most authoritatively, as if to set an example to the younger or less experienced men who have spoken before him. At other points he takes us back to the symposia of his youth, for example in book 9, where we see Plutarch as a star pupil in the skills of sympotic conversation, trumping his fellow students in front of their great philosophical mentor, Ammonius. We, too, are offered instruction, both in the prologues, where Plutarch often lays out explicit recommendations for habits of learning and speaking; and also, implicitly, in the models for action which are presented to us in the conversations themselves. If the young symposiasts are to learn from the example of watching and responding to the arguments of their elders – as Plutarch recommends repeatedly in his work *On listening* – it seems hard to avoid the impression that we are being prompted to engage with those models in similar ways ourselves through the act of reading.[30] Learning, for these young symposiasts at least, works by repetition. The recurring rhythms and gestures of sympotic conversation become ingrained in them through repeated exposure. And the repetitions of Plutarch's text invite us to share in that experience.

What, then, are the defining features of the style of speech which is on display? Most distinctively of all, it is a style of speech which aims for a variety of different explanations for each question which is proposed. The questions under discussion tend to arise from the circumstances of the symposium, as the symposiasts comment on recent events, on their surroundings, or on the running of the symposium they are attending.

[28] See Swain (1990) 130–1.

[29] Claims that Plutarch takes a back seat in this work could hardly be more wrong: e.g., see Barrow (1967) 15 and Jeanneret (1991) 167, who writes that 'the author melts into an anonymous collector and mediator'. For good examples of discussions where Plutarch makes his own contribution the climax of the discussion, see (in addition to those discussed below): *Quaest. conv.* 1.9, 5.2, 5.4, 6.4, 6.5, 6.6, 7.5.

[30] Cf., e.g., Swain (1996) 138: 'A key part of Plutarch's plan for moral improvement, with the aim of constituting one's life according to philosophy, was the observation of others', with several examples from the *Moralia*.

Plutarch and his fellow guests then take it in turns to propose solutions.[31] They quote repeatedly from their wide reading, in a way which often leads to juxtaposition of competing explanations from different authorities: Plato and Aristotle figure most often, but they share the stage with a dazzling range of other authorities. In addition, the speakers often speculate on their own account, with varying degrees of plausibility and ingenuity. Originality and ingenious, improvised speculation seem to be valued almost as much as exhaustive knowledge of earlier writing.[32] The use of alternative explanation as an interpretative strategy was of course far from being unusual. It was enshrined most influentially in a number of Aristotelian works (esp. Pseudo-Aristotle, *Problems*), which raise scientific problems in question form, and then proceed to answer them with one or more possible explanations.[33] It is also widespread within the aetiological and scientific work of many of Plutarch's contemporaries, and Plutarch himself uses variations of it in a great many of his own works.[34] There is also evidence from the Hellenistic period and later for specific association of this style of analysis with learned sympotic writing and styles of sympotic speech.[35]

What effects does Plutarch achieve through his traditional but also unusually vivid emphasis on this strategy of 'interpretative pluralism'?[36] Plutarch's use of it signals his alignment with the precedent of the Pseudo-Aristotelian *Problems*, casting his own work as a version of Aristotle's projects of systematising and advancing a great range of different areas of human knowledge. It also signals his alignment with some of Aristotle's successors. Theophrastus, for example, in his meteorological work, repeatedly accepts a variety of possible explanations for one single phenomenon. In doing so, he not only gives an impression of comprehensiveness, showing that he has

[31] Cf. Jacob (2001), esp. xxx–iii on Athenaeus' very similar use of the technique of *zētēsis* (which he suggests may date back to the symposia of the Mouseion of Alexandria (lxxii)), whereby one guest proposes a problem, which is then answered by others; the difference is that Athenaeus' speakers rely more heavily than Plutarch and his fellow-guests on quotation of texts recalled from memory, and less on the kind of ingenious personal response which so often follows on from the quotation of past authorities in the *Quaest. conv.*

[32] The *Quaest. conv.*'s distinctive style of ingenious conversation is best analysed by Frazier and Sirinelli (eds.) (1996) 177–207.

[33] Ps.-Aristotle, *Problems* are quoted and argued over in *Quaest. conv.* 1.9, 3.7, 3.8, 3.10, 6.8, 6.9, 7.5, 8.3 and 8.10; see also Boulogne (1992) 4689, n. 47, citing *Metaph.* 983b4–6, 995b1–3.

[34] In that sense Feeney (1998) 129 is surely wrong to characterise alternative explanation as an exclusively Latin technique; in doing so he mentions that alternative explanation features in Plutarch's *Roman questions*, which analyse Roman customs, but not in his *Greek questions*; that distinction ignores the prevalence of this technique elsewhere in Plutarch's work: e.g., see *Natural questions* (discussed by Harrison (2000)), *On the E at Delphi*, *On the sign of Socrates* and *On Isis and Osiris*; see also Pailler (1998) on the recurring presence of the question within Plutarch's lives of figures from the archaic period.

[35] See Cameron (1995) 71–103 (esp. 103).

[36] See Hardie (1992) 4754 for that phrase.

investigated a range of possible causes; he also emphasizes the importance of the four elements – earth, air, fire and water – for his view of the workings of the universe, offering one explanation for each element.[37] It also has some overlaps with methods of analysis which were common in Epicurean[38] and Pythagorean[39] philosophy. In addition, it was closely related to common patterns of argumentation which were ingrained in rhetorical theory, and which would presumably have come as second nature to Plutarch and many of his readers, given their likely saturation in rhetorical training.[40] And it was well suited to express the speculative, agonistic idiom which lay behind much ancient scientific reasoning, and which seems to have arisen – at least originally – from the practice whereby different experts would offer competing explanations for the same phenomenon in public contexts.[41] In the *Sympotic questions* that point takes on added complexity through the presence of individuals from a wide range of different professions, so that the variety of different responses is in a number of places represented as a vehicle for productive comparison between different professional viewpoints.[42] Perhaps most importantly, it offers the opportunity to bring different authors of the past into dialogue with each other and with the symposiasts of the present,[43] allowing for comparison between different explanations and different principles of explanation, and in the

37 See Taub (2003) 121–4; also 151 on similar techniques in Seneca's meteorological writing.

38 See Hardie (1992) 4761; Asmis (1984) 321–36. Epicurean theory holds that all explanations are equally valuable, the main aim of explanation being to remove superstition by showing that a number of plausible rational explanations exist; in some of his works Plutarch rejects that assumption, tending to hierarchise his alternative explanations according to plausibility (cf. Boulogne (1992) 4694), but the *Quaest. conv.* in places comes close to endorsing that Epicurean view, albeit for very different reasons, by the suggestion that all responses may be equally valid because of their equal capacity to inspire philosophical reflection.

39 E.g., see Hardie (1992) 4781–3, mentioning the close links between Platonism and Pythagoreanism in this period, and the influence of Pythagoreanism on Plutarch's teacher Ammonius.

40 See Schenkenveld (1997) and (1996).

41 E.g., see Barton (1994b), esp. 133–7 on medicine as a 'conjectural' skill within which guesswork played a necessary and accepted role within prognosis, and 147–9 on the centrality of the *agôn* to ancient science.

42 See Hardie (1992) 4754–6, with reference a number of examples: e.g., *Quaest. conv.* 9.14 where the guest list includes 'the rhetor Herodes, the Platonist philosopher Ammonius, Plutarch's brother Lamprias, Trypho the doctor, Dionysus of Melite the farmer, the Peripatetic Menephylus, and Plutarch himself' (4755), many of whom tailor their own answer to the question under discussion according to their own professional or philosophical preoccupations.

43 Cf., e.g., Russell (1973) 44–6: he quotes 5.3 (676C–677b) as a striking but not at all unusual example of intricate knowledge of a large number of writers within a very short passage; cf. 8.2 (718c), where one of the guests suggests making Plato a 'partner' or 'contributor' (κοινωνόν) in the discussion. That technique of introducing authors of the past into dialogue stretches back to Plato (e.g., the ventriloquising of Simonides at Pl. *Prt.* 339a–347b), but the sheer frequency of Plutarch's quotations takes it on to a different scale. For similar examples in other Imperial authors, see Vitr. *De arch.* 9, pr. 17 on the process of entering into conversation with the authors of the past.

process demonstrating the thoroughness with which one has considered the full range of possibilities. In some cases there is a sense that the explanations offered can be hierarchised according to criteria of plausibility. That impression is particularly prominent in some of Plutarch's other dialogues, where there seems to be a gradual progression from less to more plausible interpretations.[44] However, even there it is rarely the case that one single version is flagged unequivocally as the right one, and in some cases, especially in the enterprise of interpreting mythical material, that sense of indeterminacy, in the face of the secrets of the divine, is represented as necessary and even desirable.[45]

Speculative, sometimes even absurd, explanations are valued so highly within the *Sympotic questions*, as we shall see repeatedly in the section following, that Plutarch at times seems to be flaunting the unreliability of the responses he and (especially) his fellow symposiasts offer, and so making it deliberately difficult for us to judge exactly what lessons about reading and responding we should take away from this work. One explanation for that impression is the co-existence of two separate criteria for judging the value of explanations within the work. The first is the criterion of plausibility. But the second, which sometimes conflicts with that, is the requirement for explanations which conform to the requirement of sympotic harmony and entertainment – what Plutarch calls the 'friend-making' character of sympotic argument[46] (not that the two are incompatible, since for Plutarch the forging of friendship can come from measured discussion as much as from frivolous speculation). As long as one of these two criteria is satisfied, it seems, the argument in question is likely to be acceptable, although the relative significance of those two criteria is also always open to playful negotiation, and there are moments when characters are criticised for being excessively ingenious or excessively rhetorical.[47] The co-existence of these two different criteria for valuing contributions – plausibility and ingenuity – forces us to face up to the difficulty of distinguishing in practice between appropriate and inappropriate pieces of analysis. It shows us the value of ingenuity, but also underlines the need for personal experience in judging how far to take that ingenuity, or what circumstances to use it in.

[44] E.g., see Hardie (1992) 4755, making that point for *On the E at Delphi* and *On Isis and Osiris*.
[45] See Hardie (1992) 4752–4.
[46] E.g., see the prologues to books 1 (612d) and 7 (697d); see also many of the articles in Montes Cala, Sanchez Ortiz de Landaluce and Gallé Cejudo (eds.) (1999) for discussion of Plutarch's approval of moderate consumption of wine for its capacity to encourage friendly interaction (esp. Montes Cala (1999), Teodorsson (1999), Gómez and Jufresa (1999) and Stadter (1999)).
[47] E.g., see 8.4 (723f–724a).

More importantly, however, Plutarch's playful displays of 'interpretative pluralism' in this work enact his positive valuation of active reading and listening; and in the process challenge the work's readers to participate for themselves, to judge between the explanations on offer, or to come up with others which are more plausible or more ingenious. It is this skill of active reading, I have suggested, which allows us to bring coherence out of fragmentation. In that sense, the symposiasts whose answers fall towards the more speculative end of the spectrum are, paradoxically, giving an entertaining *performance* of the 'serious' philosophical requirement for personal response to discussion, where the fact of participating in the practice of alternative explanation is as important as the explanations themselves.[48] The combination of 'serious' and 'frivolous',[49] and the tension between single explanation and shared discussion where all contributions are valued equally,[50] are, of course, central to the symposium tradition. But in the *Sympotic questions* those crucial sympotic ingredients are given a distinctively Plutarchan spin, as the frivolous joys of ingenious speculation are shown to embody the most important principles of philosophical education. Not only does the text flaunt the diversity and triviality of the subjects which are used as starting-points for discussion, but its subject matter is also further fragmented, and in some cases further trivialised, by the range of pathways each discussion follows, as new speakers attempt new explanations. In other words, the diversity of the *Sympotic questions'* subject matter is itself further intensified by the action of multiplication which is central to the technique of alternative explanation, which fragments the world into seemingly independent and incompatible viewpoints. And yet this technique of fragmentation is itself, paradoxically, the starting-point for overarching philosophical understanding.

TRIVIALITY AND COHERENCE: *SYMPOTIC QUESTIONS* BOOKS 2 AND 3

How, then, does Plutarch embed these principles within the detailed texture of his work? For one thing, he regularly offers his readers or his fellow symposiasts explicit justification for the strategies of ingenious and creative

[48] Just as in the context of Roman religious interpretation the performance of multiple explanations for any single ritual may be in itself more significant than the desire for interpretative 'accuracy': see Feeney (1998), esp. 127–31.

[49] For Plutarch's justification of the mixture of seriousnesss and frivolity in the *Quaest. conv.*, see, e.g., the prologue to book 6 (686d).

[50] See Relihan (1992).

argumentation. In 6.8, for example, Plutarch records his own attendance at a public ritual designed to drive out the disease of *boulimia*; and then afterwards at a symposium gathering where the disease was discussed. First, he tells us, a number of suggestions were made about the origins of the disease's name and of the ritual which had just been performed. In summing up this first phase of the discussion Plutarch emphasises the atmosphere of co-operation in which it was conducted: 'These were the things which made up the shared *eranos* of conversation' (6.8 (694b)),[51] an *eranos* being a feast funded by the shared contributions of the participants. Discussion then proceeds to the causes of the disease. After a number of suggestions about why *boulimia* tends to afflict those who walk through heavy snow, the symposiasts lapse into silence, at which point Plutarch offers his readers a brief aside: 'When silence fell, I reflected on the fact that to idle and untalented people listening to the arguments of their elders brings a feeling of relaxation and satisfaction; whereas those who are ambitious and scholarly use it as spur to make their own attempt at seeking and tracking down the truth' (694d).[52] He then shrewdly introduces a claim made by Aristotle about the natural heat of the body, and the conversation once more begins to circulate, 'as one would expect' (ὅπερ οὖν εἰκός (694e)). This passage is typical of patterns which are repeated over and over again throughout the *Sympotic questions*: the use of past authority to provide a stimulus for present discussion; explicit recommendation of independent thought, in language which is closely reminiscent of Plutarch's work *On listening* (especially the contrast between passive filling of the mind and active kindling of it at 48c, quoted above); and use of the language of contribution to describe individual attempts at explanation. The last of those is especially frequent, and is often combined with an emphasis on the way in which Plutarch's own 'contributions' to discussion are improvised, made whether or not he is confident of having a reliable answer. In 3.5 (652b), for example, Plutarch tells us that he is reusing an argument he had come up with a few days before, when he had been forced to extemporise (αὐτοσχεδιάσαι). In 2.2 (635c), similarly, Plutarch speaks 'in order to avoid the impression of joining in the conversation without making a contribution'.[53] The requirement of being an entertaining conversationalist, and to be generous with one's

[51] ταῦτα μὲν οὖν ἔρανον κοινὸν ἐκ πάντων συνεπλήρου λόγων.

[52] γενομένης δὲ σιωπῆς, ἐγὼ συννοῶν ὅτι τὰ τῶν πρεσβυτέρων ἐπιχειρήματα τοὺς μὲν ἀργοὺς καὶ ἀφυεῖς οἷον ἀναπαύει καὶ ἀναπίμπλησι, τοῖς δὲ φιλοτίμοις καὶ φιλολόγοις ἀρχὴν ἐνδίδωσιν οἰκείαν καὶ τόλμαν ἐπὶ τὸ ζητεῖν καὶ ἀνιχνεύειν τὴν ἀλήθειαν . . .

[53] ὑπὲρ τοῦ μὴ δοκεῖν ἀσύμβολος τοῦ λόγου μετασχεῖν.

own interventions, seem to outweigh any requirement to aim for a single, correct answer.

There is also a recurring emphasis on the requirement for young men to learn from their fellow symposiasts, as I suggested earlier. Those scenes of learning contain both explicit and implicit instruction on the styles of speech and interpretation one should aim for, lessons offered both to the young symposiasts themselves and to us. Plutarch's teacher Ammonius plays a prominent role, both in book 9 and elsewhere, as if to remind us of the way in which Plutarch's own interpretative virtuosity has itself been learnt, painstakingly and gradually, in the course of a long process of development from pupil to expert.[54] In 3.1, for example, Plutarch records an occasion in Athens, at a party held after a sacrifice to the Muses, where Ammonius criticises the practice of wearing flower garlands at a symposium as an unworthy practice for serious philosophers, and so prompts the 'young men' (οἱ νεανίσκοι (646a)), at least those who do not know him well, to remove their garlands in embarrassment. Plutarch, however, knows better, as Ammonius' star-pupil should, and so sets out to refute his philosophical mentor: 'I knew that Ammonius had thrown the topic into our midst in order to encourage practice and further enquiry' (646a).[55] He seems to have grasped the need to exercise one's ingenuity, and the need to admit at least certain types of pleasure into the symposium, in contrast with the other young men who fall for Ammonius' insistence on a complete banishment of frivolity. Plutarch's impressive display then continues in 3.2, which is represented as a continuation of the conversation in 3.1. Ammonius sets out an argument for the belief that ivy is a hot plant, rather than a cold one, as is commonly believed. Once again the young men are cowed into silence. The other, more experienced, guests then urge the young symposiasts to attempt a response, and it is once again Plutarch who speaks, as soon as a promise has been secured from Ammonius not to intimidate the young men by arguing against them. Plutarch contradicts Ammonius with

[54] E.g., see Clement and Hoffleit (eds.) (1969) 95 for the point that the conversations of the *Quaest. conv.* seem to date over a period of twenty to thirty years; cf. Jones (1966) 206–7 on changing representation of Ammonius through the work, portrayed with varying degrees of authority depending on the degree of maturity in his pupils and fellow guests. In 9.15, for example, Ammonius' speech occupies almost the whole discussion, and thus stands as the closing speech of the whole work. In 9.14 Ammonius guides discussion (e.g., at 744b, where he rejects too ready acceptance of an implausible explanation; and at 746b, where he calls for more responses at the very end of his own speech, despite the fact that five guests have spoken already), and has the penultimate speech (745d–746b), with only Plutarch to follow him; in other words it is Plutarch who himself responds to Ammonius' prompting, as often elsewhere (e.g., at 8.3 (721d) and 9.2 (738a)), as if to show how Plutarch's career has itself been shaped from his teacher's encouragement.

[55] ἐγὼ δ᾽ εἰδὼς ὅτι γυμνασίας ἕνεκα καὶ ζητήσεως καταβέβληκεν ἐν μέσῳ τὸν λόγον ὁ Ἀμμώνιος . . .

reference to precisely the passage of Aristotle used by Ammonius himself. Strikingly, that technique of arguing against an opinion with reference to exactly the principles cited in support of it is used similarly by Florus in 3.4 (651c), on a different occasion, in a way which offers us the opportunity to build a cumulative picture of the lessons embedded in the text through consecutive reading of these different and apparently disjointed dialogues.

In the second half of book 3 (3.6 and 3.7) – as often elsewhere – Plutarch switches from description of his early philosophical training to description of occasions much later in his life, after he has reached a position of authority.[56] In 3.7, for example, Plutarch's aged father proposes to 'the young men who were interested in philosophy' (τοῖς φιλοσοφοῦσι μειρακίοις) (655f) a discussion of why sweet new wine is the least intoxicating kind of wine, while the experienced Plutarch looks on. On this occasion there are several contributions, in contrast with the silence of 3.1 and 3.2.[57] Plutarch sums up these contributions with praise of the young men's ingenuity and readiness to speak, although he also points casually to two very obvious explanations they have missed, one of them taken from Aristotle, as if to remind us – and them – that ingenuity on its own is never enough, unless it is supplemented by exhaustive reading. The contrast between the young Plutarch and the old Plutarch, at the beginning and the end of book 3, seems deliberately pointed, reminding us of how the day-to-day experience of philosophical speculation can contribute to lifelong education and philosophical self-improvement.

The *Sympotic questions'* many scenes of learning are thus threaded through the work in a way which prompts us – as well as the young symposiasts of the dialogues themselves – to draw lessons from them. We undergo a repeated process of exposure to common patterns of argumentation, dialogue after dialogue, just as the young men must learn night after night, and gradually we begin to build up a sense of how we can make the different dialogues fit together with each other. There is space here to discuss only one other example of that kind of patterning, from book 2. Here

56 There are several similar instances in other books of older men setting an example for their younger fellow-guests, although their authority is not always unchallenged: e.g., at 1.2, Plutarch's father begins the discussion by playfully criticising Plutarch's brother Timon for his seating of the guests; Timon disagrees, and Plutarch and his other brother Lamprias then argue themselves over the dispute; in 5.3 (676e–677b) a learned rhetorician impresses the younger men, but Plutarch and his friend Lucanius make a point of disagreeing.

57 That silence has also just been mirrored in 3.6 (653e), another of the dialogues set later in Plutarch's life where the young men are reduced to silence when one of the older guests contradicts their claim that Epicurus should not have introduced discussion of the best time of day for sex into his *Symposium*.

again recurring themes and principles of argument cluster together in a way which gives a shadowy impression of coherence and progression to the book. 2.1, for example, is a discussion entitled 'What are the subjects about which Xenophon says people, when they are drinking, are more pleased to be questioned and teased than not'. This is the longest dialogue in the whole of the *Sympotic questions*, but the problem also spills out beyond the end of 2.1, into the repeated scenes of teasing with which the rest of book 2 is saturated, as if to emphasise the need to supplement theoretical discussion, however exhaustive, with personal experience: it may not be enough to have theoretical knowledge of teasing; in addition one must learn by seeing teasing in action. In 2.2 (635a), for example, in the course of a discussion on 'Why men become hungrier in the autumn', Plutarch's brother Lamprias is teased for his gluttony; at 2.10 (643e), a totally separate occasion, Lamprias acknowledges his own gluttony but accuses Hagias of the same; in 2.3 (635e) Plutarch is teased by Alexander for not eating eggs, but then teases Alexander in return (635f); and Soclarus is teased in 2.6 (640b) for the strangeness of the plants which grow in his garden, an observation which then leads into erudite scientific/horticultural discussion on techniques of grafting.

Equally prominent in book 2, though perhaps less obvious, since it is not the subject of explicit discussion at any stage, is a recurring interest in the dangers of misattributing causes in analysing remarkable natural phenomena. In 2.7, for example, there is a long discussion of a type of fish called the 'ship-holder' (ἐχενηΐς), which is said to have the power to slow down ships, despite its tiny size, by attaching itself to their hulls. At the end of this discussion, Plutarch debunks a whole series of explanations for that remarkable power by suggesting that the ships are held back not by the fish, but by seaweed, which is precisely the thing which attracts the fish there in the first place. In other words, he rejects the possibility that the presence of the fish is the cause of the ship's slowness, pointing out that the presence of the fish and the slowness of the ship may instead be common symptoms of a third phenomenon, the seaweed. That strategy of argument is closely matched in the two *quaestiones* which follow. In 2.8 Plutarch rejects the explanations offered for the belief that horses bitten by wolves tend to be unusually spirited, by suggesting that it is only the spirited horses who escape from the wolves in the first place. And then finally in 2.9, we hear a discussion about why sheep bitten by wolves have sweeter flesh. Here, however, there is no explicit attempt to draw the obvious conclusion – not that they have sweet flesh because they are bitten, but rather that they are bitten because they have sweet flesh to begin with. As so often, Plutarch seems to be leaving us to make that conclusion independently, drawing out for ourselves the lessons of the two preceding dialogues.

Plutarch thus repeatedly emphasises the requirement that the philosopher should be able to use any conversation as a starting point for philosophy, by applying his or her own distinctive skills of reading. In that sense it should not matter whether we read things disjointedly and out of context or not. And yet at the same time he weaves complex thematic continuities into his work, challenging us to draw these out for ourselves and so to experience the way in which disparate material can begin to resolve itself into unity if only we read carefully enough – not only ethical unity, but also narrative unity, for example in this intricately developed progression of examples of particular types of argumentation in Book 2, which between them tell a carefully structured story about how we can hone our own skills of analysis.[58] I do not mean to suggest that the *Sympotic questions* aspires to any kind of overarching and continuous narrative coherence. As we have seen, it is a work which values single, disjointed facts and specific occasions very highly. The *Sympotic questions* aspires to unity only through its attention to the specific, which we must put into shape for ourselves. But it does, I suggest, frequently gesture towards thematic connections and progressions between its different parts, as if to give us a faint and preliminary glimpse of the kind of coherence we can expect to emerge from our own readings of Plutarch's work, and of the world, if we are only willing to put the effort in.

Those impressions throw a provocative light on Plutarch's deceptively simple programmatic statements of intent. Here I look briefly at the most often quoted of those, in the prologue to book 2. There Plutarch catches breath to look back at the subject matter and arrangement of the previous book. He distinguishes, first, between 'sympotic' subjects (συμποτικά) on the one hand, which consist of debates about the proper way to run a symposium and to behave at one; and 'symposiac' subjects (συμποσιακά) on the other, which form suitable topics of symposium conversation, but without having any direct connection with the symposium setting; and he categorises the *quaestiones* of book 1 retrospectively according to that scheme. And then in the closing sentences of the prologue he proclaims the randomness which underlies his principles of composition – 'These things have been recorded haphazardly and without being put in order,

[58] In that sense Gallardo (1972) 189 seems wrong to say that the gaps between adjacent *quaestiones* in the *Quaest. conv.* prevent a coherent reading: 'Dada la estructura de la obra, resulta imposibile hacer un estudio de los personajes o de la progresión de la acción, pues esta última no existe.'

but rather as each of them came to my memory' (629d)[59] – before finally explaining to his addressee, Sosius Senecio, that the reader is likely to come across some of Sosius' own contributions to discussion in what follows. In both of these claims, I suggest, this prologue is more complex than it might initially appear, and than is usually assumed. First, in the light of the work's constant saturation with didactic material, of the kind we have already glimpsed above, the stated distinction between 'sympotic' and 'symposiac' subjects begins to seem disingenuous. In fact all of the discussions of the *Sympotic questions* are 'sympotic', in the sense that each of them is equally concerned with exploring the question of how best to speak and behave in the symposium. All of these dialogues are equally didactic. Secondly, the claim to have composed haphazardly breaks down, as we come to perceive the *Sympotic questions'* shadowy overtones of patterning emerging from the mass of disjointed detail. Plutarch's claim to have composed 'as each thing came to my memory'[60] looks, on closer inspection, not like a statement of the work's randomness, but rather like an attempt to equate the ordering of the work with the retrospective patterning which memory inevitably imposes. The statements Plutarch makes in this prologue might at first sight be taken to support stereotypes of artlessly structured miscellanistic writing. But on closer inspection we can see that Plutarch actually plays along with those stereotypes knowingly, while at the same time ultimately resisting them, at least for those who can read closely and creatively enough to spot the trick.

LOCAL IDENTITIES IN THE *SYMPOTIC QUESTIONS*

What, then, of the political world, the world of the Roman Empire? The first thing to note is Plutarch's representation of geographical distinctions. The overriding impression the *Sympotic questions* projects is one of men with a shared Hellenic education talking as equals. However, this stress on the cultural homogeneity of the speakers is regularly nuanced by references to the local origins of individual participants, who are drawn from a range of cities across the Greek world. In much the same way, Plutarch pays frequent attention to their differences of philosophical persuasion or of profession, explaining, for example, that one is a rhetor, another a grammarian, another a doctor and so on.[61] These sporadic reminders of difference contribute to an impression of cosmopolitanism in their conversations, conjuring up a

[59] σποράδην δ' ἀναγέγραπται καὶ οὐ διακεκριμένως, ἀλλ' ὡς ἕκαστον εἰς μνήμην ἦλθεν.

[60] Cf. n. 3, above, on parallels for this claim (not always justified) in Aulus Gellius and others.

[61] Cf. n. 42, above.

picture of a network of cohesive elite hospitality stretching through the cities of the Greek world, while also foregrounding the possibility that differences of background might contribute to differences in their conversational contributions. There is a similarly frequent though sporadic attention given to the geographical settings of these conversations: some are introduced with no specific setting, while others are carefully grounded in specific occasions and specific cities. In 2.2, for example, we hear of a gathering at Eleusis: 'At Eleusis, after the mysteries, at the height of the festival, we were having dinner at the house of Glaukias the rhetor. When the others had finished eating, Xenokles the Delphian as usual began to tease my brother for his Boeotian gluttony' (2.2 (635a)). The phrase 'as usual' (ὥσπερ εἰώθει) lets us in on a world where ease of travel and widespread sharing of a common literary heritage oils the wheels of elite guest-friendship across the Greek world. The stereotype of Boeotians as gluttonous reminds us, however, that it is a world where local difference is still conspicuous, even if frivolously treated. In 7.7, we see Plutarch hosting a similarly cosmopolitan gathering in his home city:

> When Diogenianos the Pergamene was visiting in Chaironeia, there was a conversation over drinking about types of entertainment, and we had trouble fighting off a bearded Stoic sophist who brought up Plato's criticism of those who listen to flute-girls in their symposia but who are unable to entertain themselves through conversation. Philip of Prusa, even though he came from the same philosophical stable, told us to forget about those guests of Agathon's, who spoke more pleasantly than any flute or lyre. (7.7 (710b))

The effect of these repeated patterns is to draw attention to the way in which shared Greek culture is formed from Panhellenic diversity. And the world it conjures up, where a cosmopolitan, Hellenic philosophical identity will often be combined with political engagement in specific local contexts, is one which Plutarch himself was committed to throughout his life, in his devotion to his home city of Chaironeia.[62]

Rome can be made a part of this world, as I suggested earlier.[63] The addressee of the *Sympotic questions*, Sosius Senecio, is a Roman politician probably from the west.[64] Plutarch suggests that Senecio can use the text as a

[62] See Jones (1971) 3–64 on Plutarch's career, including detailed discussion of the identity of the many friends mentioned in the *Quaest. conv.* and elsewhere.

[63] See Jones (1971) 48–64 on Plutarch's western friends; Swain (1990) 129–31 on Roman participants learning Greek styles of speech.

[64] See Swain (1996) 426–7 on Senecio's western identity; even if Senecio did not come from the west of the empire, as Swain claims, he nevertheless 'presented himself consistently as a Roman, and held high positions in Trajan's administration', as Stadter (2002) 23, n. 27 points out.

spur to his own progress in philosophy, using it to reconstruct entertaining and edifying conversations he has been involved in. As that suggestion implies, Senecio is present at many of the discussions Plutarch describes, trying his hand at Greek styles of ingenious, philosophical speech. The same is true of a number of other Romans, most conspicuously Mestrius Florus. Both Florus and Senecio can hold their own in the philosophical banter of the Plutarchan symposium, although both are represented as having things left to learn. At one point, for example, Florus, in a fit of ostentatious Hellenic role-playing, objects to the inclusion of Egyptian material in a discussion; Plutarch takes him to task for failing to realise the capaciousness of the Greek interpretative tradition, its capacity to accommodate other cultural traditions (5.10 (684f–685a)).

Moreover, it is striking that many of these conversations are set at local festivals, at sympotic gatherings which have a semi-official flavour. Often the symposium hosts are festival officials or local priests, holding small banquets for friends and local notables in their own homes – a common convention in festivals in the east, where the dividing line between the private symposium and the large-scale civic banquet was far from clear. The implication, as I aim to show in the examples which follow, is that the philosophical conversations of Plutarch and his guests are somehow equivalent to the social interaction of sacrificial banquets, and the skills of agonistic competition, though Plutarch also hints at differences between them, representing his *quaestiones* as more elevated versions of those festive activities. If we follow the implications of that parallel, sympotic conversation is to be seen as a performance of cultural memory just as much as the processions and sacrifices which traced their way through the city streets of the Greek east so frequently.

Approximately 25 per cent of the *Sympotic questions'* conversations are explicitly set at specified festival occasions[65] (it is difficult to give an exact figure, since in some cases the text makes it difficult to be sure about whether consecutive chapters are held on the same occasion). Of those, I give just two examples. 5.2, first of all, is a discussion of whether or not the poetry contest is the most ancient component of the Pythian games. Plutarch starts the dialogue as follows: 'At the Pythian festival there was a discussion about whether the newer competitions ought to be abolished' (674d). He then he sets out some of the main arguments used on either side. It sounds at

[65] The number of conversations does not correspond to the number of chapters, since some conversations are spread over more than one chapter. I count a total of fifty-seven conversations; of those, I count fourteen which are set at specified festival occasions: 1.10, 2.2, 2.4–5, 2.10, 3.7, 4.1, 5.1, 5.2, 5.3, 6.8, 6.10, 7.5 and 9.1–15.

first as though the discussion which Plutarch summarises is a symposium discussion, but we then learn in the next paragraph that these opening sentences actually refer to a discussion of that topic in a Pythian Council meeting, where Plutarch himself had spoken against making any change to the festival programme:

During the Council meeting I argued against those who wanted to change the established programme and who criticised the contest as if it was a musical instrument with too many strings and too many notes. And then at the dinner which Petraios the agonothete hosted for us, when the same topic of conversation came up again (ὁμοίων λόγων προσπεσόντων), I once again defended the musical arts; and I demonstrated that poetry was not a late and recent addition to the sacred games, but that it had been awarded victory crowns even in the ancient past. (674e–f)

The learned and ingenious styles and topics of speech which we find in Plutarch's symposium conversations seem to be useful and authoritative for more public, official contexts too, in the sense that they contribute to highly publicised decisions about festival programming. Plutarch hints here that there is no clear dividing line between frivolous private speech and authoritative public pronouncement. The phrase ὁμοίων λόγων προσπεσόντων – 'since similar topics of conversation happened to come up at dinner' – backs up the impression of links between the two different types of speech. Plutarchan symposium conversation, by that standard, has political significance, whatever its surface appearance of frivolity.

My second example comes from 8.4:

When the Isthmian games were happening, during the second of Sospis' spells as agonothete, I avoided most of the dinners, when he entertained together all the foreign visitors, and often all the citizens as well. Once, however, when he invited his closest and most scholarly friends to his home, I was present too. When the first course was cleared away, someone came in bringing a palm-frond and a woven garland to Herodes the rhetor, sent by a famous competitor who had won a contest in the encomium contest. He accepted them, then had them taken away again; and then he said he had no idea why different contests have different types of garland, while all of them alike give palm-fronds as prizes. (723b)

They then proceed to discuss that problem at length. Here we see several characteristic features. For one thing Plutarch hints that the conversational skills which Plutarch and others – like the rhetor Herodes – are displaying are connected with the skills on display in the festival's contests. It is as if Herodes' style of speech – which is in a very loose sense encomiastic, in the sense that his answer to the question under discussion involves him in praising the palm tree – has been the model for his pupil's victory in the competition. The opening lines also imply that the banquet Plutarch attends is part of Sospis' official duties as agonothete. Plutarch thus represents the conversation as an episode which falls within the boundaries of festival time, although he also insists on his own discriminating dislike of banquets where too many people are present. In that sense the conversations he records are not direct equivalents of general festival practice, but instead are represented as more elevated versions of festival banquets and festival contests as they are most commonly done.

This grounding of philosophical discussion within particular social occasions, even within the bounds of festival time, is, of course, not new for the literary-philosophical symposium tradition.[66] We need only look back to the fourth-century BCE *Symposium* of Xenophon, which shows Socrates and friends relaxing at a banquet held to celebrate the victory of the boy Autolykos in the *pankration* at the Panathenaic games, to see that; or the *Symposium* of Plato, in honour of Agathon's victory at the Dionysia. But Plutarch's text takes this structural feature to a new level by recording a whole range of symposium conversations, set at many different social occasions, and many different festivals, and so driving home the point that the true philosopher can do philosophy in any setting. In choosing that structure for his work, Plutarch offers us glimpses of conventions of festival feasting which we see from a very different angle in the many Imperial-period inscriptions which record provisions made by benefactors for sacrificial banquets.[67] Many of the common features of the banquets those inscriptions record are replayed in a more elevated, philosophically inflected form within the conversations of the *Sympotic questions*.

For readers familiar with the conventions of sacrificial feasting, that resemblance may well have enhanced the sense that the philosophising of Plutarch and his fellow guests was an activity particularly appropriate to festival time. For example, the presence of young men learning from their elders in Plutarch's sympotic dialogues picks up the common motif of young men attending banquets together with their fathers, preparing themselves, presumably, to take up their roles as full citizens.[68] Moreover, banquet

[66] That said, there were models available – which Plutarch chooses not to follow – for almost entirely non-contextualised portrayal of erudite sympotic conversation, perhaps most famously in the sympotic writing of Epicurus, which seems to have conspicuously neglected any detailed attention to dramatic setting: see Usener (ed.) (1887) 115, with reference to several passages of Athenaeus (177b, 182a, 186e, 187b); Plutarch's own familiarity with Epicurus' *Symposium* is clear from a number of references, including, in the *Quaest. conv.*, prologue 1, 3.5 and 3.6.

[67] See Schmitt-Pantel (1992) 255–420 on sacrificial feasting in the Hellenistic and Imperial periods, and 471–82 on the way in which Plutarch's *Quaest. conv.* engages with the realities of public banqueting.

[68] See Schmitt-Pantel (1992) 396–7.

inscriptions often stress the presence of foreign visitors in festival banquets, often with particular attention given to Roman visitors, just as Plutarch's text fosters an atmosphere of cosmopolitan hospitality, where Greek cultural tradition can forge unity across local and even Greek/Roman boundaries.[69] Most importantly of all, perhaps, inscriptions celebrating banquets – like inscriptions celebrating festival benefaction or agonistic victory – tend to be very much aware of their part in a series. In some cases, for example, we find inscriptions for individual benefactors recording a whole string of different banquets and distributions spread across the year, each one slightly different from all the others.[70] In other cases, large numbers of almost identical banquet inscriptions seem to have been put up very close to each other within Greek cities, recording each new event through familiar, formulaic language, adjusted only to take account of variations in the identity of benefactors or setting.[71] These inscriptional series conjure up an impression of the recurring rhythms of festival time as something which structures the life of the city. Plutarch draws on those patterns of representation in his *Sympotic questions*, showing us how the recurring rhythms of sympotic conversation are both framed by and equivalent to – though also elevated above – the repeated patterns of the local and Panhellenic festival calendar.

In what sense, then, does Plutarch politicise his patterns of textual organisation? He does so, I have argued, above all by showing us that the gesture of combining the specific and the universal, of drawing universal significance out of fragmented detail, is not only an abstract, intellectual one. It is also, as the framing passages of his dialogues make clear, a process which is central to social and political interaction throughout the Greek east, where Panhellenism always requires an awareness of local specificity. Plutarch repeatedly characterises the speech of himself and his fellow guests as an elevated equivalent of festival competition and display, carried out within a philosophical version of cosmopolitan festival commensality. In making that equation, he brings an added dimension to his portrayal of the techniques of active reading by which, so he suggests, the diversity of the world can best be understood. It is not only that these techniques find a productive breeding-ground within the cosmopolitan, elite society of the Greek city; they are also, Plutarch suggests, performances which can match the central role played by festival performance

[69] See Schmitt-Pantel (1992) 389–96.

[70] E.g, see the long inscription in honour of the banquets and distributions funded by Epaminondas of Akraiphia at *IG* VII, 2712, with Oliver (1971).

[71] E.g., see the series of second-century CE banquet inscriptions from Syros in *IG* XII, 659–67.

and festival commensality within Hellenic cultural self-definition. The text hints, moreover – by repeated juxtaposition of locally specific dramatic framing with philosophical discussion – that the technique of allowing universal knowledge to emerge from engagement with the smallest and seemingly most insignificant details of argument may be related to the fundamentally Greek instinct, ingrained within centuries of civic ritual and political engagement, of constructing Panhellenic unity through attention to local diversity. I began this chapter with claims about the potential for both thematic and ideological unity to be encoded within the random accumulations of miscellanistic compilation. Plutarch's unwieldy collection of scientific, literary, historical conversations, I have argued, is powerfully, and paradoxically, imbued with both, not only through the complex narrative patterns which lie beneath its surface, carefully designed to provoke response from us as readers if only we can read with proper philosophical attention; but also through the way in which Plutarch imprints those patterns with political resonance, foregrounding their link with the civic and religious rhythms of his contemporary world.

CHAPTER 3

Galen and Athenaeus in the Hellenistic library

John Wilkins

INTRODUCTION

The ordering of knowledge was a major issue for two writers of the late second and early third centuries CE. Athenaeus in his *Deipnosophistae* (*Sophists at dinner*) and Galen in a number of nutritional and pharmacological treatises set out to review the state of knowledge in the domain of food, nutrition and culture. The Galenic treatises under discussion here, although only a small part of his output, reflect important aspects of his research methodology; of the works of Athenaeus, only the *Deipnosophistae* survives.[1] The *Deipnosophistae* and Galen's treatises on nutrition and pharmacology have in common not only a subject matter based on foods, drinks and medical treatises, but also an aim to research previous technical works and a strategy for cataloguing complex data. The data could be found by library-based research, and tested by experiment and personal experience. Once gathered, this data could be ordered in a list or catalogue, or other format. In this respect, Galen and Athenaeus resemble, for example, the lexicographer Pollux, some of the Hellenistic doctors, and others who produced catalogues in their own and previous centuries, although they apply those techniques rather differently from each other. Galen prefaces his works with chapters or whole books on methodology, while Athenaeus combines cataloguing with anecdote (for which compare Aulus Gellius and Aelian) and sympotic forms that resemble Plutarch's *Sympotic questions*.[2]

The domains of food and pharmacology had become vast and complex by the second century CE. This was partly due to the diversity of practice found in the many Greek and non-Greek cities of the Mediterranean. A second factor was the stimulation to the flow of foods and related goods from Asia into the Mediterranean area that followed the expeditions of

[1] For recent discussions of his treatises on the kings of Syria and a comedy of Archippus see Braund (2000b) and Wilkins (2000).

[2] Cf. Chapter 2, above.

Alexander. The third and crucial factor was the centralising power of Rome in the Mediterranean world and beyond, which drew all things to itself. As Aelius Aristides put it, all things came to Rome, so there was no need to travel to the areas of production in order to enjoy regional specialities. Rome was an international centre of exchange for many goods: the two that concern us here are foods and intellectual products. How was an intellectual to reflect the vastness and complexity of the Empire? Galen as a doctor and an intellectual was drawn to Rome from Pergamum and became court physician to Marcus Aurelius, while Athenaeus, whether or not he travelled to Rome personally, set his *Deipnosophistae* at a series of meals at the home of the rich Roman magistrate Larensis.

In addition to the great supply of foods and drugs that became available to Rome and the other major Mediterranean cities from all quarters of the Empire, there was an accompanying expansion in nomenclature as local names were no longer necessarily attached only to their own place of origin. At the same time there had been, from the fourth century BCE. onwards, systematic attempts to classify and order the relationships between certain classes of plants, animals and foodstuffs. Examples relevant to this study are *Regimen I–IV* of the Hippocratic doctors, medical treatises of Diocles of Carystus and Mnesitheus of Athens, and the zoological and botanical works of Aristotle and Theophrastus. Over a period of several centuries, attempts to systematise and classify were often at odds with each other and sometimes internally contradictory. So it was that when Athenaeus and Galen came to review these works as an aid to their own classifications, many apparent errors needed to be cleared out of the way before work could proceed.

These two authors thus face many of the same challenges. As we shall see, they also respond to many of them in similar ways. However, there are also important differences between their objectives and the ordering principles they employ. Charting some of those similarities and differences is one of the main aims of this chapter. Galen was writing treatises for the use of doctors, patients and on occasion a wider educated audience. *On the properties of foods* (henceforth *Food*) and *On the properties and mixtures of simple medicines* (henceforth *Simples*) are technical texts,[3] which complement other works of Galen on physiology and the digestive processes, such as *On the natural faculties, On mixtures* and *On the opinions of Hippocrates and Plato*. Galen's aim, for example in *Food*, was to classify foods

[3] For the reader's convenience, since I use them throughout my discussion, I have adopted these English titles for texts that are abbreviated elsewhere in the volume as *SMT* and *Alim. fac.*

coherently and authoritatively in order to assist a branch of medicine that he declared to be particularly useful (1.1). The aim of Athenaeus seems to be less straightforward. He appears to review the evidence for food and eating practices over a millennium of Greek culture, from Homer to his own day. This evidence is presented in the form of a discussion over food and wine in the tradition of sympotic prose literature.[4] Athenaeus adds a variety of forms, one of which might invade another. An alphabetical list of items might, for example, be temporarily suspended to allow room for a number of speeches from the comic stage, as happens in book 7. Or a list of foods might be interrupted by the dramatic arrival of a character into the dining room, such as the cook in book 9. Such invasion of one form by another might suggest poor writing, as some have suggested of Athenaeus. But it might also reflect a culture which was so familiar with literary forms of ordering and classification that the disturbance of one by another provided particular pleasure to the reader. There is a certain destabilisation of hierarchy and canonical texts in Athenaeus, which raises important questions for the reading of texts in the second and third centuries CE.[5]

Galen and Athenaeus concentrate on the Greek-speaking cities of the East. Yet both are aware of the demands of Rome. Galen has a great geographical range that is Mediterranean-wide and clearly brings the Hippocratic *Regimen II* up to date, geographically speaking. Athenaeus, meanwhile, sets his work at a series of lavish meals in Rome at the home of a host who is said to have the best library in the Greco-Roman world. His semi-fictional diners also on occasion cite Latin authorities. Neither Galen nor Athenaeus, however, draws on obvious Roman authorities such as Pliny the Elder, Celsus or Columella, who might have contributed to their efforts to identify clearly particular species of plants or animals. The extent of their engagement with Latin authors as well as with the power of Rome will be discussed further below.

These two authors are of particular interest in the current century since they are first-rate witnesses to the intellectual preoccupations of their times and were objects of the greatest interest in the Renaissance and later centuries. Yet they are less read in our own time. Technical treatises are often relegated to the cupboard containing such topics as the history of medicine and the history of science,[6] while the *Deipnosophistae* is sometimes dismissed as

4 For which see Romeri (2002).

5 This is not the place to discuss the nature of the *Deipnosophistae*, but I return to this topic in concluding remarks at the end of this chapter.

6 There are naturally many works on Galen that take a broader perspective. See for example the essays collected in Nutton (1988) and Kollesch and Nickel (eds.) (1993)

incoherent, even, in its current form, as a pale and fragmentary image of the imagined original of Athenaeus.[7] This chapter argues against those judgements for a more sensitive acknowledgement of the richness and cultural interest of both.

FORMAT

The presentation of the results of research is largely determined by the format of the final product. Galen set out to write treatises that would catalogue foods and drugs in a coherent order. The items in *Foods* are arranged in three books, one of cereals and pulses, one of other plants (herbs, fruits and vegetables), and one of meat and fish. Galen takes account of how earlier treatises had been ordered (2.1; see further below), and of his target readership. In both treatises, previous medical scholarship is examined at length, and in *Simples*, methodology is given considerable attention. So, for example, the desirability or otherwise of an alphabetical order is considered at some length.[8] In pharmacology, more than foods are at issue. In addition to plants, animals and fish, metals and other products are also found to have pharmacological properties. After the five books introducing methodological problems in pharmacology, different categories are located in six different books. The ordering of *Simples* indicates more than one principle at work internally, and also a different ordering principle from that used in *Foods*. Galen has then composed *Foods* and *Simples* as two cataloguing treatises, to provide a resource for doctors and other readers, and for cross-referencing within Galen's own writings. So *Simples* is often referred to in *Foods*, while both treatises are referred to in *On the preservation of good health* (*Hygieina*). Galen almost certainly has different readerships and different purposes in mind. *Simples* is earlier; devotes more space to methodology; has fewer anecdotes, and is often referred to for definitive statements on properties (or *dunameis*). *Foods* may then be comparatively less technical and more accessible to the general reader. It is certainly a richer hunting ground for the social historian.[9] Here Galen gives good evidence that he has his whole output in good order. Galen's own commentary on this is well known in the short treatises *On my own books* and *On the*

[7] On the supposed version of the *Deipnosophistae* in thirty books, and the processes of epitomisation it has undergone, see Kaibel (ed.) (1887–90), Arnott (2000) and Rodríguez-Noriega Guillén (2000).

[8] Galen discusses alphabetical order at *Simples* 11.792 K (following Pamphilus, on whom see below), 12.2 K, 12.209 K. Alphabetical order is used in *Simples* books 6, 7 and 8 (the plant books); is not used in part of 9 (stones), but is used in most of 9 (metals); is not used in 10 or 11 (animals and fish). See further Barnes (1997) 10–11.

[9] See Garnsey (1999), Wilkins (2003).

order of my own books (see Flemming in this volume). Galen makes clear the perils that threatened his books, which include fire, plagiarism and the passing off of false works as his. It is also clear how far his interests extended beyond medicine, narrowly defined, into philosophy, lexicography and literary concerns.

Galen the writer of medical treatises thus shares with Athenaeus certain features, among which lexicography and philosophy play a large part, as do the nature of texts and the use of medical texts in any account of food in antiquity. Both Galen and Athenaeus are major sources for the Hellenistic doctors such as Mnesitheus of Athens, Diocles of Carystus, Philistion, Praxagoras, Phylotimus and Diphilus of Siphnos. In fact, these authors depend heavily on Galen and Athenaeus, together with the late compiler Oribasius, for their survival, albeit in fragmentary form.

I shall return later to the content of the *Deipnosophistae*. In format, it is curious. Formally, it is a literary symposium in the tradition of Plato's *Symposium*, with a series of diners and symposiasts speaking on a set theme, in this case of meals and foods and the literature and other cultural products that accompany them. As in Plato's model, lively dialogue on certain issues is interwoven with material in a complementary format. Where Plato writes beautifully crafted speeches in praise of love, Athenaeus has lengthy literary quotation and lists of items, sometimes in alphabetical order, sometimes not. Overall, the material is organised in the order of a meal, with starters, fish and meat dishes and then the dessert or 'second tables' to accompany the symposium. Within that principle of organisation, some of the foods and mealtimes described are presented by a speaker, such as Plutarch in book 6 on parasites. Sometimes Athenaeus gives a summary of what was said, as in the list of fish in book 7: 'I shall record for you what the deipnosophists said about each fish. They all brought to the common table their contributions from books, whose names I shall omit because of the vast number' (277b–c). Sometimes Athenaeus seems to discard the dialogue form with speakers altogether, and to present material of his own, as in the twelfth book on luxury. However we interpret the ordering of Athenaeus' work, it does not follow the clear and systematic alphabetical order of the modern dictionary or encyclopaedia. I shall return later to the effect that Athenaeus appears to create. I note for the moment that he seems to be writing for a reader who is familiar with different modes of processing a great deal of varied material and who is able to sustain the subversion of certain forms such as the alphabetical or non-alphabetical list.

Certainly, the reader of Galen and Athenaeus would appear to belong to the elite strata of Greek and Roman society. Galen clearly speaks of peasants being different from 'us' (the author and reader, *Food* 1.2). Often he seems to envisage a Roman elite audience, for example in *On prognosis*, where his voice is that of the competitive doctor who can meet any challenge in the rhetorical displays beside the sickbeds of Rome. In the treatises in question, however, his evidence is more likely to be drawn from Asia Minor and Egypt than Rome (and even less so from the Greek mainland). Athenaeus by contrast sets his semi-fictional meals at the home of the Roman Larensis, who is said to provide lavish meals (on which see the concluding remarks below) and to possess the finest library ever seen. He thus appears to offer meals of unparalleled splendour and a research library with unparalleled resources. Poorer citizens are sparsely represented in the *Deipnosophistae*.

RESEARCH

The *Deipnosophistae* is a literary thesaurus, full of wonderful discoveries in rare books and little-known texts. It artfully presents research in the Hellenistic library. The model of the *Deipnosophistae* as a virtual library has been developed by Too (2000) and Jacob (2000) and (2001). The diners at the meals of Larensis rely on reading and memory, with only a little personal experience, such as, for example, Larensis' reports on Moesia in book 9 (398b–399a), or Athenaeus' references to his native Egypt (such as 312a–b, for which see n. 23). Galen, on the other hand, combines research in the Hellenistic library with case studies, experiment and personal autopsy (see below, and Wilkins (2003)). Despite different aims, formats and methods, however, Galen and Athenaeus share a number of features, which highlight the intellectual preoccupations of their time.

Attention to detail

Let us consider what each author says of the pistachio nut. Athenaeus first (649c–e):

At this, the Syrian,[10] who had been tested and bitten very hard,[11] said, 'well, pistachios have been served on our table as well. If you tell me which author mentions the word pistachio, "I will give you" not "ten gold staters" (in the words of the idler from Pontus)[12] but this drinking cup here.' When Democritus remained silent, Ulpian said, 'since you don't know, I'll tell you. Nicander of Colophon

[10] Ulpian, the symposiarch, is in the middle of a debate.

[11] Terms appropriate to food are frequently transferred to the deipnosophists themselves. Thus, for example, they are abused as pigs, or they are in danger of becoming fish.

[12] Heraclides of Pontus the Younger, who wrote an anecdotal book entitled *Leschai*.

mentions them and says in the *Theriaca*, "pistachios (*phittakia*) on branches like those of the almond". But there is also the reading, "pistachios (*bistakia*) have appeared, almond-like". And Posidonius the Stoic writes as follows in Book 3 of the *Histories*: "Arabia and Syria grow the persea-fruit and the so-called pistachio (*bistakion*) [there follows a description of the fruit] . . ." But the brothers[13] who wrote the *Georgics* write as follows in Book 3: "and the manna-ash and the terebinth which the Syrians now call pistachio (*pistakia*)." So these two call pistachios *pistakia* with a pi, while Nicander has an aspirated *phittakia*[14] and Posidonius *bistakia*.'

This might initially sound like a laborious and pedantic journey through variant readings, with mild jokes on the identity of the Syrian speaker and the Syrian origin of the pistachio. A more positive and nuanced reading might emphasise the geographical and linguistic range of speakers and cited authors, and the range of dates and literary forms used. In addition, Ulpian draws on special data on the pistachio from both predictable authors such as Nicander and the Quintilii and from Posidonius whose work is normally directed at other agenda.

Galen's note on the pistachio shares with Athenaeus a geographical interest, but is otherwise more narrowly focused on medical properties. He says (*Food* 2.30) that pistachios grow in Alexandria and Beroea in Syria; that they are not very nourishing; that they are bitter, astringent and aromatic, and neutral on the stomach. When discussing carobs, however, Galen comes closer to Athenaeus' interest in spelling and terminological accuracy.

The carob (*ceration*), whose third syllable is spoken and written with the letter t looks nothing like the cherry (*cerasia*) with an s since it is a food that is woody and full of bad juices; consequently it is difficult to digest, for nothing that is woody is easy to digest. Since it does not pass through the body quickly, it is furnished with considerable bad qualities. So it would be better if these fruits were not exported from the areas in the east where they grow. (*Food* 2.33)[15]

Here, we find that Galen has added to nutritional and geographical criteria a lexical concern (the danger of confusion of terms). The potential for error is created not by a botanical confusion of species but by the similarity of name; this apparent similarity is disproved by their appearance (*ouden eoike*). Note that Posidonius in Ulpian's quotation referred to the appearance of the pistachio tree. Galen has a further comment, that the carob would be better omitted from the diet. The inclusion of a non-approved item suggests he

[13] Quintilii. See Canfora (ed.) (2001) 1686, Zecchini (1989) 16.

[14] On manuscript inconsistencies in this passage see Gulick (1927–41) *ad loc.* and Canfora (ed.) (2001) *ad loc.*

[15] Translations from *Food* are taken from Powell (ed.) (2003), in some cases with minor adaptations.

aims at inclusiveness above medical acceptability. More important for the moment is the concern for lexical accuracy.

Matching the lexicon and the science

This concern reappears in Galen's entry on wild chickling (*arakos*: *Food* 1.27):

I have found the last syllable of the word *arakos* written with a k in the *Merchant Ships* by Aristophanes where it reads: 'wild chickling, wheat, pearl barley, groats, one-seeded wheat, darnel and fine flour'. The seed is similar to that of chickling (*lathurai*), and some people do not think they are separate species, for the general use and power of wild chickling is similar to that of chickling, except that wild chickling is harder and more difficult to boil and so is harder to digest than chickling. Where I live there is a wild variety that is round and hard, smaller than vetch (*orobos*) and found in cereal crops. It is called *arachos*, pronounced not with a k in the last syllable but a chi. Harvesters throw it away, just as they do with axeweed (*pelekinos*).

For Galen there is a lexical and a botanical problem. There is clear confusion between chickling and wild chickling, and the botanical confusion between the seeds can be established by comparing similarity and difference. Principles of analogy and relationship are at issue. For chickling and wild chickling, the term Galen uses is *paraplesios*.[16] The lexical confusion however is more insidious. To clarify the issue, Galen goes back to Aristophanes. We might note his works on 'political' terms in Aristophanes, Eupolis and Cratinus, which do not survive but are listed in *On my own books* 48. In these treatises, Galen probably explored just such words as the present one, which could help clarify terms in the medical lexicon, and to give an Attic identity to them.[17] Similar concerns also inspired the pseudo-Galenic text on Hippocratic glosses. This is not a merely pedantic concern that finds its way into a grammarian's or antiquarian's work such as the *Deipnosophistae*. It is also at the heart of Galen's concerns.

These concerns are best worked though in Galen's chapter on the primitive wheats. This is an extensive section (1.13) on *tiphai, olurai, zeiai* (species of emmer, einkorn and possibly spelt). Galen finds that his medical textbooks are inadequate:

Mnesitheus placed *tiphai* third after the naked wheats and barley. Diocles discussed it rather superficially since he preferred brevity in writing to exactness (*to akribes*) in exposition. At any rate, that is how he wrote, abbreviating the discussion about

[16] See p. 77 below on the terms used by Theophrastus. [17] See p. 85 below on Atticism.

wheat, barley and much else. Praxagoras and Mnesitheus wrote about them a little more fully than Diocles, but they also omitted some things. Phylotimus wrote at length about some but inadequately about others, and some, like *zeia* he totally forgot. It is plain that his teacher Praxagoras did likewise. For while Phylotimus ignored nothing that Praxagoras spoke of, he works over it and adds much. One can wonder that the compiler of the Hippocratic *On Regimen*, whoever the ancient author was, did not even mention the name of *zeiai*.[18]

Galen's review of the medical sources shows not so much error as omission. He next subjects Diocles to close analysis, which includes consideration of manuscript variants in the text. Next he considers Mnesitheos, who has found a type of cereal in northern regions that he calls *zeia*. Galen does not praise Mnesitheos' geographical range, but tests his data. Galen cannot find any corroboration of this in visits to northern peoples, whether, he notes, the grain is spelt *zeia* or *zea*. He considers a further possibility, namely that *zea* is a Greek term for what the Thracians and Macedonians call by another name. A visit to the region reveals that the main cereal there is called *briza* (Galen spells it out in the nominative and accusative case); that it is 'very similar' (*homoiotatos*) to what people in Galen's area of Asia Minor call *tiphê*; and that it makes a malodorous black bread. Conclusion: if Mnesitheos had mentioned the black bread then Galen would tend to believe *briza* is what Mnesitheos called *zeia*. Further comments follow, which are drawn from Theophrastus and Dioscorides, with Homer added for good measure. Many other similar (*paraplesia*) seeds are considered that fall somewhere (but it is not possible to say where precisely) between wheat, barley, *zeia*, *olura* and *tiphê*. This use of analogy resembles the approach of Theophrastus who declares *zeia* and *tiphê* to be very similar, *homoiotata*, to wheat (*Food* 1.13).

In this passage, Galen attempts to bring some order to a confused picture. It is worth noting that the picture at the end of the twentieth century remained confused. Sallares 1991 discusses the difficulty of identifying 'by-forms' of wheat; two recent translations of Galen *On the properties of foods* do not agree with each other; and an encyclopedia of food notes many different modern terms for the primitive wheats.[19] Galen for his part uses autopsy and consultation with peasant farmers to discover how the plants in the field match the medical and botanical accounts. Note that Galen concerns himself with peasant names for plants, and not only with names

[18] The manuscript text of Hippocrates *Regimen II* 43 was emended by Wilamowitz, followed by the *CMG* text of Joly and Byl, on the strength of Galen's comment here. *Zeia* is preserved in the Hippocratic tradition.

[19] Kiple and Ornelas (eds.) (2000), Grant (ed.) (2000), Powell (ed.) (2003).

in high-status botanical and medical texts. He searches for the greatest possible accuracy and clarity, and notes the absence of these qualities in most of his predecessors, including the Hippocratic *Regimen II*. He might well claim implicitly to have the most thorough, accurate and geographically wide-ranging account available. His techniques in verifying taxonomy include those highlighted by Athenaeus: spelling, dialectal and other linguistic variation, geographical considerations and analogy. Comparison between Athenaeus on pistachios and Galen on primitive wheats indicates a shared interest in reading widely, spelling, definition of meaning, and textual variants. These are familiar concerns in Hellenistic scholarship, to be found in the analysis of texts from Homer to the Hippocratic corpus (indeed in Galen's own commentaries on the latter).

Athenaeus, too, provides testimony on the primitive wheats. In book 3, the deipnosophist Pontianus says

> Tryphon of Alexandria in the book entitled *Plants* sets forth the kinds of bread, if I remember rightly, as leavened (*zumites*), unleavened (*azumites*), with fine flour (*semidalites*), with rough flour (*chondrites*), with unsifted flour (*sunkomistos*) – this he says is more laxative than bread made of refined flour (*katharos*) – bread made from *olurai*, bread made from *tiphai*, bread made from millet. Bread made from rough flour, he says, comes from *zeiai*, for rough flour cannot be made from barley. (109b–c).

After several other deipnosophists have responded to Pontianus, Galen (Athenaeus' semi-fictional version of the Galen under discussion in this chapter) files a research report on what medical writers have to say (115c–116a). He quotes Diphilus of Siphnos,[20] Philistion of Locri, and Andreas (on whom see below), and then comes to Mnesitheus: 'Mnesitheus says that bread is more digestible (*eupeptoteron*) than barley cake and that bread from *tiphê* is more sufficiently nourishing because it is digested with no great effort. Conversely, bread made from *zeiai* if eaten to excess is heavy and difficult to digest. The result is that those who eat it do not enjoy good health. And you must know that cereals that are not parched or ground generate wind and heaviness and cramps and headaches.'

What overall treatment does Athenaeus give on these primitive wheats? His first reference to these cereals is oblique, coming at the end of a series of different kinds of bread. Bread is under discussion since the bread stage of the meal has been reached. The authority chosen is Tryphon of Alexandria, a grammarian of the first century BCE who wrote on pronunciation

[20] Athenaeus tells us that Diphilus was physician to Lysimachus. Interestingly, Athenaeus quotes this Hellenistic doctor extensively on foods but Galen ignores him in *Food*.

and dialect forms. He is apparently an ideal source for a deipnosophist, effortlessly supplying a list of recondite words (cf. 109e, 114b, 114e) that can be refashioned from a grammarian's manual into service at a literary dining table. Both Pontianus and Cynulcus (who is also involved in the discussion but not quoted above) are listed among the philosophers rather than the grammarian deipnosophists, but as we shall see, nearly all deipnosophists display to some extent the interests of the grammarian. Galen, who represents the medical interest, draws exclusively on medical texts, but other doctors in the *Deipnosophistae* also display concerns within the domain of the grammarian (e.g., 3.116f–118d).

Comparison between the interests of the deipnosophists and those of Galen might reveal completely different working methods, to match the different aims of Galen and Athenaeus. To some extent this is so, and is illustrated by the discussion of wine by the deipnosophist Galen at 26c–27d, which differs from the work of the real Galen.[21] Galen in his own treatises presents case studies of patients,[22] and (in the case of cereals) goes out into the fields to discover what the peasants are growing and what their names for each cereal might be. Athenaeus is more concerned with the finished product, bread, than the cereals in the field. But Galen, too, mentions finished foods (as did the Hippocratic author in *Regimen II* 42 and 44). So, on bread he says, 'After wheaten breads the best are the ones from *olura* when it is a good strain; and those from *tiphê* are second. But the latter in no way falls short of the *olura* breads when the *olura* is of poor quality. When the *tiphê* is very good quality, warm breads from it are much stronger than those from *olura*' (*Food* 1.13). He goes on to note heaviness in the stomach. We might have expected a link to Mnesitheus here, since the views are similar to those quoted in Athenaeus, but there is none.

[21] See Nutton (1995a) 368–9. Athenaeus' presentation of Galen is limited. He is introduced at 1.1e as 'Galen of Pergamum who published so many philosophical and medical texts that he exceeded all who went before'. But none of these works is quoted in the *Deipnosophistae* and Galen only speaks twice, on wine and on bread. The descriptions of Italian wines seem to follow different criteria from those in Galen's own works: Brock (2000). On bread, as noted above, Athenaeus' Galen draws on the same medical texts as does Galen in his own works. In particular, the same text of Mnesitheus of Athens is quoted at Ath. 115f and *Food* 1.13. The relationship of Athenaeus with Galen may be compared with the treatment of Plutarch, whom Athenaeus quotes only twice, but who seems to be adapted in a complex way into the deipnosophist Plutarch of Alexandria. Berra (2005) explores Athenaeus' conversion of the Chaeronean Plutarch into Plutarch of Alexandria. As he shows, Athenaeus seems to take little from Plutarch by way of intertext but to engage with him seriously through a fictional character based on Athenaeus' own centre of scholarship, Alexandria.

[22] Notable examples are the young vegetarian in Alexandria (*Food* 1.25); the young man who ate mushrooms (*Food* 2.67); wet-nurses, who ate wild herbs when hungry (*Food* 3.14).

Autopsy and local knowledge

Galen places particular emphasis on autopsy. This is found too among the deipnosophists, as noted above.[23] But it is less of an issue than in Galen. Consider two striking passages. The first comes from chapter 1.13 of *Foods*, the chapter quoted above:

> It is a matter for wonder that Mnesitheus was unaware in what way *olura* differs from *tiphê*. For each occurs in quantity in Asia, especially in the hinterlands of Pergamum, since the country people always make bread from them because the wheat is taken down to the cities.[24]

Again we see that Galen makes claims based on apparently authoritative evidence drawn from Asia Minor rather than mainland Greece or Italy. This might reflect the claims of Greek speakers against the claims of the centre in Rome;[25] or it may reflect Galen's implicit claim to have an empire-wide overview. His geographical frame of reference in *Foods* extends from Spain to Syria. Furthermore, in this context, Athens for Galen represents a dominant dialect within a linguistic system and not a geographical reference point for foods. Rome, by contrast, is referred to for its foods.

Autopsy, for Galen, can clinch a difficult problem of identification, and perhaps reinforce his claim to authority. His correction of Mnesitheus would appear to prove that the doctor who knows his cereals is superior to the doctor who has only read about them in books. Galen makes the case for autopsy particularly strongly in *Simples*: in the introduction to book 6 (XI, 793–4 K) he says that

> Pamphilus is clearly a man who assembled a book on plants and did so as a *grammatikos* from the books he had written. He had neither seen the plants he wrote about nor tested their properties, but he believed all those who wrote before him without testing them.

Later Galen adds, (XI, 796 K) 'Pamphilus has not even seen in his dreams the plants whose form he tries to describe.' For himself, Galen declares

[23] The entry on the *latos* in the catalogue of fish (Ath. 311f–312b) is a particularly good example. Athenaeus recalls the fish of the Nile even after many years' absence from Egypt. The Nile *latoi* are similar (*paraplesioi*) to the *glanis* of the Danube. Athenaeus attempts to comment on many different species, but without reference in this case to zoological authors, which is unusual for the fish catalogue.

[24] Galen's rhetoric of wonder could be turned against him, for he seems unable to distinguish rye from wheat in Thrace: see above.

[25] See Swain (1996) and Flemming in this volume.

I censured those who first of all described the forms of plants, thinking it best to be an *autoptês* learning in the presence of the teacher and not to be assimilated [to the study of plants] by guides in books. (XI, 796–7 K)[26]

Galen champions autopsy, but not at the expense of books. Discrimination is needed in the study of books. Galen appears to find Pamphilus to be a bad botanist but a good lexicographer. Good books in pharmacology, for example, that he can recommend (proemium to book 6) are by Dioscorides of Anazarbus, Sextius Niger,[27] one of the Heraclides[28] and Crateuas.[29] Galen in his treatment of Pamphilus thus follows the lexicographer in his ordering of terms alphabetically and rejects his botanical observations on the grounds of too little personal knowledge of plants. He has further complaints against Pamphilus (book 6 (XI, 792–3 K)), who, he says, turned to the stories of old women and some nonsensical mumbo-jumbo from Egypt, mixed up with incantations which they murmur when they pick the plants.

Reading technical works

Galen's range of reading is thus not completely different from that of the deipnosophists. Athenaeus is as likely to quote Diocles of Carystus and Mnesitheus of Athens as is Galen. The deipnosophists have also read Pamphilus widely; Andreas too,[30] whom Galen also censures, and possibly Heraclides of Tarentum also (though there is confusion between several medical authors of this name). A difference is, however, discernible between Athenaeus and Galen, in discriminating between books read. Athenaeus' diners accept almost all books with relish, the rarer the better, the more

26 We might contrast Galen's censure of the research techniques of Pamphilus (for which see Jouanna and Boudon (1997)) with his adoption of alphabetical order for the plant section of *Simples*, explicitly after the example of Pamphilus.

27 An important botanist also used by Pliny and Dioscorides. Scarborough and Nutton (1982) 206 note that Dioscorides often uses Sextius without acknowledgement. Sextius is an important example of a Roman author who wrote in Greek and was used by Galen. Roman authors are admitted to *Simples* (as they are quoted occasionally in the *Deipnosophistae*) but Galen does not cite Roman authors in *Foods*. It may be that Galen thought Roman botanists had more to offer than farming and medical authorities; Galen may have been influenced by disciplinary conventions between pharmacology and nutrition; or the distinction between Greek authors and Roman authors writing in Greek may have had some other significance in technical fields at this date.

28 Possibly the Empiricist Heraclides of Tarentum who was taught by Mantias (of whom Galen also approved) and whom Dioscorides also mentions in his preface (1), on which see below.

29 The doctor of Mithridates VI of Pontus, 'the most celebrated pharmacologist of antiquity' (Scarborough and Nutton (1982) 204).

30 Another Hellenistic doctor, from the court of Ptolemy IV. He wrote, among other things, on bites and on false belief.

canonical the better. Pamphilus the grammarian; Homer or Plato; the rare Matro are all embraced, with the critical comments on text and manuscript traditions that Galen also uses. Both Galen and Athenaeus, for example, comment on the Hippocratic authorship of the treatises known as *Regimens* (e.g., *Food* 1.1; *Deipnosophistae* 45e–f). Athenaeus has little more interest in Hippocrates than that. Galen has enormous interest in Hippocrates and takes on the interpretation and improvement of Hippocratic thought in many different forms, ranging from commentaries to comments in passing. With reference to books he has read in general, Galen rejects some and relies on others, according to doctrine and to the principles of order, which vary from book to book.

Crucial questions about order are raised for Galen by Dioscorides, whom Galen quotes with approval in both *Food* and *Simples*. On the vexed matter of the primitive wheats and their correct names, Galen (*Food* 1.13) turns to *On the Materials of Medicine* and finds that Dioscorides has identified two kinds of *zeia*, and compares them with *olura*, not so much as plants but in bread and porridge preparations. As plants, they are distinguished by being either single grained or two-grained. Here, he simply quotes Dioscorides and does not comment further, possibly because this level of botany (albeit medical botany) is beyond his competence.

Dioscorides will help us to place Galen and Athenaeus in this difficult area of ordering materials that have very technical applications. In his striking preface to *On the Materials of Medicine*, Dioscorides sets out certain principles.[31] He aims for a survey that is complete (1) and looks for more than written sources. He aims for accuracy (both precision and comprehensiveness),[32] which is lacking, he says, among recent writers (2). Predecessors have been inconsistent in ordering material, some being guided by properties, others by an alphabetical order: Dioscorides will group material by properties (3–5).[33] Much attention is given to season, parts of the plant and location, for which Dioscorides stresses his widespread travels which have given him extensive opportunities for autopsy. The virtues of Dioscorides are largely replicated in Galen, where autopsy, reading and seasonal and botanical details are valued. Galen turns to more literature than Dioscorides might wish: Herodotus and Homer are quoted in the section on primitive wheats, alongside Theophrastus and Dioscorides (Homer at 1.13.9; Theophrastus at 1.13.11; Herodotus at 1.13.12; Dioscorides at 1.13.13–17). We might

[31] On the preface see especially Scarborough and Nutton (1982).

[32] Compare Galen's complaints about the doctors' comments on primitive wheats, quoted above.

[33] Scarborough and Nutton comment on the manuscript tradition of Dioscorides which sought to rearrange his material into alphabetical order.

then place Galen somewhere between the technical author Dioscorides and the literary author Athenaeus.

Ordering the results of research

Once the results of research have been gathered how are they to be ordered? Here too, Galen can help us to understand what Athenaeus is trying to do. In *Food*, Galen says (2.1) that he has followed a number of earlier works in dividing his books into cereals and pulses, other plants and finally fish and meat. He also follows those who place the most useful cereals first in their section, and the most nourishing meats first (starting with pork) in their section.[34] Alphabetical order is thus not used, and for ease of reference the reader is urged to refer to a particular book for a particular class of food. Within books, related plants are often put together. Thus wild chickling (see above) follows chickling because (1.27) 'the seed is similar to that of chickling and some people do not think they are separate species'. In *Simples*, by contrast, as we have seen, Galen uses alphabetical order in some books but not in others.

To order by affinity or alphabetically? Athenaeus and his semi-fictional diners consider both approaches to ordering the vast literary materials they have been researching. In some books of the *Deipnosophistae*, the dialogue form is used and a speaker reports on his reading at some length. Examples of this, with some listing in addition, may be found in book 14 on cakes and book fifteen on garlands. Elsewhere, a list is given: vegetables and fruits in book 2, breads in 3 (see above), fish in 7, meats in 9, cups in 11. Lists often begin with a reference to an authoritative glossary, such as Callimachus (643e), who, in his *Pinax* of miscellaneous writings, recorded books on cakes. At 676f, Ulpian is urged to show more originality in listing garlands than mere reference to a standard work, the *Garlands* of Aelius Asclepiades. These lists in Athenaeus may or may not follow alphabetical order. They may refer to other orders. They may also consider affinities. Because of the grammatical and lexical interests of the work, affinities are often determined lexically rather than botanically or zoologically. This is best seen in the extensive list of fish in book 7. On the *muros*, for example (312e), Athenaeus quotes Aristotle, in book 5 of *Parts of animals*,[35] who is reported as saying that the *muros* 'differs from the *smyraena*'. In the next entry (313a), on the *mainides* or sprats, Speusippus the philosopher

[34] A different order is followed in *On the Thinning Diet*, where the list begins with the most efficacious foods for thinning the humours, namely alliums.

[35] A mistake: in fact book 5 of *History of animals*.

is quoted for his book *Similars*, which declares that, '*boakes* and *smaridae* are similar to the *mainis*'. Speusippus is widely used in book 7, and quotation from his *Similars* helps to locate the bewildering world of fish with their many local names and apparently countless species. To some extent the division between zoologist and lexicographer is artificial, since Aristotle, for example, finds difficulties with names[36] and Speusippus on the *mainis* is clearly trying to compare similar fish rather than similar names.

CONCLUSION

Both Galen as a doctor, then, and Athenaeus as the orchestrator of a literary symposium have stepped into the very technical areas of botany, zoology and many others. Galen does not seem to challenge the working methods of botanical writers such as Theophrastus and Dioscorides in the same way that he takes on medical writers such as Mnesitheus and lexicographical writers such as Pamphilus. From a pharmacological perspective, Scarborough and Nutton judge that Galen's catalogues of Simple and Complex Medicines do not compare well with Dioscorides for accuracy, excellence and in particular pharmacological properties.[37] Again, (if this judgement is correct) Galen has entered into a technical area in which organisational skill, botany and medical skills are all needed, and his botany is not strong enough. Nonetheless, he uses autopsy and travel, as Dioscorides would wish, along with seasonal and regional factors and is also greatly concerned with properties or *dunameis*.

However, what is at issue in this chapter is not Galen's technical expertise as a pharmacologist, but rather his approach to ordering the material he has gathered. One of the striking things about that approach is the way in which it is marked, as is also the case for Athenaeus, by engagement with distinctively Imperial intellectual preoccupations, many of which were themselves vehicles for exploring relations between Greek and Roman culture, or between Greek past and Greek present. Galen's writing includes, as we have seen, literary sources such as Herodotus and Homer, which puts him firmly in line with the literary preferences of other Imperial Greek

[36] See for example Aristotle on crabs, *Hist. an.* 525a 30–525b 6. He considers close resemblance (*paraplesion*), names and varieties that are too numerous to describe and too small to have a particular name.

[37] Scarborough and Nutton (1982) 191. They note that Galen did not harmonise the comments from the many sources he had read.

authors. It also includes material arising from arguments over Atticism, on which he has much to say, often of a dismissive nature. Generally he seeks clarity and easy understanding in place of the complexity and obscurity that Atticism might foster. At the same time, as we have seen, Galen was quite capable of checking a term with Aristophanic usage.[38] Atticism is also an issue for Athenaeus, since Ulpian the symposiarch favours pure Attic precedents (e.g., 94c, 347c–e, 366a, 368c) while Cynulcus and others argue for clarity and contemporary relevance, not least in regard to Latin usage. Ulpian in fact is not simply an Atticist, but a Syratticist, an Atticist from Syria, on which a number of Ulpian's fellow diners make satirical comment. Certainly attention is often drawn to Ulpian's poor understanding of Latin (e.g., 97d–98e). This anti-Atticism is close to Galen's position in *Food*, where he notes Latin terms as required. Thus on *smyrnion*, or Cretan Alexanders, Galen notes 'In Rome now, everybody usually calls this vegetable *olisathron*, not *smyrnion*' (*Food* 2.51).

Atticism is one distinctive intellectual concern of the Imperial period. Fascination with geographical range, which perhaps owes something to the awareness of the geographical scope of the Mediterranean world as it is united under Roman rule, is another, discussed above. A third is the use of Latin texts by Greek authors. Athenaeus cites some Roman authors (the brothers Quintilii, cited above, seem to be the most recent). Galen uses no Roman authors in *Foods*, but does use Sextius Niger and others in *Simples*.[39] Dioscorides in his proemium clearly had no difficulty in using Bassus, Sextius Niger and other hellenising Romans. Galen and Athenaeus are typical of the Greek technical writing of the Roman world both in engaging with Roman authors but also in holding them at arm's length, as a relatively small part of the network of sources they rely on.

Just as important for our cataloguers as these issues of Greek and Roman cultural interaction was the project of bringing technical writers from a special domain to a broader one. Both Galen and Athenaeus read and brought medical texts, botanists and zoologists into their own texts, though Galen's practice varies according to treatise. These general principles highlight some of the range of possibilities open to a compiler in the period. Compiling is not a term of abuse (as it is often applied to Athenaeus), in the mind of Galen at least, since he, the cataloguer with utility in mind, clearly sees it as vital for medical practice. Part of his job is to give a comprehensive account as

[38] On Galen's Atticism see further (among others) Swain (1996) 59–62, Wilkins (2003).

[39] Nutton (1988) 316 notes that Galen is not quoted in Latin medical sources until the fifth century CE.

accurately and as precisely as possible. Galen's claim to authority embraces interpretation of Hippocrates, the best methodology for classifying foods or drugs, breadth of reading in scientific texts, literary texts and lexicographical texts and autopsy.

The *Deipnosophistae* seems to have been composed for readers familiar with these research techniques. Massive reading is expected; consultation of library catalogues or the *pinakes* of Callimachus (for example) on cakes or fish is normal practice. There is some, but limited autopsy. There is criticism of texts in order to establish authenticity (as there is in Galen). When we consider the presentation of research, Athenaeus seems to subvert expectations. Alphabetical ordering of massive amounts of data is sometimes used and sometimes not (as in Galen, who is not himself always consistent even within one book). Standard works are sometimes welcomed and sometimes dismissed. The invasion of the catalogue by foreign material occurs on so many occasions that we should probably explain the phenomenon as part of the design of the work. Athenaeus takes us by surprise in this respect, as he does in his setting up Homer as the architect of the symposium and the founder of the genre of sympotic prose, and Plato as the main threat to the genre. The *Deipnosophistae* is a literary product full of paradox and surprise. This extends to its form as much as its content, which can present everything from fish sacrifice to the foundation of the Olympic Games by a cook. That Athenaeus did not seek to set out his researched data as Pliny does, with sources set out within books ordered by topic, suggests that lexicography and cataloguing were sufficiently familiar to non-specialists to be suitable for subversion within the sympotic genre. Such a picture conforms with Athenaeus' apparent subversion of other forms of power and authority, not least that of Larensis the Roman host, who preaches restraint but presides over a most lavish table in Rome.[40]

We are told at the beginning of the *Deipnosophistae* that Larensis has gathered together diners who are skilled in special areas of *paideia*. Comparison with Galen's methods show how knowledge is gathered in this period. Athenaeus follows much of this method, and gives it a special twist. This can even be seen in relation to Galen's watchword, utility. Athenaeus engages with utility, while also often subverting it. The Epitome (the summary version of the *Deipnosophistae* which is the only text to survive for books 1, 2 and part of 3) at 1a says that Athenaeus introduces the uses, names and explanations of all his subject matter. At 185a, Athenaeus declares that

[40] See for example Whitmarsh (2000), Wilkins (forthcoming).

he is going to introduce the most useful parts of the symposium, for which Homer is the best witness. And at 694c there is a discussion of songs, which give useful advice for life. These claims to utility are not frivolous. They help the reader to see that Athenaeus has blended methods that belong to technical writing into a sympotic form.

CHAPTER 4

Guides to the wor(l)d

Andrew M. Riggsby

Most of the articles in this book focus on the ideologies implicit in particular means of packaging and processing information; the present essay has a more technological focus. I do not mean to suggest that the technological and ideological approaches are, in the last instance, separable, nor even that such a separation would necessarily be desirable even to the extent that it might be possible. But, while one might be able to postulate that, say, British world imperialism and the industrial revolution that gave it much material support had shared ideological underpinnings, it is clearly not the case that ideologies can simply produce any resources that might further them. While I will discuss the role of technology in the broader culture, I will approach the question first by surveying just what technology is available. In the first section I will treat several general issues having to do with tables of contents in general (beginning with their definition) and several commonalities of the four existing Roman examples. The next sections examine the distinctive uses to which tables of contents are put in those four texts (taken in chronological order). Then, I return to a more general issue of how the potential and actual uses of tables of contents interacted with their functions in individual texts and with each other. Finally, I will make a few remarks on causal connections between ideology and technology in this case.

TABLES OF CONTENTS

I begin with two stipulative definitions. A 'table of contents' ('TOC' through most of this chapter) is a summary of the contents of a work by means of listing its contents in abbreviated form and in the order of the text. An 'index' is a summary list ordered with reference not to the particular work but to some more broadly available principle (typically alphabetisation today) and then keyed back to the work by some diacritic feature (typically page or section numbers today).[1] A table of contents may also include such keying, but it is not definitionally necessary. I introduce the distinction primarily to limit the present study to tables of contents; to the best of my knowledge, and despite more casual use of the term in discussing several of the texts below, there are no indices in classical Roman texts. A table of contents or an index can be placed at the beginning or end of a work (as varies today by national custom) or at the beginning or end of large segments of an extended work or at both points. In this last case the point is normally to divide the work up at larger and finer levels of detail (a common approach in modern cookbooks). In principle, this can be done with many different levels of detail and, given appropriate typographical devices, without creating separate tables for every level of observation. Such multiple-level organisation will be one of the main technical issues considered below.

Now, Doody[2] has objected to the use of terms such as 'table of contents' and 'index' as anachronistic in Roman contexts. In particular, it falsely naturalises centuries of the eventual evolution of the book and of scholarly methods. The underlying concern here is significant, but need not deter us from using the modern terminology as long as we keep the potential problem in mind. Doody, in the specific context of Pliny uses *summarium* (taken from the author himself) but neither this nor any other Latin word I am aware of has classical authority as a general term for the four lists of contents I will discuss below. Yet, as we will see, they share numerous similarities, and it is clearly fair to take them as a group. I prefer to use the modern terms, but to emphasise that the definitions offered above say nothing about use, and more generally to keep in mind throughout the discussion the issues raised in Doody's paper.

Let me also introduce a distinction between two uses of the table of contents. (The two are not exclusive, nor, of course, do they exhaust the possibilities). One is reference. A TOC may be used to allow the reader to skip more or less directly to a specific part of the text to learn about any chosen topic. The other is segmentation. A TOC may give shape to a work which the reader may still read linearly. The works under study here all comprise fairly miscellaneous collections of 'facts'. These need to be gathered together in bunches to have much importance. The very order of a text helps create these segments, but they will be clearer to the reader

[1] There are other possibilities even beyond replacing numbers with letters or the like. Chapters are often marked by colour as well as number. Within such a system, one could use variation in intensity instead of page numbers.

[2] Doody (2001) 2.

if signalled somehow. Thus the TOC helps set up the intended context for individual observations. A TOC might be judged 'useful' by a reader if it could perform either of these tasks, even if it did not achieve the other.

How common was the TOC in Latin writing? A few manuscripts of Suetonius' biographies of grammarians and rhetoricians open with a crude table of contents, a list of the persons treated in the order of treatment in the text. However, even editors who feel this is ancient assume it is not Suetonian.[3] Nonetheless, the device does clearly exist in classical Latin texts. It seems, however, to have been quite rare. The individual books of Columella's *On country matters* (*Rust.*) also begin with such lists, although these can be shown from Columella's own text and the manuscript tradition to have been duplicated from their grouping at the end of Book 11 to the beginnings of the respective books by the ninth century.[4] The manuscript traditions of Pliny's *Natural history* (*HN*) show similar transformations.[5] In addition to the three just cited, the only other examples I can find are in Scribonius Largus' *Compounds* (*Comp.*), a mid first-century CE collection of medical remedies and in Aulus Gellius' *Attic nights* (*NA*). One might imagine, if they were common only a few centuries later (and even readers at that time were apparently comfortable with inserting their own, as in the Suetonius example above), that the apparent lack at earlier times is due to a gap in the evidence rather than actual practice. When Pliny mentions the device to Titus in his prefatory letter, he seems to allude to its rarity by citing a now-lost precedent:

> And thanks to you no one else will have to read through all this, but, when anyone wants something, he will seek precisely that and know where to find it. Valerius Soranus (in Latin, at least) did this before me in the books which he entitled 'Enlightened'. (*HN* pr.33)[6]

This Soranus appears to be the Q. Valerius Soranus (tribune of the plebs in 82 CE) cited by, among others, Cicero on linguistic and antiquarian matters.[7] However, while we can point to intervening examples, they are hardly from prominent texts.

The actual rarity of the table of contents is also suggested by the fact that not only Pliny, but also the other three surviving authors preface it with an explanation:

[3] Kaster (1995) 41–2. [4] Henderson (2002b) 111–13.

[5] Pliny: Doody (2001) 3–9; similarly on Gellius see Holford-Strevens (2003) 31.

[6] *Tu per hoc et aliis praestabis ne perlegant, sed, ut quisque desiderabit aliquid, id tantum quaerat et sciat quo loco inveniat. Hoc ante me fecit in litteris nostris Valerius Soranus in libris quos* ἐποπτίδων *inscripsit.*

[7] Cic. *De or.* 3.43, *Brut.* 169; Gell. *NA* 2.10.3.

First, then, I have added [a list] below of what problems the recipes are calibrated and fitted to, and have numbered it, so that it is easier to find what one seeks. Then to the ailments I have added the names and proportions of the medicines of which the combinations consist. (Scrib. *Comp.* pr.15)[8]

But since it often happens that our memory of what we have learned fails, and it must be frequently revived from notes, I have added an outline of all my volumes, and when necessary, it will be easy to find what is to be sought in each one and how each thing must be done. (Col. *Rust.* 11.3.65)[9]

Here I have laid out all together the headings which go with each entry so that it will immediately be clear what can be found in each volume. (Gell. *NA* pr.25)[10]

That is to say, the authors do not just supply a list of topics, but also explain that it is to be used by the reader to jump to a particular topic. Note also that none of the four even has a word for 'table of contents' or even 'list.' They all name what is to be listed: *capita rerum* ('headings' or 'chief points') for Gellius, *argumenta* ('outline') for Columella, and indirect questions for Pliny and Scribonius. One might even go so far as to note some verbal similarities between the four prefatory passages and their promise of a book-by-book accounting:

Pliny	Gellius	Scribonius	Collumela
quaerat	quaeri	quaeretur	quaerendum
inveniat	inveniri	inveniatur	[reperior?]
subiunxi	–	subiecimus	subieci
singulis libris	qui . . . libro	–	quoque
[indirect questions]	capita rerum	[indirect questions]	argumenta

Now, none of these usages is individually striking given the general sense of the passages. In the aggregate, however, we might wonder why there is not more small variation such as Columella's use of *reperior* for *invenire.* It is possible – though certainly no more than that – that all four passages have a common source. (For a more specific argument about the relation between Pliny and Gellius, see below.)

[8] *Primum ergo ad quae vitia compositiones exquisitae et aptae sint, subiecimus et numeris notavimus, quo facilius quod quaeretur inveniatur; deinde medicamentorum, quibus compositiones constant, nomina et pondera vitiis subiunximus.*

[9] *Quoniam tamen plerumque evenit, ut eorum, quae didicerimus, memoria nos deficiat eaque saepius ex commentariis renovanda sint, omnium librorum meorum argumenta subieci, et cum res exegisset, facile reperiri possit, quid in quoque quaerendum et qualiter quidque faciendum sit.*

[10] *Capita rerum quae cuique commentario insunt, exposuimus hic universa, ut iam statim declaretur quid quo in libro quaeri inveniri possit.*

Finally, we should note the relative complexity of the syntax of most of the items in these tables of contents. A great many are indirect questions ('whether the universe is finite', Plin. *HN* 1.2.1), indirect statements ('that a young vine must be pruned before the winter wind', Col. *Rust.* ind. 4.11), or at least prepositional phrases ('on ancient frugality and on the sumptuary laws of old', Gell. *NA* 2.24). Only a few entries are simply nominal. Goody[11] has noted that, though in some respects a quite primitive device, the list is nonetheless a fairly arbitrary one, far from the forms of normal oral discourse; most distinctively in this respect lists are characterised by their discontinuity. The Roman tables of contents resist this formalisation, by pointing to a more normal discursive context. (That book incipits in Latin take the form of full sentences suggests a similar lack of abstraction.) The specialised usage has not yet been internalised by our authors.

SCRIBONIUS LARGUS

Scribonius' *Compounds*, apparently datable to 44–8 CE,[12] comprises a substantial prefatory letter to one C. Iulius Callistus, followed by a table of contents, then the bulk of the work – a list of formulae for various remedies. Though it is the earliest and the shortest (117 Teubner pages) of the works under consideration, it has arguably the most sophisticated table of contents. The entries are brief and normally begin with a prepositional phrase with *ad* giving the ailment to be treated; occasionally this is preceded by a description of the form of treatment (for all headaches, 4; *catapotium* for a productive cough, 87). The most novel feature of the TOC is that each of these entries is then followed by an index number, keyed to the individual paragraphs of the main text, each containing a recipe (occasionally more than one) for the ailment in question.[13] Thus this TOC is quite well adapted to a reference function.

Scribonius' TOC also seems to show two-level organisation. This is slightly less apparent from the text than the numeration, but would quickly become apparent to the user. The text itself is organised by symptoms. Different treatments for the same or similar ailments are generally found together. Moreover, the text is subdivided by over 100 lemmata at fairly short intervals.[14] The TOC does not show the same subdivision by lemma, but

[11] Goody (1977) 81–2, (1986) 54–5. [12] Sconocchia (1983) vi–viii.

[13] One of the manuscripts omits the numeration, but the prefatory epistle (15) explicitly directs the reader to look for them.

[14] Some of the later lemmata for plasters give the type and inventor, but not the function of the treatment. These may not have added much use value.

subcategories are frequently reflected in the wording of individual entries.[15] Many of those entries have a form something like 'for the same', 'another for the same', or 'also for the same, but made of multiple ingredients'. Whatever the precise form, the terms *alius* ('another', referring to the treatments) and *idem* and *item* ('the same' and 'also', referring to the complaint) nearly always signal the continuation of a group, as might be expected just from the meanings of the individual terms. However, the groups defined by this wording in the TOC do not cross the lines between groups in the text (as defined by lemmata), and in several cases correspond exactly to them. Subdivision, legible in both the TOC and the main text, further increases the reference value of the work to the reader. The overall ordering does the same as well. From the point of view of the user, the important issue will ordinarily have been to be able to find the treatment for particular diseases as they actually present themselves. Hence, that becomes the main parameter of arrangement. (Within groups, there is a strong tendency to start with the simplest treatments, which might be considered a similar convenience).

There is one set of exceptions to the signalling function of 'the same'/'also' (*item*/*idem*) noted above, and it does point to one limitation of Scribonius' organisational scheme. Each is used a few times to signal that a given treatment is effective for more than one ailment.[16] It has no good way to deal with such treatments. Now, in a modern drug manual this could be catastrophic. In that case, where there are more categories of information to deal with (contraindications, drug interactions, side effects), it is important to be able to gather the whole pharmacopia together to optimise treatment. Yet since all of Scribonius' compositions are equivalently effective (or are treated as such), the ancient physician was little handicapped. At worst, his list of potential treatments for any particular malady came in slightly short, and that only if he did not have time to think through his entire repertoire (rather than just this specific section of the text). And over the course of extended treatment, he did not need to depend so much on the TOC.

PLINY

The first book of the Elder Pliny's *Natural history* consists of a table of contents, listing the various contents in order and divided by books. (When referring to the text of the TOC below, I will cite the entry

[15] There are apparently lemmata between sections 18/19, 31/32 and 200/1.

[16] Numbers 1, 25, 26, 81–2, 92, 121, 126, 213.

corresponding, say, to book 5, chapter 12 simply as 5.12. I trust context will disambiguate chapters of the main text from their TOC entries.) It is preceded by the preface, so the work has essentially the same structure as Scribonius'. There is also a fair amount of non-summary material interleaved into the TOC. Pliny offers counts of the 'facts' recorded and lists of the authorities used in each book. None of this information is correlated with the topics.

I need to preface further discussion of Pliny by saying a few words about the chapter numeration of modern editions. By working back and forth between the TOC and the main text, it is generally possible to divide the latter into segments that must have been intended by Pliny. The list of topics and their actual execution generally line up, and difficult cases can often be resolved by establishing the borders of surrounding chapters. There are a few problems with the modern numeration. In some cases a single chapter is made to correspond to what look like two topics in Pliny's index (e.g., 8.61). And in some books (e.g., 19, 28) there are long swathes of numbered sections that do not correspond to anything in the TOC. This seems to indicate a lapse in Pliny's TOC. Most reconstructible chapters are of fairly modest length, and there are seemingly obvious changes of subject within the potentially unbroken sections of book 19. There are also occasionally problems of ordering in the TOC. So, for book 25 we read:

on the greatest pain (25.7)
discoverers of noble plants (7–39)
  moly III (8)

At first, it might seem that editors should have divided what they call chapter 7, but the discussion of discoverers comes up before that of pain in the chapter. Nonetheless, it seems fair to take these chapters as original units, at least compositionally, unless there are specific reasons not to in individual cases.

In the process of lining up TOC and text, modern editors have seen that some entries in the TOC seem to indicate higher order categories than the rest. That is, they correspond not to a single chapter, but to a whole set of related chapters, and so ranges of chapter numbers are recorded next to them. Mayhoff's Teubner text further emphasises such groupings by indenting the sub-entries; Beaujeu's Budé prints the higher order entries in bold-face. The close agreement between editions as to these groupings seems to indicate something originally Plinian, just as in the basic chapter division. Take, for instance, this segment of the TOC for book 7:

examples of the greatest reverence (7.36)
those outstanding in the arts (37–39)
astrology, medicine (37)
geometry, architecture (38)
painting, sculpture in bronze, in marble, in ivory, carving (39)[17]

Someone reading the previous sections will note that *excellentes* here is substantive: 'those outstanding in'. The substance (and grammatical case) of the entries for the next three sections pick up on *artibus* ('in the arts') and that binds the grouping together. This ends with the new substance and syntax of the entry for 40. I take it these editorial decisions can also largely be taken for granted unless specific circumstances warrant. So, for instance, Naas[18] is justified in taking them as evidence for a greater degree of structure in Pliny's working methods than some have wished to concede. I noted in the introduction that such multi-level structure would be one of my main interests in this paper, but what I am particularly interested in is a more audience-oriented approach. Would such a structure have been useful or even visible to an ancient reader of Pliny? In fact, we will see that there are numerous problems.

First, there are actually two slightly different situations which modern editors treat identically. In the passage just cited, the heading *artibus excellentes* ('those outstanding in the arts') is purely conceptual; it can only be seen as an organisational device. There is no such topic in the main text. But there are many examples of the following sort:

wonders of the sea (2.101–5)
what powers the moon has over land and sea (102)
what the sun (103)
why the sea is salty (104)
where the sea is deepest (105)

Here 'wonders of the sea' could refer to a single chapter, rather than a whole grouping. The two configurations are roughly equally common.[19] And in a few cases of this sort, the opening section of a group in the main text is itself nearly just a lemma:

[17] *summae pietatis exempla* (7.36)
artibus excellentes (37–39)
astrologia, medicina (37)
geometria, architectura (38)
pictura, scalptura aeraria, marmoraria, eboraria, caelatura (39)

[18] Naas (2002) 189.

[19] Ignoring groupings of only two chapters, I count 43 where the purported super-category also corresponds to a chapter and 53 where it was purely organisational. Editorial differences might make for small changes in these figures, but would not affect the general picture.

wonders of fire and water together (2.107–110)
  on maltha (108)
  on naphtha (109)
  places always on fire (110)

The segment of the main text that could correspond to the proposed lemma ('wonders of fire and water together') reads: 'Let us add certain wonders of fire (the fourth element in nature), but first some from water' (107).

Now, frankly, it would have been hard for a reader to see much of this without use of modern typographic conventions, such as the indentations I have used above. Similar devices were not unknown by Pliny's time (notably in inscribed legal texts), but there is no evidence of any papyrus text relying so heavily on them.[20] (And note that with something like the Budé's bold-face headers, one can find the ends of sections only by reference to printed numbers, which seem clearly not to have been original.) This combined with the underlying formal inconsistency (pure lemmata in TOC or main text or neither) may suggest that what structure there was in the *Natural history* was not for the direct benefit of the reader.[21] The same is suggested more strongly by some other devices that could have been used to structure the TOC. There are entries in the TOC which are introduced with terms like 'their' (*eius/eorum*), 'in which' (*in quibus*) or 'in the same way' (*item*) that could point out groupings. For instance, Pliny's famous treatment of elephants is recorded as follows:

On elephants (8.1–11)
  on their senses (1)
  when they were first harnessed (2)
  on their docility (3)
  wonders among their deeds (4)
  on the nature of beasts in grasping their own danger (5)
  when elephants were first seen in Italy (6)
  their combats (7)
  how they are captured (8)
  how they are trained (9)
  on their birth and the other matters (10)
  when they are born. Their enmity with dragons (11).[22]

[20] Moatti (1997) 222–3.

[21] The same is also true at a higher level at which Pliny gives some books, but not all, individual titles.

[22] *de elephantis* (1–11)
  de sensu eorum (1)
  quando primum iuncti (2)
  de docilitate eorum (3)

This passage illustrates a number of potential strategies. The frequent use of 'their' (*eorum*) tells the reader to refer back, suggesting a continuing sequence. Alternatively, one might repeat a key term to emphasise ongoing sequences, as here at (6), but also more extensively at, say, 13.24–6. Terms meaning 'the rest' like *cetera* (e.g., 6.5) or *reliqua* (here) not only suggest membership in a sub-section, but also the end of that section;[23] here, however, Pliny has placed *reliqua* (10) somewhat sloppily. More abstractly, one might note the ellipsis of the subject (in all cases = 'elephants') in 2, 8, 9 and 11. Yet all of these strategies are used only intermittently. The segmentation they do create does not seem to fight against the one reconstructed on a purely topical basis, but it almost never adds anything either.

The issue of ellipsis may be of further interest here. This would be an extremely subtle signal if it were meant for benefit of the reader. Instead, it seems more like a symptom of the composition of the TOC based on the linear text of the *Natural history*. To repeat, I do not wish to deny that Pliny's composition involved structural sophistication; I suggest that it is not always obvious in the work itself (nor necessarily even in his ultimate outline).[24] If Pliny indeed composed the TOC in this fairly close-to-the-ground fashion, then he could easily have lost much of his own structure. This would also explain a phenomenon such as the following. At 8.17–21 we are given the heading 'on lions', followed by five reasonably straightforward sub-headings. However, if one reads chapter 8.17, it actually introduces not only lions, but the other big cats discussed in the following chapters (panthers and tigers). I noted above that the reader would have difficulty knowing how long a section might go on even with knowledge of when it started. If Pliny were composing the TOC linearly and close to the ground, he could well have the same confusion, and that would explain why the lemma for 8.17 is too limited. In such a large work as Pliny's *Natural history*, a multi-level hierarchy would have been valuable both for reference and for segmentation. Pliny may have had some of this

mirabilia in factis eorum (4)
de natura ferarum ad pericula sua intellegenda (5)
quando primum in Italia visi elephanti (6)
pugnae eorum (7)
quibus modis capiantur (8)
quibus domentur (9)
de partu eorum et reliqua natura (10)
ubi nascantur. discordia eorum et draconum (11)

[23] See already on these terms Doody (2001) 22, n. 20.

[24] Naas (2002) 192–3 argues that some numerical information in the TOC is based on an intermediate text or set of texts, rather than the main final text. This is possible, though not certain.

to start with (though perhaps inconsistently), but he passed little of it on to his reader.

Let me also mention one other organisational oddity here, which will be taken up more analytically in the following section. Books 29 and 30 both give animal-derived remedies. Both books, however, include almost exactly the same set of animals (and animal parts). Those in the former book are 'medical' as indicated by the head-matter on the history of medicine; those in the latter seem 'magical' on similar grounds. The order in each book is roughly parallel and based on the animal source of the remedy. Each rubric in either TOC lists the number of remedies based on the given animal in each book individually and then in total.

COLUMELLA

Columella's TOC comes not at the opening of the work, but as a postscript to his book 11. It is, however, introduced and explained by the immediately preceding text. Most of the lemmata in the TOC appear in the main text as well. This seems to have been an original feature of the text, but it is not signalled in the introduction to the TOC. It is hard to make generalisations about the entire TOC. The lemmata take on many forms, including all those attested in any of the other TOCs of this chapter (single and multiple entries, all syntactic constructions, long and short, use of pronouns like *is* and *idem*), and even show some unique features. Some of them address the reader in the second person (3.8, 9; 5.1). At the level of the individual book, however, we may be able to make more of this diversity. More specifically, the choice of forms for different books seems to suggest a choice of orientations that the reader might take toward those books. Books 2 through 4 lay out basic procedures for preparing the land (book 2) and growing vines (books 3 and 4). These seem to be imagined to provide step-by-step instructions to the reader. The lemmata are primarily in the form of indirect questions (how? how many? how much?). Key terms such as *vitis* (vine) or *malleolus* (a technical horticultural term usually translated as 'slip') are repeated in successive lemmata, so that they may each stand alone (here Columella is much more consistent than Pliny). They are arranged roughly in the order they would need to be carried out through the year. Once the text even points to the sequence:[25]

[25] One might also point to *deinde* at 2.3, though that is arguably a logical connector, more than a temporal one here.

How meadows are made from fields
How meadows are cultivated once made.
(1.27–8)[26]

Here the 'once made' (*facta*) in the second line warns the reader that the 'are made' (*fiant*) of the first line needs to be carried out earlier, imagining not a timeless set of procedures, but a particular execution of them which does require an order.[27] Twice the verb even falls into the second person, suggesting a specific audience.[28] Contrast books 6 and 7 on various kinds of farm animals (as well as 8 on birds). Here the lemmata are almost always in the form of 'on' (*de*) plus a word or short phrase. They also show some segmentation. The main sections are arranged by type of animal: 'on horses', 'on mules'. Subtopics to these follow immediately and are frequently marked by lack of the preposition and/or use of a pronoun (usually 'their') to refer to the section topics. So, 'their treatments' (*medicinas eorum*, viz., horses) shows both markers (again, when Columella does this at all, he does it more consistently than Pliny). Now, this distinction between a sequence of orders and the more encyclopedic approach is not strictly required by reality. Horses, for instance, could perhaps be treated chronologically by working though their life cycles. Nonetheless, the formal differences are clear, and seem to suggest that Columella imagined his reader consulting different books in different ways.

We might then suggest that Pliny was doing something broadly similar in his elaborate co-ordination of books 29 and 30. He has certainly expended more energy on organisation than we have seen previously. Now, he has not adopted a physician's work process as his model. If he had, he would have arrived at something more like Scribonius' organisation (i.e., based on symptoms). He may, however, have in mind a reader who is differently oriented toward these books than toward the others (whether or not they happen to be the same person). To specify that orientation is more difficult. One could imagine a 'scientific' or 'theoretical' reader, who might want to compare or contrast the domains of magic and medicine, who could use

[26] *quem ad modum ex arvo prata fiant*
quem ad modum facta prata colantur

[27] For a parallel phenomenon, see Riggsby (2003) 176–7 where I argue that Pliny's time-deep familiarity with his own villas vitiates his sense of their spatial order, whereas someone else's villa is more likely to be presented linearly. In Columella the question is temporal rather than spatial order.

[28] Here again we can see a parallel phenomenon in Pliny; see Riggsby (2003) 175. In de-ordered contexts, most people tend to fall out because they have no consistent place in the author's experience. There are, however, categories of functionally equivalent persons in some cases, and it might be argued here that the addressee of advice is such a person (or, rather, place-holder for many people).

the parallel lists as controls. Or it might be useful to a person who has one method in hand and wants to compare others (this, we noted, was a potential weakness of Scribonius' version). At any rate, we should keep in mind this kind of book-by-book targeting of longer texts.

GELLIUS

Gellius' chrestomathy, the *Attic nights*, has the same general structure as the *Compounds* and the *Natural history*: prefatory letter, table of contents (foreshadowed by the letter), then main text. Though modern texts usually print the lemmata at the head of the individual sections, the introductory letter and the practice of the manuscripts do not really support their placement anywhere except in the TOC. In general, the syntax of lemmata in the four texts is not as standardised as in modern TOCs. As noted above, they include bare nouns (or noun phrases), prepositional phrases (usually with *de* 'about'), indirect statements and indirect questions. The form of Gellius' lemmata is distinctive in several respects. First, he is relatively sparing of *de* (and the equivalent *super*); instead, he prefers indirect questions or (most notably) a variety of constructions dependent on nouns which refer to types or segments of discourse: 'words', 'account/history', 'epigram', 'consultation', 'conversation' and so on. And even when he does use the *de*/*super* construction, the ultimate reference is still normally to text[29] or to a word[30] rather than the thing itself. For example, 'On the constellation which the Greeks call ἅμαξαν and we call *septentriones* and on the theory and origin of either word' (2.21) tells us more about words than constellations. I will return to the significance of this fact shortly. Secondly, there is occasionally more than one lemma per section, such as the tripartite:

> That great honours were granted of old to advanced age, and why afterwards the same was granted to husbands and fathers, and certain things on this point in the seventh chapter of the *lex Julia*. (1.3)

In itself, this would prevent the lemma from segmenting the work; that is, you cannot tell from the lemmata how many sections there are or what their topics are without referring to the actual text. Some diacritical feature, such as indentation, line-breaks, or literal rubrication would be required for that purpose, and we noted above that this would have been unlikely. (Fortunately, the radical changes of subject between sections guarantee that the standard section divisions are correct.) However, different questions within

[29] E.g., 2.7, 22, 24; 3.3; 4.7, 18. [30] E.g., 2.21; 4.16, 17.

a lemma are always joined by *et* or *-que* ('and'). This never happens between lemmata. Third, the information in the lemmata is shaped differently from that in the sections of the main text. A disproportionate number of the sections open with the name of the source on which Gellius is drawing. The easiest thing to do would be to reproduce the same pattern in the lemmata, too, and this is common enough. However, in a number of cases the name is delayed until the reader has some sense of the substantive topic (or even omitted altogether).[31] So, for instance 15.10 opens:

Plutarch, in the first of his books 'On the Soul' . . . said that many Milesian maidens have hanged themselves to death.

while the lemma for the same section is:

On the wilful and wondrous death of the Milesian maidens.

Here there may be a functional distinction. In the text, Gellius immediately confirms for the reader the authority and interest of the text in question. In the lemmata, he more often puts forward the topic, that is, a criterion by which the average reader is more likely to be searching.

Beyond the structural similarities of all four works under consideration here, however, Gellius' work has particularly close ties to Pliny's. Both, of course, are large collections of information on a wide variety of topics, but substantial overlap in the topics of their prefaces has convinced many that Gellius is specifically responding to Pliny.[32] Most important of these themes for present purposes is a particular economy of time noted by Henderson and Ker. For instance, Henderson[33] shows how Pliny (followed in part by his like-named nephew) emphasised a need to maximise one's return on available time (that is, time not already devoted to *negotium*) by means of *studium*. Ker[34] notes an additional constraint that is less well defined, but thereby the cause of potentially dangerous strategic negotiation. This is the proper use of night time, that is the 'borrowing' of limited amounts of darkness to supplement the day, without simply inverting the normal, healthy opposition of night and day. Both Pliny and Gellius respond to these imperatives with ferocious acts of compression. Students of Pliny have seen this as a gesture of power in his work; cataloguing the things of the world claims actual authority over them.[35] It seems natural then

31 Switched: e.g., 2.10, 20; 4.9; 5.7 Omitted: e.g., 1.14; 2.6; 3.1, 5, 6, 9, 13; 4.14; 5.17.

32 E.g. Astarita (1993) 20–3 and Holford-Strevens (2003) 165–6 on adversarial/ironic aspects of the relationship.

33 Henderson (2002a) 80–4, 88–9.

34 Ker (2004) 232–6.

35 Naas (2002) 421–3, 435–6 with much secondary literature.

that Pliny's TOC recapitulates the gesture made by the work as a whole.[36] Reduction of the whole world to thirty-seven books asserts control of that world; reduction of those thirty-seven books to one does the same thing. (The quantification of facts and authorities is perhaps then a third-order compression.) While I have suggested above that the TOC is not entirely symbolic, its reading in these terms must be at least partially correct.

What I want to suggest here is a parallel reading of Gellius' TOC. The terms of Pliny's lemmata are overwhelmingly things in the world. As I noted above, Gellius prefers the terms of discourse. Pliny was part of the state apparatus and wrote notionally for the emperor. His work is part of the project of empire and accordingly looks outward from centre to subordinate periphery.[37] Gellius' dedication, if any, is now lost, but remains of his preface do not appear to address a single, specific reader. While Pliny's writing was distinct from his 'true' duties, it was still meant to be useful to the same emperor. Gellius' is the opposite of duty, being a matter of recreation (pr.1). Its function is the enhancement of liberal education in the most general sense (pr.13). His is a private, individual project. Gellius, then, is attempting to master not the real world, but the world of discourse, reducing it first to twenty books of notes, then further to a single book of lemmata. His location may even be relevant here. Athens could perhaps stand as the symbolic centre of this world of texts, even as Rome was the political centre of the material world.[38]

CONCLUSIONS

What I have offered above is a provisional sketch of the table of contents in Latin literature. Though certain themes came up across discussions of different authors, I hardly developed either a historical narrative or a single conclusion. But this diversity in its own right suggests certain conclusions about why we do not see more of the device. Obviously, certain technologies rely on the pre-existing development of other technologies. The steam engine, for instance, could not exist until certain developments in metallurgy (among other things) had taken place. Less obviously, some technologies require their own broad availability to succeed. Telephones, for instance, are of little use if too few people have them. Hence, there can be a huge lag between the invention of a device, and the boot-strapping needed for it really to take off. I noted in the first section that Doody had

36 Ker (2004) 233 already adumbrates this claim.

37 See further Naas (2002) 416–17 on the geography.

38 For a different, though not contradictory, reading of the location see Ker (2004) 237.

recently raised the question of utility. The formal differences between the ancient and modern tables of contents almost always render the former much less useful to the reader looking for certain kinds of information. We might then wonder whether they were meant to be utilitarian at all. Perhaps their main value was somehow symbolic. On the other hand, all four texts stress some added utility to the reader of these devices.

To pursue the paradox, we might turn to another, more subtle observation of Doody's. To wit, different approaches to a given TOC can be seen as 'with' or 'against the grain'. That is, while certain information is hard to extract with the help of one of these tables of contents, other questions are easily answered. (The precise distinction varies considerably from text to text.) Individual texts are designed to be approached in certain ways, and thus in some sense are for certain audiences. Scribonius' text had the best defined audience of the texts considered here: doctors, and more specifically doctors whose patients presented with certain systems, and needed to prescribe appropriate treatments. Scribonius developed a version of the TOC that would accommodate them nicely, and whose theoretical weaknesses were of little practical consequence. The fact that he actually carried out this plan suggests we should not dismiss claims to utility quickly nor look for a radical cultural mismatch in that notion.[39]

It also explains a second technological oddity – that the 'best' surviving TOC was also the first. Scribonius had the most specific audience and so the easiest task. The various instances, moreover, were so scattered that each author was essentially working in isolation rather than benefiting from a growing tradition. Contrast here, say, the regular division of long literary works into standardised books, a development which was even retrojected once developed.[40]

In Columella and (perhaps) Pliny, however, we saw a somewhat different version of shaping the information of a particular work with a specific audience in mind. Recall that Columella seems to have expected different books of his work to be consulted in slightly different ways, and thus shaped their tables of contents differently. The relatively full syntax of both is a further indication of this audience-specificity. Recalling Goody, cited in the introduction, we can see this further as an indication of imperfect abstraction. But this kind of innovation can have bad long-term consequences. The better these texts are for their ideal readers, the harder they are for others. Moreover, if tables of contents do not work in a consistent way, they will be difficult even for users who may notionally be in the target

[39] On the utility even of Gellius, see Holford-Strevens (2003) 36–44.

[40] Moatti (1997) 223.

audience. Here the syntax just noted above as an 'indication' may have been historically important. Being able to fall back on it to some extent may have kept Roman authors from learning to let the overall form of the TOC do its work. At least some of these Roman tables had significant symbolic functions, yet at the same time they were meant to help readers find information. The fact that they were specifically targeted to certain kinds of practical users implies this. For the former purpose, they were generally well designed. For the latter, less so, but the cause of this was not so much conflict between functions, as difficulties in developing the technology which are not necessarily obvious to modern readers so accustomed to modern data design.[41]

BACK TO IDEOLOGY

Earlier in this chapter I suggested that there is a homology between the relationship of Pliny's *Natural history* to the outside world and the TOC's relationship to the main text of the work. The double compression is a double gesture of authority. Trevor Murphy has made a very similar claim and furthermore set it in an important context. Having shown the essentially 'triumphal' nature of the whole work's 'demarcation' of the world,[42] he continues:

> The index is a textual analogy to the *Natural History*'s use of the Roman triumphal procession as heuristic and analytical metaphor: both triumph and index make a thing known while at the same time signifying both the total availability and the total objectification of what is known.[43]

Given the nearly seamless fit of the analogy text:world::TOC:text and thus the parallel gestures made by both, it would hardly be surprising if the TOC (especially given its rarity overall) had been developed specifically for the same ideological context as the main text itself.[44] Yet, as far as we can tell, that was not the case. It is unfortunate that we know so little about the work of Valerius Soranus that Pliny cites as precedent, but his attested linguistic and antiquarian interests suggest a programme more like those of slightly later writers like P. Nigidius Figulus and M. Terentius

[41] Cf. Bodel (1995) on consular Fasti, whose form shifts from the Republic to the Empire. There is an ideological shift here in the nature of the consulship, but the new form is still well-designed to display the newly important information (succession patterns of suffects, rather than dates in office).

[42] Murphy (2004) 145–64, 214. [43] Murphy (2004) 214.

[44] I should make it clear that Murphy does *not* make this claim.

Varro. That is, to the extent that we might suggest a particular ideological background, it is more likely an inward look at 'Romanness' rather than a view outward to empire.[45] And at any rate, Soranus' and Scribonius' works were backed up by the expertise of the author, not (as in Pliny's case) just his diligence. Our earliest clear parallel for Pliny's TOC arises, recall, in Scribonius' narrowly occupational context. That Pliny would be drawing on a tradition originating with works by scholars like Soranus or professionals like Scribonius is reasonable enough. As noted above, the more restricted and predictable an audience, the easier it is to produce a TOC.

If this is in fact more or less the history, it provides a lesson in connections between ideology and technology alluded to at the beginning of this paper. I do not propose to offer here (nor do I think there can ever be) a general solution to the question of causal relationships between the two. I can, however, point to a particular case not often taken into account. Pliny's TOC is an instance of what, in the biological sciences, Gould and Vrba have labelled an 'exaptation'.[46] That is, a feature which carries out one function in response to one set of external pressures, yet arose originally with a different (or no) function in response to different pressures.[47] (Wings, which in their early stages could hardly have been useful for flight, are a standard example.) So, for instance, a structured TOC would not only have been easier for a Scribonius to invent, but it might have met specific needs of his.

To describe what those needs might have been, I need to digress slightly on the subject of the technical handbook in the first century CE Scribonius' work is reminiscent of two others, somewhat better known – Vitruvius' on architecture and Quintilian's on rhetoric. Architecture, medicine and rhetoric were areas of marginal status. They provided useful, intellectual knowledge, but excessive specialisation in any one of them would be inappropriate to the 'omnicompetent' citizens of the political class. They had been, as Cicero says explicitly of the first two and of education in general, 'honourable for those whose status is fitted to them' (*Off.* 1.151). Since the late Republic, however, this kind of intellectual specialisation became more

[45] I refer here not to the supposed adoption of Greek methods by Varro, but by the substance of his research which famously made Cicero think he had previously been wandering Rome as an 'alien' or 'guest' (*Ac.* 1.9).

[46] Gould and Vrba (1982).

[47] I do not want to make a general claim here for the applicability of evolutionary theory to cultural features, but the notion of 'exaptation' is largely descriptive and does not, I think, import any particularly problematic assumptions.

respectable in certain areas, most notably the law.[48] But individual 'professions' largely had to construct their own legitimacy. The three handbooks just cited share various strategies for doing so. First, two claim that their arts presuppose other, not obviously related disciplines: geometry, drawing, optics, history, philosophy, music, medicine, astrology (Vitr. *De arch.* 1.1.4–11), music and geometry and the whole 'encyclion paedian [sic]' (Quint. *Inst.* 1.10.1). The authors' own fields are made out as not so narrow after all and therefore more suited for a gentleman. Secondly, and at the same time, two make a point of establishing boundaries with related (perhaps lesser) disciplines: medicine vs. surgery (Scrib. *Comp.* preface), grammar vs. rhetoric (Quint. *Inst.* 1.2.4–6). There is a real expertise involved in creating these handbooks, which only a few could claim. Thirdly, all three establish their ties to the emperor (Vitr., *De arch.* 1.pr. 1–7, Quint. *Inst.* 4, pr. 2; 6 pr. 1, Scrib. *Comp.* pr. 13).[49] As with Pliny's *Natural history* (pr. 4–12), utility to the emperor becomes an absolute standard. If it's good enough for him, it's good enough for you. Wallace-Hadrill has already pointed out the value to the emperor of appropriating (the practitioners of) these 'professions' as they emerge as relatively independent loci of authority.[50] Here we see there is also the value to the technical experts: their professions are ennobled by being appropriated. At any rate, the value of the TOC to Scribonius is easy to explain in this context. First, the reader's ease of access illustrates the utility of the text. Secondly, the structure the TOC provides shows both the sytematicity of the field and Scribonius' own mastery of it. Once the multi-level TOC had come into being in a work like this, the device could more easily be borrowed for other purposes, such as Plinian nationalism.

In this case the causal connection between ideology and technology is at best indirect, but should the process be described as ideology-free? No. If we take a modified Althusserian approach to ideology, this kind of borrowing is easy enough to account for. Individual subjects are configured by recognising themselves in various positions offered to them by their societies; they respond to various 'interpellations'.[51] So Pliny the triumphalist writer identified himself not just as a Roman, but specifically as part of the imperial apparatus. And there can be multiple identifications. Most famously and explicitly, Cicero could recognize himself as both Roman and Arpinate

48 For the underlying notion of the separation of spheres, see Habinek's (1990) reading of Cicero's *De amicitia*. For the specific history of some fields, see Wallace-Hadrill (1997).

49 Rhetoric was apparently a problematic area from its inception as an area of knowledge, and so there had long been counter-arguments for its utility. What is striking about Quintilian is the appeal not to a generic audience, but specifically to the imperial family.

50 Wallace-Hadrill (1997).

51 See Gunderson (1996) 117–19.

(*Laws* 2.6). But other identities might line up less with local and national communities and more with, for instance, age, class, gender or occupation. Scribonius, whatever his national identity, also self-identifies as a physician of a new sort. Gellius' retreat to the world of texts suggests the intersection of the Romanness of a *subject* (rather than Pliny's imperial *agent*) and an elite whose leisure gives it time to focus on nice distinctions of ethics and aesthetics. New ideas can appear in any of these contexts, but once they have come into being, they are available for use in entirely new areas.

CHAPTER 5

Petronius' lessons in learning – the hard way

Victoria Rimell

Petronius' *Satyricon* looks like the joker in the pack. Not only is it the one text in this volume to have made the twentieth-century big screen – it's also the only chunk of prosimetric fiction, and perhaps the only work regularly read for fun, or even read much at all. If we can't swallow Petronius' toxic disordering and perversion of systems and 'facts', we can decide it's really off our map (and what did we expect from a hyper, Neronian pantomime anyway?). Moreover, questions about how the *Satyricon* embroiders, applies and tests knowledge, especially when set against the fetishisation and codification of Roman learning in the first century CE, will always jigsaw with debate about what we (can) know about, or learn from, the text itself. As Conte warns of the *Satyricon* in his *History of Latin Literature*: 'Few masterpieces are as shadowy as this . . . We would do well to keep in mind the extent to which our knowledge and the hypotheses built on it are limited and partial'.[1] The *Cambridge History of Classical Literature* makes similar claims: 'No Latin writer excites more lively interest. Unfortunately it is not always accompanied by due recognition of our ignorance'.[2] Slater classifies the fiction as 'singularly uninterpretable',[3] Sullivan concedes that it 'presents more puzzles than any other ancient text',[4] while Rudich ranks it 'the most controversial text in all of classical literature'.[5] These comments are all referring, to a greater or lesser extent, not only to the mutilated and probably jumbled state of the extant text (a problem which haunts all close readings of the *Satyricon*), but to the interpretative stumbling blocks posed by a hybrid, opaque anti-narrative which is notoriously difficult to follow and categorise, apparently undertaken, as Zeitlin writes, 'with the deliberate intention of defeating the expectations of an audience accustomed to an organising literary form'.[6]

[1] Conte (1994) 454. [2] Kenney and Clausen (eds.) (1982) 139. [3] Slater (1990) 250.
[4] Sullivan (1968) 21. [5] Rudich (1997) 186. [6] Zeitlin (1971) 635.

From where we stand, this text's extraordinary density makes for an overthickened, engulfing broth: as Connors writes, the *Satyricon* is 'flooded' with literature,[7] and Sullivan is convinced of its mammoth original scale (perhaps over 400,000 words).[8] Yet as the *Cambridge History* goes on to note, 'Petronius writes for a highly literate audience, able to recognise widely scattered allusions',[9] and I have argued elsewhere that we should not allow our own Aristotelian biases to shroud complex verbal, thematic and imagistic designs in the text.[10] The *Satyricon* is both grotesquely learned (demanding to be read as a whole, rather than a pick-and-mix of Menippean tales) and anti-intellectual: it celebrates 'learned' Nero while sending him up, and lays an exhilarating assault course for educated readers whilst also satirising our nerdy, Encolpian curiosity.

Conte is keen to stress that the *Satyricon* represents an 'encyclopedia' of labyrinthine artistry, a label which reflects a (controversial) view of the text as more comprehensive than fragmented:[11]

> we may recognise as a unifying feature of the work the fact that Petronius has collected, reinterpreted and parodied all the literary genres and cultural myths of his day (Homer and Virgil, tragedy, elegy, history and philosophy) as well as popular literature (sentimental novels, short stories, mimes, declamation and sensational stories of witches, magic and werewolves): Petronius may be studied as a shrewd depicter of customs and also as the author of a kind of literary encyclopedia of imperial Rome. Nor is this encyclopedia surprising in a period that opens with Ovid's *Metamorphoses* and was to have, on the constructive, institutional side, its Pliny the Elder and its Quintilian.[12]

Yet while the *Satyricon* is clearly a text of peculiar complexity, it is also clear that it cannot be said to systematise and inform in the manner of a reference work (although it also gets us thinking about how totalising, solid and even any 'encyclopedia' truly is). It cuts and pastes, pulverises and rehashes a vast body of literary, medical, zodiacal, culinary and physiognomic knowledge, as well as less assortable savvy on how to make your way in a big, bad world. In many ways, as Conte hints (and this stands up, I think, even when we try to estimate the impact of a fragmented text), the *Satyricon* has more in common with Lucan's immersion in disorder and discontinuity, or with the daedalean, anti-Lucretian universe of Ovid's *Metamorphoses*, than with

[7] Connors (1998) 145.
[8] Sullivan (1968) 34–8. See further debate in Walsh (1970) 73–6 and Reeve (1983).
[9] Kenney and Clausen (eds.) (1982) 140, cf. Harrison (ed.) (1999) xxi: the *Satyricon* 'appeals to a specific repertory of literary knowledge'.
[10] Rimell (2002) 7–9 and *passim*.
[11] See also Connors (1998) 145, and Rimell (2002), esp. 7–9.
[12] Conte (1994) 464.

the fastidious technicalities of Frontinus' *On aqueducts*, the reader-friendly patterning of Valerius Maximus' *Memorable deeds and sayings*, or even the fragmentised compilation of Plutarch's *Sympotic questions*. In particular, Ovidian didacticism and Roman satire, with their emphasis on the obfuscation and unreliability of knowledge and truth, set the immediate stage for the experiment of first-person narration in the *Satyricon*. Yet its megalomaniac scope and hybridity, as well as its exploration of the relationship between knowledge (and different types of expertise) and power/freedom make it a fascinating text to include within the parameters of this project. The *Satyricon*'s Neronian over-consumption might be construed as a further symptom and expression of the universalising drive epitomised by Pliny's *Natural history* or Seneca's *Natural questions*.

In the *Cena*, for example, the site of learned (or 'learned') discussion which forms the backbone of the text as we have it (*Sat.*, 26–78), Trimalchio's Neronian reign (both of the microcosmic dining room and his empire-like estate)[13] is contingent on his command and display of knowledge: he boasts of Greek and Latin libraries, employs a clerk to enumerate daily happenings on his property, recites poetry off the cuff, offers dishes laden with mythological reference, and decorates his walls with scenes from Homer. But, as I'll argue, Trimalchio's performance also pressures Foucault's power/knowledge formula, buttressed so effectively by many of the other texts dealt with in this volume. Or rather, it re-highlights, with different emphasis, the equivocality of knowledge: for whether Trimalchio really is in the know has little bearing on his dominion, and the notion that his properties are so innumerable he is unsure where some of them are (*Sat.* 48.2) is of course an index of his power. Under his rule what counts as erudition, and what distinguishes pattern from mess, constraint from liberality, gets tainted and confused, mirroring the challenges of reading what often seems a slyly tyrannical text. Dis-ordered knowledge, Petronius reminds us, can always hamstring rather than empower reader response, just as deranging rather than regulating text and world might equally flaunt an imperative to control.[14]

One of the things this essay will stress is that the *Satyricon* is a disturbing fiction on many different levels, perhaps never more so: to some extent we all want and need to believe that literary texts embody knowledge and are

[13] Trimalchio's properties are such that if he wants to go to Africa, he can do so by travelling only through his own land (48.3).

[14] Compare Jason König in this volume, who argues that the surface fragmentariness of Plutarch's *Sympotic questions* urges readers to forge their own consistent view of the world. Also see, in particular, Riggsby on Pliny, and Alice König on Frontinus, where ordering knowledge is construed, at least in part, as a gesture of control.

useful, enriching and categorisable. At the same time, we have embraced the fracturing, de-totalising critiques of post-modernism (which themselves shadow shifting models of knowledge born of twentieth-century advances in theoretical physics),[15] while our late-capitalist world shortcircuits a Foucauldian reciprocity between power and knowledge before our eyes, and reshuffles a shaky hierarchy of specialisms. The dominant cult of celebrity rates looks, wealth, self-promotion and being in the right place at the right time over talent and skill, and we live in an era where plumbers can earn more than university professors,[16] big drug companies govern medical research, and where access to and speed of information have helped fuel a zeitgeist characterised by anxiety and fear.[17] The most extreme impact of a postmodern unpacking of encyclopedism is that those who profess to know most, and to know the most authoritatively, end up the least trusted (take politicians, journalists and, increasingly, conventional doctors). Yet this same cultural vibe has rated Schott's *Miscellanies* runaway bestsellers.[18]

Similarly, Petronius' rambunctious novel illustrates how, in the accelerated culture of high empire (then, as now), knowledge can become a kind of madness even as it reassures and entertains, an (addictive) recipe for paranoia and a threat to the self. We can't forget what we (think we) know about Petronius, the politician and *Arbiter elegantiae* at Nero's court, a figure of expertise always presumably at risk of not knowing enough, or conversely being a little *too* good at his job, who was finally forced to suicide in 66 CE.

[15] E.g., Heisenberg's articulation of the uncertainty principle, the discovery of wave-particle duality, or, more generally, the emerging model of a holographic universe. These new (and not so new) ideas have yet to make their way (at least consciously) into the cultural imagination at large, but are beginning to make themselves felt through the New Age, or Post-Secular movement.

[16] All of which recalls scenes from Petronius and Martial (e.g., Ascyltos' enemy tells him school fees are a waste of money, as teachers these days are not worth twopence (*Sat* 58), and Martial 9.73 lashes out at an illiterate cobbler who got rich through sheer luck, while *he* slaved away at his grammar).

[17] Thanks, largely, to advances in information technology, in particular the internet, which has become an uncontrollable breeding ground for everything from hypochondria to conspiracy theories and hard-core pornography.

[18] Schott's *Original Miscellany* (2002) and his *Food and Drink Miscellany* (2003), Plinean collections of trivia on everything from glove sizes to public-school slang and euphemisms for offal, have been publishing sensations. In many ways these books charm because they offer readers complete guilt-free control over how to sample what are self-consciously useless minutiae. They allow the literary equivalent of bored internet surfing (which has turned us all into high-speed sorters and amateur lexographers), offering fast-food for info-addicts but at the same time a nostalgic respite from the daily onslaught of the type of information which *must* be assimilated 24/7, or which is traumatic and hard-hitting. Similar follow-up publications include (with same retro school-book-style paper cover) Peter Bowler's *The Superior Person's Book of Words* (2002), a dictionary of eccentric sounding, neglected words for readers to 'rediscover', Bill Bryson's *A Short History of Nearly Everything* (2003) and Michael Cook's *A Brief History of the Human Race* (2004), a more complex narrative than ever thanks to scientific advances like DNA and carbon-dating.

In pained contrast to the many kinds of orderings we've seen so far, Petronius' labyrinthine rehash of classical literature (like many other Neronian texts which parade creativity and anxieties of influence in grotesque bodies) presents knowledge as a problem of personal identity, and of physical and psychological, as well as intellectual, management. For us, especially, it upsets lingering presumptions that disembodied objectivity is the only form which the acquisition of knowledge can take.[19] This is an era, or literary system, in which, as Eumolpus argues at *Satyricon* 118, one's mind must be 'flooded in a vast river of literature' (*ingenti flumine litterarum inundata*), to achieve intellectual self-actualisation.[20] To tackle today's literary trends, exemplified by the heady entanglements of civil war poetry, you must be 'full' of literature (*plenus litteris*), the same image Agamemnon conjures when he describes the student's ideal trajectory in his poem at *Satyricon* 5: 'full of the learning of the Socratic school' (*Socratico plenus grege*) (line 13); 'thus, full up, you shall pour out words in a swelling river from a heart the Muses love' (*sic flumine largo / plenus Pierio defundes pectore verba*) (lines 21–2). Petronius depicts (and performs) a paradoxical literary cosmos in which knowledge, like other commodities, is demanded in excess, but where that excess is ultimately too much to swallow or sustain, or at least to enjoy in any straightforward way. In Quartilla's brothel, curiosity is cruelly punished ('A man cannot look upon forbidden things and go free', Quartilla warns at 17.4), while in the *Cena*, excess food and entertainment turn an orgy of hedonism into a nauseating death-trap. The culinary spectacles stage the cutting open and tasting of dishes as discoveries, enticing the guests to play detectives (or soothsayers, poring over fake entrails). Yet the more they 'know', the less they desire to know – discovery becomes sickening ('the whole event was getting really nauseating'; *ibat res ad summam nauseam* (78.5)), even a torture ('we could have put up with this, had a far more fantastic dish not driven us to prefer death by starvation' (69.7)).

Like Persius' *Satires* and Manilius' *Astronomica*, the *Satyricon* turns the imperial knowledge project inwards, to explore deeper, darker realms of body and mind: its narratives open up a nexus of 'inner' spaces, from Trimalchio's stuffed pig, staged within the underworld-labyrinthine cavity of the dining room itself, to the bellies of the dinner-guests, of trainee

[19] That is, despite feminist critiques of the disembodied, Cartesian knower, of which Jaggar and Bordo (1989) is a good example. Also see Merleau-Ponty (1962), who posits a subject who knows because the body knows.

[20] In this respect the *Satyricon* looks quintessentially Neronian. As Elsner and Masters (1994) 7 put it: 'Neronian culture exploded in a glorious and ultimately rejected orgy of transgressive experimentation. The result was excessive texts like the *Satyricon*, excessive arts like the *Domus Aurea*, and excessive acts like the liberation of Greece.'

orators (licking up the teacher's bait, *Sat.* 3.4), of the Trojan horse (in the *Capture of Troy* (*Troiae halosis*)) and of the cannibals at Croton, imagined finally at *Sat.* 141 ('Just shut your eyes and imagine you are devouring a million sesterces instead of human flesh', Eumolpus instructs, 141.7). When the guests are served a zodiac platter in the *Cena* (35), made up of foods representing all twelve signs and accompanied by Trimalchio's 'urbane' lecture (39), they take spoof astrology lessons in understanding the self by literally (and perhaps disturbingly) putting that knowledge inside them.[21] Yet whereas Manilius' drive to grasp the *praecordia* of the sky and how they relate to our inner worlds takes us on a journey of glorious, satisfying progress, Petronius' delving within, we shall see, is shadowed by literal and metaphorical *intestinum bellum* ('inside', i.e., 'civil' war) and leads not so much to clarification and empowerment but to guilt and angst, even horror. The trauma surrounding inside-bodies is perhaps epitomised by the Trojan horse in the *Troiae halosis*, opened up to spur forth violence, and shameful defeat. Meanwhile, the acceleration, concentration and re-evaluation of literary education discussed in the *Satyricon*'s narratives and dramatised in the fiction as a whole, work to oppose and destabilise a set of educational ideals to do with the objectives, pacing and exclusivity of learning – yet these are precisely the ideals to which its audience necessarily refers in attempting to untangle this text. This basic irony will surface several times in the readings that follow, and I will be stressing the extent to which investigations of how this novel deals with and represents imperial knowledge turn a critical spotlight on readerly perspectives.

THE WORLD ACCORDING TO ENCOLPIUS

I want to start, then, by reviewing the 'problem' of first-person narration in the *Satyricon*. For like the guests in Trimalchio's dining room, myopic second-guessers bound to be caught out (that's part of the show), Petronius' readers experience the world of the text via an impenetrable first-person narrative told by an 'unreliable narrator par excellence',[22] the drop-out student Encolpius. Although he claims to be learned ('both you and I know literature'; *et tu litteras scis et ego* (10.5)), and (during the *Cena* and in the picture gallery, for example) is intellectually inquisitive, Encolpius looks dizzy and naïve, making discoveries by chance rather than by deduction, skipping the smallprint ('I couldn't take them all in at once' (30.1); 'there

[21] On hints of cannibalism in the *Cena*, see Rimell (2002) 49–59.

[22] Rudich (1997) 186–7. The concept of the unreliable narrator is taken from Bakhtin, (1981), e.g., 402.

were, like, six hundred of these jokes, which have now escaped my memory' (56.10)), wandering haphazardly and blindly rather than planning his route, even (in *Sat.* 6–7, 79) going round in circles and forgetting where his lodgings are ('"Excuse me Madam" I said, "do you happen to know where I live?"' (7.1)). As protagonist he reports more than once that he is drunk and so unable, presumably, to see things clearly ('we were also drunk, and our ignorance of the area meant we'd have got lost even in the daytime' (79.2)). His literary knowledge, such as it is, is apparently rarely used to his advantage: at *Sat.* 29.9, he does not recognise that Trimalchio has two of the most famous texts of the ancient world, the *Iliad* and the *Odyssey*, painted on the walls of his atrium. As Odysseus ('much praised Encolpius', *polyaenos Encolpius* (*Sat.* 127.7)),[23] he forgets his antidote to Circean magic and his resulting 'epic' performance inevitably flops. Later, on board Lichas' ship, his Odyssean 'escape-from-the-Cyclops'-cave' plot is desperately crude: he advocates splashing poet Eumolpus' ink all over, like cheap cologne, to disguise himself and his fellow stowaways as Aethiopian slaves (rather than use it to 'write' branding marks – Eumolpus' more sophisticated plan, designed to revamp Odysseus' self-revelatory scarring as a strategy for disguise). When the gang happen upon a cloak which has been stolen from them in *Sat.* 12, Ascyltos convinces Encolpius to ignore his knowledge of official legal processes and legal rights in favour of dissimulation, bribery and brute force ('what's the point of laws when money rules?', he argues in verse at *Sat.* 14.2).

Yet the problem is that Encolpius' outlaw perspective, fogged by drink, drugs, inanity or paranoia, is ours too: we cannot, as Auerbach, Sullivan and Conte propose, ally ourselves securely with a sophisticate Author (Petronius) and condescend to a buffoonish Narrator (Encolpius) from a position of objectivity and superiority 'outside' the text:[24] it is ultimately impossible to disentangle narrator from author, or even narrator from protagonist,[25]

[23] Cf. the epithet πολύαινος used by Homer for Odysseus, e.g., at *Od.* 12.184.

[24] See Auerbach (1953) 47, Sullivan (1968) 258 and Conte (1996) 72 and *passim*. Laird (1999) 210 has recently criticised this view.

[25] The concept of Encolpius' split (or coherent) 'personality' has been the source of much debate in criticism of Petronius. Sullivan (1968) 81 argues that Encolpius' character is disorganised and fragmentary, not because he is at odds with himself, but because he displays 'those traits which are appropriate responses to the demands of the particular episode'. Against this, Beck argues (1973) that Encolpius is two distinct personalities, the wise, retrospectively self-critical narrator and his former wild, idealistic, foolish self. In his later article (1975), Beck reaffirms this idea, stressing the need to 'disentangle' narrator from protagonist: Encolpius' foolishness is to be taken 'at face value', as is his distanced, knowing commentary on it. Compare the counter-argument from Veyne (1964) 303–6, who sets out claims for Encolpius' 'fausse naïveté'. Yet the point is rather that we never can tell when Encolpius is telling it straight or pretending, is being naïve, or ironic, any more than we can tell (as

to decipher whether Encolpius is simply a half-witted figure of fun or whether he is sometimes or always posing as a clown, empowered by ironic self-mockery (the stand-up's stock tactic) and by his audience's inability to tell when they are being manipulated or fooled. So when Encolpius says, in typical form: 'I kept quiet as if I didn't know what the story was about' (92.13)[26] we confront several potential takes on his narration: is this a privileged point where fausse naiveté is transparent, where this *idiot savant* nods and winks in our direction, or does it placard our narrator digging himself a deeper hole, unaware even of what he *doesn't* know? Or how do we read his claim on glimpsing Trimalchio's latest culinary riddle – 'Of course, being pretty smart, I immediately realised what it was' (*Sat.* 69.9):[27] as guileless self-inflation, or heavily ironic put-down, with wry stress on *scilicet*? This is the central dilemma and joke of this text: the *Satyricon* tempts and dares us to laugh, to boast we're in the know, yet the 'joke's on you' threat, as in Roman satire, looms large and loud. The characterisation of *Satyricon* as *merely* farce, pantomime, or light entertainment, a *jolly* experiment in Neronian excess,[28] has been contingent on ignoring or muffling the complexities of reading this first-person narrative, with its unsettling, satiric momentum.

Lack of authorial signposting of any kind (in the text as we have it), and the unusual opacity of this narrative, are mirrored in Petronius' hellish or comically surreal cosmos. We are to imagine a world that could not be farther removed from Pliny's idea of the universe as a complex but comprehensible whole governed by divine foresight. Whereas in Ovid's *Metamorphoses* and *Fasti*, divine agents are flawed characters inhabiting a relativistic universe and offering conflicting, subjective 'truths' (as critics have argued, Ovidian encyclopedism repeatedly stages crises of legitimation),[29] in the *Satyricon*, gods, like mythic heroes and heroines, are merely glamorous stage-names, men in costume, actors ('Indeed, this place is so

Beck himself argues ((1982) 208) whether a line or passage in the *Satyricon* is 'authorially privileged'. The dilemma of Encolpius' 'personality' is by definition insoluble. Indeed as Jones points out ((1987) 811), Beck's separation of narrator and protagonist is already shaky when examined on its own terms, as there are several instances in which our narrator gets 'so involved in his recollections that he loses his ironic distance'. Similarly, George (1966) finds that despite arguing that Petronius would never wish to identify himself with the effeminate, subtle-as-a-brick Encolpius, 'the dissociation between author and Encolpius is not complete' (355).

26 *utcumque tamen, tamquam non agnoscerem fabulam, tacui.*

27 *ego, scilicet homo prudentissimus, statim intellexi quid esset.*

28 In the work of Auerbach (1953), Slater (1990), Sullivan (1968) and Panayotakis (1995), for example.

29 For further discussion of poetry and knowledge in Latin literature, from Lucretius to Ovid, see Schiesaro (1997) and (2002). On epistemological crises in Ovid's *Fasti*, see Newlands (1995) and Barchiesi (1997).

full of divine spirits that it's easier to meet a god than a man' (17.5); 'for no-one believes the gods are gods' (44.17)); heaven is Trimalchio's ceiling ('we were wondering what was going to be announced from the heavens next' (60.2)), and the zodiac fits on a plate (35). The *Satyricon*'s all-penetrating mode of theatricality, the constant collision and confusion between the real and the artificial – most hard-hitting at the level of the narration itself – effectively takes a sledgehammer to epistemics.

DOWN WITH SKOOL: CARNIVAL, SEX AND THE UNIVERSITY OF LIFE

Yet this is also a text which, as I have said, seems to elicit, demand and contain an ultra-sophisticated level of literary knowledge. Many of the stories and scenes that make up *Satyricon* are themselves concerned with learning, education and the status of knowledge, and I want now to consider these more closely. The main characters – Encolpius, Ascyltos, Giton, Eumolpus – are educated men, or at least playacting as such (they are 'just like students'; *tamquam scholastici* (10.6)), our narrator an adventuring researcher whose curiosity gets him into trouble and propels the events that constitute this fiction. The vocabulary of law, rhetoric and literary criticism, as critics have recognised, seeps through the work as a whole,[30] while the performances of myth (whereby characters act out versions of well-known plots in the guise of mythical characters, such as the scene of *polyaenus* Encolpius' meeting with Circe at *Sat.* 126), often look like (bungled) school *suasoriae*.[31] The diners at the *Cena* jostle for intellectual position, swopping witticisms, put-downs, horror stories, sermons: in this pressured milieu, Niceros says he's reluctant to recount his own adventure because he fears being mocked ('I am afraid your clever guests might laugh at me'; *timeo istos scholasticos, ne me rideant* (61.4)), while the young scholar Ascyltos is attacked for presuming that he is more educated than the lowly freedmen, who all have degrees from the University of Life. Meanwhile, Trimalchio sells himself as a learned symposiarch, setting his guests a menu of riddles cooked up by the masterchef Daedalus, boasting of his libraries and command of facts ('I do nothing without a reason'; *nihil sine ratione facio* (39.13)), dependent on an entourage of 'experts'.[32] But his learning

[30] E.g., Barchiesi (1996) and Wooton (1976).

[31] Shelton (1998) 117 comments on the speeches practised by the rhetor's pupil: 'The topics of their speeches were much more akin to prose fiction than to legal actions or court cases'.

[32] At 76.11, Trimalchio reports that he was encouraged in his work by an astrologer called Serapa: 'he knew my own insides, and only fell short of telling me what I had had for dinner the day before', the

is obviously posed: he acts the Pythagorean at *Sat.* 56.4–5,[33] preaching the evils of eating lamb while also wearing sheepskin, whereas at *Sat.* 47.10, he was Pythagoras' nemesis, boasting of his cook's ability to outdo 'Pentheus' mincemeat'. There are several points at which he could almost be parodying (or badly impersonating) 'official' knowledge-orderers like the late-Augustan Manilius or the sober Tiberian Celsus. His lecture on astrology and the *orbis* of the zodiac dish at *Sat.*39 (Smith comments on its 'pedantic accuracy'[34]) is potentially a mish-mash of many ancient astrological writers: compare this passage with Manilius' account at *Astronomica* 1.672–80:

> The circle is held by Cancer at the top, by Capricorn at the bottom, and is twice crossed by the circle which balances day and night, whose line it cuts in the signs of Aries and Libra. Hence the curve of the round is drawn through three circles, and covers its straight path by its downward slope. Nor does it escape the sight of the eye, as if it were just to be perceived in the mind, even as the previous circles are perceived by the mind, but throughout its enormous circuit it shines like a star-studded baldric and lights up heaven with its broad outline.

As Smith comments, Trimalchio has the zodiac dish in front of him as he gives his exposition, and 'is perhaps imagined as turning it round as he proceeds, as if it were a celestial sphere':[35] Manilius' orrery becomes Trimalchio's edible platter, which also 'draws the eyes' of his audience ('its novelty turned the eyes of everyone'; *novitas tamen omnium convertit oculos* 35.1, cf. *Astronomica* 1.677). At *Sat.* 47, similarly, when he plays doctor and makes a speech on bowel health after returning from the bathroom, Trimalchio looks as though he has been reading his Celsus. He says (47.2) that for constipation he has found pomegranate rind useful, and 'pinewood boiled in vinegar' (*taeda ex aceto*): Celsus 2.29 gives a long list of such remedies. Or at 50.5–6, when Trimalchio boasts of his knowledge of Corinthian bronze ('and lest you think I'm some kind of ignoramus, I know perfectly well how Corinthian plate originated' (50.5)),[36] he plays on (and mocks?) a background of Roman connoisseurship of the metal which culminates in Pliny's entry in his *Natural history* 24.6.12:[37]

joke being, perhaps, that this 'amazing' fortuneteller can only inform Trimalchio of what he already knows, and 'prophesy' what has already happened.

33 Compare Ov. *Met.* 15.116–21.

34 Smith (1975) *ad loc.*

35 Smith (1975) *ad loc.*

36 *et ne me putetis nesapium esse, valde bene scio, unde primum Corinthea nata sint.*

37 See also Vell. Pat. (1.13), discussed by Connors (1998) 20–1, who uses Corinthian bronze to distinguish between two conquering generals, Scipio Aemilianus, who is cultured (*elegans*) and Mummius, who is uncultured (*rudis*), and who therefore can't tell the difference between true and fake Corinthian bronze. For Connors, Trimalchio's retelling of the story (he alone has real Corinthian bronze because he obtained it from a craftsman named Corinthus) displays, on the one hand, his 'foolish and ignorant pretensions', and on the other symbolises the *Satyricon*'s strategies of refashioning epic.

But although it is agreed that there are no lampstands made of Corinthian metal, this name is nevertheless often attached to them, because although Mummius' victory destroyed Corinth, it caused the dispersal of bronzes from a number of the towns of Achaia at the same time.

The *Satyricon* as we have it begins with the speeches of Encolpius and Agamemnon on the crisis in contemporary education outside the school of rhetoric where Agamemnon teaches (*Sat.* 1–6), and similar speeches are given by (poet and 'teacher') Eumolpus at *Sat.* 88, before the *Troiae halosis* poem, and at 118, before the recitation of the *Bellum civile* (*Civil war*). These diagnoses of infected Roman intellectual culture are played out in the *Bellum civile* itself: Eumolpus' epic sample imagines the seeds of civil war sown in the flesh of those who shun the moral codes and strict curricula of traditional education to guzzle up the *satura* of experiences, ideas and material goods that is *imperium romanum*. Likewise, in the cityscape of Croton, the population of cannibalistic legacy hunters – a hungry new breed of *human*ities students – vilify all *studia litterarum*, all *eloquentia* (116.6). Chrysis knows nothing of astrology, but can expertly read character in a man's face and walk (126.3), while Oenothea's scholarship in impotence and its cures is unsurpassed (134.10).

Indeed, throughout the text knowledge is focused on, or accessed through, physical and sexual desires. Teachers are corrupt paedophiles: at *Sat.* 85–7 Eumolpus tells Encolpius the story of how he convinced a pretty boy's parents into letting him teach their son, only to exploit his position and seduce the boy with bribes (masters of oratory, who bait their hooks with tasty titbits to pull the crowds, are accused of similar crimes in *Sat.* 3). In Croton, Philomela (herself, elsewhere, a figure whose myth connects the birth of writing and elegiac poetry with rape and incest)[38] entrusts her two children to the poet Eumolpus for 'instruction' ('he was the only man in the whole world who could teach kids a wholesome philosophy on a daily basis' (140.2)), which turns out to be a crash course in sex ed. The girl's talent in this area (sex and/as legacy hunting) is described as an *artificium* (140.8) which can mean trick, device or work of art as well as talent, craft, profession or education.[39] The term perhaps recalls Philomela's artistic trick, the telling of her violation in a tapestry, but also, more obviously, the discussion of the role of education and learning among the freedmen, where *artificium* is used twice to mean something like 'learning', for example at (46.7–8):

[38] Kilgour (1990) 33 argues that the Philomela myth shows how 'poetry is produced by the disorder of relations and the confusion of identity represented as incest, cannibalism or civil war'.

[39] OLD *s.v.*

'I want him to have a trade . . . literature's a treasure, and education never dies'.[40] The second example comes at 58.14, where it is used ironically to mean, approximately, 'cunning' by a freedman who discredits traditional education as 'nonsense': 'yes, I thank the gods for my "education" – it's made me what I am'; *ego, quod me sic vides, propter artificium meum diis gratias ago* (58.14). *Artificium* with its range of meanings, perhaps encapsulates the freedmen's boast that knowledge can be faked: in their world, 'education' is always also a scam or artifice. Used in the Philomela episode, it hints that we may not be sure who is teaching who within a system where both learning and teaching are reducible to seduction, where knowledge is (just) turning tricks, and where true intellect lives below the waist ('It's so much more profitable to rub groins than minds', as Encolpius puts it at 92.11). Yet while these images comically degrade the pursuit of knowledge, they also extend ideas about knowledge as a bodily as well as intellectual practice, and spell out the sexual dynamic in all epistemophilia. After reading Petronius, it is hard to turn back to Pliny's *Natural history* or Frontinus' *On aqueducts* without feeling there is something narcissistic (and/or repressive) about the obsessive, controlling diligence of their compilations.

More generally, the *Satyricon* makes frequent reference to the decline of eloquence and de-valuing of education (a theme familiar to us from Seneca, Columella, Quintilian, and the Roman satirists). In the dialogue outside the rhetorical school at *Sat.* 1–6, Encolpius argues that pupils these days are turned into complete fools by their education, that colleges teach nothing but vapid cliché; students are all fed on the same banal diet, schools resemble stuffy kitchens, a stinking environment which stalls all refinement of the senses/intellect (*sapere* = to taste/to know, 2.1). Oratory, ruined by the flatulent Asiatic style, has gone to the dogs, and no literature today matches the Greats of old – Sophocles, Euripides, Demosthenes, Thucydides. Indeed, Encolpius' image of clichéd *sententiae* as 'honey-balls of phrases, every word and act sprinkled with poppy-seed and sesame' (1.15),[41] gets served up by the king of bad taste, Trimalchio, for his very first course, 'dormice rolled in honey and poppy-seed' (31.10).[42] The teacher Agamemnon is apparently also one of the guests, and finds himself censured by the chattering freedman Norbanus for thinking he is above dinner-table banter, and accused of being 'mad with learning' (*scimus te prae litteras fatuum esse* (46.2)), just as the students are *insanientes* at *Sat.* 3.2. When the freedmen guests talk about education, they reject learning for its own sake and are suspicious

[40] *destinavi illum artificium docere . . . litterae thesaurum est, et artificium nunquam moritur.*

[41] *mellitos verborum globulos et omnia dicta factaque quasi papavere et sesamo sparsa.*

[42] *glires melle atque papavere sparsos.*

of elite systems of knowledge: qualifications are esteemed for their vocational (i.e. cash) value. Denied a conventional education, these ex-slaves boast certificates in survival and (night-class) diplomas in *Sales & Marketing*. Norbanus tells Agememnon about his young son, who never lifts his nose from the slate; he can do simple division, is relishing Latin, and has made a decent start on Greek, but doesn't want to earn a living (*non vult laborare* (46.6)). His other son is no scholar, but is *curiosus*, and can 'teach you more than he knows himself' (*plus docet quam scit* (46.6)). In the *Cena*, knowledge, or the types of knowledge to be privileged, are being redefined, but now Norbanus implies that expertise can be simulated, given a bit of enthusiasm and the gift of the gab (and you can't teach that): it's not a system, it's a *lingo*. Norbanus argues that his boys have dipped quite enough into literature – it's time to cash in the CV points, for it's not what you know but what you do with it that counts: ('law has bread and butter in it'; *habet haec res panem* (46.7)). Similarly, in *Sat.* 57–8, a freedman explodes with rage at Ascyltos' condescending attitude; having paid for his own freedom, this ex-slave didn't need an education ('no, I never learned geometry, criticism, or other such nonsense' (58.7)), and claims that Ascyltos' father wasted his money on private school fees. He thanks the gods for the practical education he received as a slave, which taught him basic manners and how to look after numero uno; now he's not just street smart, he's loaded ('let's go to the forum and borrow cash; then you'll see that my iron ring commands credit' (58.11)). Out of all the surviving episodes of the *Satyricon*, the *Cena* in particular gets us thinking about the ranking and validity of different kinds of knowledge (practical versus theoretical, worldly-wisdom versus conventional expertise). It also draws attention to a cultural context in which freedmen could rise to magisterial positions, but where apparently the prestige of teachers (many of them freedmen) was at an all-time low, while for slaves, education usually meant only the power to serve their masters in a particular way. Trimalchio's head-chef Daedalus has a 'very good mind' (70.3), but is a slave in his own labyrinth, and when Habinnas voices his beliefs in 'practical' education at *Sat.* 68, he is referring to a 'hopelessly clever' slave, who is a 'servant to his talents' (68.7), and turns out to be '*nequissimus*' (worthless, 69.4).

We might say, more broadly, that the *Satyricon*'s irreverent take on the relationship between (traditional) knowledge, authority and power is indicative of its 'topsy-turvy' world[43] and wobbly, Saturnalian universe. So in Agamemnon's speech at *Sat.* 3–5, roles are reversed when teachers are

[43] Holzberg's phrase (1995) 63.

parasitic on their pupils, feeding them only what they want to hear. In the *Cena*, the dominant mode of acting and theatricality confounds hierarchy, and slaves and professors, freedmen and knights alike neck Falernian wine by the gallon. At *Sat.* 57–8 the freedman Hermeros knocks *eques* Ascyltos off his high horse, lashing him with the satirist's favourite quip ('you see the lice on others, but not the flea on yourself' (57.7)),[44] before turning on Giton, first for laughing out of turn at the outburst ('Merry Saturnalia indeed – what is this, December? When did you pay five percent on your freedom?' (58.2)) and then for his obsequiousness ('I'll bring down the wrath of Athena on you, and that guy who first made you his slave' (58.7)). Hermeros confronts his opponent with a series of riddles ('What part of us am I? I come far, I come wide – solve me!' (58.8)),[45] a suitable conclusion to a speech which seems contradictory and potentially self-implicating. Hermoros' tirade enacts the transgressions of the Saturnalia (ex-slave puts aristocrat in his place and slurs higher education), but he goes on to berate Giton first for not acting according to his subservient status, and then for not standing up to his master (hence they are both tarred with the same brush: 'like master, like slave' (58.3)). Like Giton (with his long curls: he is a *cepa cirrata*, 'curly-headed onion' 58.2), Hermeros himself was once a *puer capillatus* ('a boy with long curls') who was devoted to his master, and while he was a slave for forty years, nobody could tell whether he was a slave or free (57.9); he is (or was) both the effeminate, sycophantic slave, *and* the arrogant, aspirational gent he attacks – for, as he says himself, it is in rotten flesh that worms will breed (57.3). His speech seems to blur, more than reverse, the roles of master and slave, *eques* and freedman, typifying the angry satiric persona which ultimately undercuts its own authority.

NOT WHAT YOU KNOW . . .: KNOWLEDGE, SATIRE AND ROMANISATION

However, almost all critics have argued for reading anti-intellectual posturing in the *Cena* as straightforwardly, farcically comic. Implicitly, then, these rough-and-ready speeches function as an anti-model for Petronius' readers, cueing mockery of these ex-slaves' crass, low-life, dumbed-down perspectives. While this is one valid reading, I would suggest that the humour of (anti-)intellectualism, in the *Cena* and throughout the *Satyricon*, is potentially much more layered and satirical, and might be framed differently as a creative reordering of hierarchy, a mischievous take on regulating scientism,

[44] *in alio peduclum vides, in te ricinum non vides.*

[45] *qui de nobis longe venio, late venio? solve me.*

of the kind we see in Varro, Celsus, Seneca or Pliny. In the speeches berating education at *Sat.* 1–5, the images of endemic decline, of the loss of precision, clarity, originality and moral muscle in an era of conspicuous consumption, solipsism and greed, the consequent degrading of the teaching profession, the new generation of obnoxious, spoon-fed students, the nostalgic throwbacks to schoolrooms of old, are very familiar – but do they affirm the current moralising vibe, or is this nauseating, textbook cliché? Is Encolpius' performance at *Sat.* 1–2 sincere or self-mocking, does its clumsiness cynically or unconsciously enact the inadequacies of those it purports to attack, and is he voicing his own opinions or simply following a formula dictated to him by a teacher, who may or may not be his respondent, Agamemnon? And is Agamemnon in agreement only because he is being forced to do precisely what he bemoans the teacher of oratory must do to stay in business – flatter his students and offer them seductive bait? As Encolpius' and Petronius' audience, our inevitable uncertainty over how to read these scenes situates us right inside the hothouse school of *Sat.* 1–2, where we experience the same smothering of sensibility.

Similarly, as much as they enact stereotyped 'decline' in their obsession with credit ratings, the freedmen's attitudes seem also to revive the ideals outlined by Agamemnon and Encolpius as lost – whereby a boy learns by practical experience and by observation of his respected elders, and grows up with his feet firmly on the ground,[46] rather than being pushed aggressively through the artificial school system and pressured to reach the highest levels (the image of contemporary education painted by Agamemnon at *Sat.* 4: 'they drive the unripe schoolboy into the lawcourts, and thrust eloquence, the noblest of callings, on young boys who are still kids' (4.2)). Or, as well as performing (self-consciously or not?) as butts of ridicule, are the self-taught freedmen also discomforting, subversive figures whose lectures from the University of Life undermine traditional educational values and ways of systematising knowledge?

Trimalchio himself, who boasts of his sophistication and learning but apparently exposes his deficiencies when he misremembers myths, is usually taken to model both the transparent witlessness of the freedmen and our privileged, editorial perspective on their flawed knowledge. Thus at *Sat.* 52.1, Trimalchio announces that he owns four-gallon cups engraved with the image of Cassandra killing her sons (error: that was Medea), in which the sons are dead but are painted so realistically you would think they were still alive. Here, he may well be presenting a crude/funny muddling

[46] See, e.g., Tac. *Dial.* 34.1–6.

of knowledge gleaned from Pliny's account of realist artworks in book 35 of the *Natural history* (according to the grammarian Apione, Pliny tells us, when experts saw the faces of men depicted by the great Apelles, they were able to divine not only how old they were, but how long they had to live),[47] the joke being that Trimalchio confuses realness with aliveness. The symposiarch also has jugs on which you can see Daedalus shutting Niobe into the Trojan Horse (52.2). Howler: the wife of Amphion had nothing to do with Troy, and we all know Daedalus made a wooden cow, not a horse. And in *Sat.* 59.3–5, when performers act out a mythical story in Greek, Trimalchio reads out a Latin translation from a book, saying,

> 'So do you know what story they're doing? Diomede and Ganymede were brothers. Helen was their sister. Agamemnon kidnapped her and sacrificed a deer to Diana in her place. So here Homer is telling the tale of the war between Troy and Parentium. He won of course, and after that he married his daughter Iphigenia to Achilles, which drove Ajax bonkers – that bit's coming up in a second.'

Trimalchio obviously perverts these well-known narratives. On *Sat.* 50–2, Smith comments, 'Petronius will unfold Trimalchio's absurd ignorance of history and mythology, as well as his pretensions to good taste.'[48] Slater emphasises the 'dense comedy' of these 'typical Trimalchian confusions of mythology'.[49] On *Sat.* 59, Smith states the accepted view that, 'Trimalchio's wild version of the story is entertaining precisely because each detail distorts some identifiable part of the normal version':[50] thus he supplements Castor and Pollux with the Greek warrior Diomedes and the Trojan boy Ganymedes, whose names merely *look* related. Agamemnon replaces Paris as Helen's abductor, and Helen is crossed with Iphigenia when, as in the Iphigenia story we know, a deer is sacrificed in her place. The false promise that Iphigenia will marry Achilles if she comes to Aulis is realised in Trimalchio's version, but the marriage occurs at the end of the war, not the beginning. Instead of being maddened when the arms of the dead Achilles were given to Ulysses rather than to himself, Ajax is enraged with sexual jealousy at Achilles' marriage, whether we are meant to imagine his lust fixated on Iphigenia or on Achilles himself.

Yet we shouldn't be too quick to laugh down Trimalchio's buffoonish 'confusions'. In the case of the four-gallon cups, we read Trimalchio's commentary through Encolpius' sloppy, tipsy narration, and we're blind to the correspondence (or lack of it) between the engraving and its description. Similarly at *Sat.* 59, there is no comment from Encolpius to confirm or deny that what Trimalchio says he is reading from the book in Latin corresponds

[47] Plin. *HN* 35.88. [48] Smith (1975) 134. [49] Slater (1987) 168. [50] Smith (1975) 165.

to the dining-room drama that is being enacted in Greek. Equally, we might read Trimalchio's 'mistakes' as a provocative gag, an attempt to take his role as deceptive and manipulative 'author' of the *Cena* to an extreme, loosening myths as he preaches lax bellies and bladders. If so, then he could hardly have picked more appropriate mythical characters to personalise these tales than Cassandra (the revealer of truth who is never believed, to whom Trimalchio compares Fortunata in *Sat.* 74.15), and Daedalus (the labyrinth architect and Trimalchio's talented head chef). Indeed, his wild 'imagination' is reminiscent of, for example, Dio Chrystostom's sophistic revision of Homer (the fall of Troy at *Or.* 11, for instance, where he argues the city was never taken, the Greeks came home defeated and Hector killed Achilles), or Dictys of Crete's *Diary of the Trojan war*, which gives us, among other things, a startling new version of Homer's Achilles.[51]

Meanwhile Trimalchio's game of character swapping in *Sat.* 59 reflects the topic of acting as metamorphosis at work in the *Cena* and throughout the *Satyricon*: characters are either named after mythical figures (Agamemnon, Menelaus, Circe, Psyche, Ganymede, Bacchus, Dionysus, Philomela, and so on)[52] or playact as them (Giton plays Ganymede in *Sat.* 92 and Ulysses in *Sat.* 97, where Lichas is imagined as a Cyclops; Encolpius adopts the epithets of Odysseus in *Sat.* 127; Eumolpus is Aeneas reciting the fall of Troy as in *Aeneid* 2 in *Sat.* 89, and so on). His novelistic imagination becomes a dramatisation of acting and theatricality run wild, a lesson in the transformative strategies and re-writings of the *Satyricon* as a whole. Just as he owns land, slaves and luxury goods, breeds his own pedigrees from exotic imports and relabels his human possessions with mythical names, so he can monopolise and concoct mythic plots. In the megalomaniac imagination, (even) knowledge can be bought. The possible political undertones of this are obvious: those in power dictate and delimit belief, gloss the 'facts' any way they please. His audience will (be forced to) publicly marvel and applaud, whether we read Encolpius' comment at 60.1 ('we were not allowed very long to admire these elegant performances'[53]) as naïve sycophancy, a sarcy joke, or real praise of innovation. Yet there may be restrictions, too, on Trimalchio's (or any autocrat's) ownership of knowledge: he risks accusations of ignorance and foolery in asserting his 'control', while the idea that knowledge has a

[51] For discussion of this and other Greek revisions of Homer's Trojan War, see Merkle (1994).

[52] Slaves were often given mythological names, perhaps, as Fitzgerald suggests (2000) 5, because it 'allowed their masters to share in the civilized world of which Greek culture was the most precious fruit'. For Trimalchio, naming is always about creative control, about playing at being a (Greek) poet thinking up new names to fit his fictional worlds.

[53] *nec diu mirari licuit tam elegantes strophas.*

price tag must by definition entail its destabilisation. Literary structures, systems of belief, are up for grabs in Trimalchio's kitchens, when even the tyrant's perspectives can be outbid (if only in the imagination).

The notion that Trimalchio reads a 'translation' in Latin rather than follow the 'pretentious' Greek ('while the reciters conversed in Greek, in that conceited way of theirs, he intoned Latin from a book' (59.3)) raises further questions about the politics of (re)ordering knowledge in the *Satyricon*. Might we assume that he doesn't understand Greek (and that his Greek library is just for show), so uses the Latin book as subtitles? At one point Encolpius hints that he doesn't know Greek either (Plocamus 'whistled out some offensive stuff I didn't catch – he declared afterwards that it was Greek' (64.5)). But we might also read Trimalchio's 'translation' as an act of cultural imperialism, whereby he exploits the kudos of Greek education while also publicly appropriating and Romanising it (compare 38.3: civilised Athenian bees are inbred with Roman bees, and relied upon to improve the natives). At *Sat.* 53.13 similarly, he tells guests that he once bought a Greek comedy company, but preferred them to perform Atellan plays (i.e., native Latin comedy), and told his pipe player to have Latin songs. His proclivities resonate with what we hear of Nero's egomaniacal philhellenism, whether or not we imagine Trimalchio to be a recognisable caricature of the emperor (and whether or not we read them more as an anti-Neronian rejection of Hellenistic influence than as a mirroring imperialistic strategy). These passages may prompt us to think more broadly about the politics of cultural hybridity in the *Satyricon*: Petronius transfers and Romanises the milieu of the Greek novel, and explores a colourful Roman literary culture, steeped in pungent satire and monumental epic, through the adventures (in Greek or half-Greek landscapes) of non-Greek characters bearing Greek names.

KNOWING TOO MUCH, TOO FAST

Yet Petronius and his characters do more than creatively undermine, redefine and (re-)appropriate knowledge. The narratives of the *Satyricon* also regularly imagine the threat, shame or sordidness of literary learning. At *Sat.* 56, for example, Trimalchio estimates the professions of medicine and money-lending to be the most difficult after writing, because the doctor gets to know what poor men's guts really look like, just as the money-lender sees the copper under the silver. This bogus philosophising ('He was just throwing the philosophers out of work' (56.7)) is bang in line with Trimalchio's 'expert' jurisdiction over interiors, from his conveyor-belt of layered dishes carved or bitten to reveal their unexpected (and in the case of the pig, gut-wrenching) insides, to his cushions, stuffed (as only he appreciates) with feathers in princely purple (38.5). As the *Cena* progresses and the opera of tyranny crescendoes, the nauseous guests are less and less keen to be party to this inside-info, as horror and trepidation replace surprised glee. In Quartilla's brothel, the gang stumble upon the secret rites of Priapus and are tortured for their discoveries, afterwards swearing that the 'horrid secret should die with the two of them' (*inter duos periturum esse tam horribile secretum* 21.3): in any tyranny, knowing too much is always far more dangerous than knowing too little. Encolpius describes Eumolpus' recitation of poetry as a 'disease' (*morbus*, 90.3), while in the *Bellum civile*, conflict is sparked (and is imagined as a 'disease' metastasising in Roman bones) when conquering Rome 'held the whole world, both sea and land, and the course of sun and moon' (lines 1–2) yet is not satisfied (line 3) even with the thrill of constant discovery and invention. And in Lichas' tale at *Sat.* 111–12, the widow of Ephesus is corrupted by snippets of Virgil, the narrative of her downfall propelled by (our) rereading of the story of Dido, whose spectre suggests that this post-Ovidian seduction cannot rule out an ominous fate for chastity's fallen queen.

Knowledge and discovery in the *Satyricon* seem to foil and be opposed to pleasure, even while they also equate to (imperial) power, just as readings of this text's riddling complexity as entertaining farce seem dependent on concomitant conclusions of readers' inescapable (and enjoyably escapist) 'ignorance' or oblivion. And on the one hand, the weirdness and episodic architecture of this fiction, together with its Saturnalian rhythms and the rejection of traditional education both by the guests at Trimalchio's *Cena* and our narrator himself, seem to invite an experience of reading that is more sensual than intellectual, more ludic than learned. On the other, the *Satyricon* is also incredibly demanding, immersed in and descriptive of dense, high-pressure systems of education, and a text which imagines a symbiosis of corporeal and intellectual knowledge. I have already highlighted the imagery of literary consumption used by the two teacher-figures of the *Satyricon* – Agamemnon (in his poem at *Sat.* 5), and Eumolpus (in his speech before the *Bellum civile* at *Sat.* 118): today's students and poets must almost drown themselves in learning, be full to the brim with literary knowledge, in order to reach the higher echelons of scholarship (and, it seems, of *empire*). Eumolpus even threatens that the poet will sink under the burden if he attempts a trendy new poetic topic like civil war, unless he is stuffed with knowledge (*plenus litteris*), so as to command an artillery of allusions, great thoughts coloured by mythology, and strokes of vatic inspiration (118.6). Civil war comes to exemplify the poetry of excess, the chaotic landscape dreamt up in flashes by the *liber spiritus*; this is the kind of war Eumolpus envisages being sown, like a biological attack, in the marrow of Rome's insatiable citizens, just as the poet himself must fulfil an insatiable appetite for literary knowledge in order to write about civil war. The portrait of conspicuous consumption and unquenchable greed at the beginning of the *Bellum civile* is immediately reminiscent of the scenes from Trimalchio's *Cena*, and not only in the activity of eating to the point of nausea/moral corruption; the tyrant's estate, with its armies of slaves and endless imports of exotic products, is a microcosm of Roman empire. Indeed civil war is a dominant image throughout the *Satyricon*, from the ongoing 'Theban' love triangle between (sexual 'brothers'/comrades in arms) Encolpius, Giton and Ascyltos, to the civil war scenes on board Lichas' ship at 108–9; civil war epic infects this fiction from beginning to end, sending the heavyweight plot-model of Odyssean wandering into a tail spin.

The culmination of literary expertise in the *Satyricon* is a physical ordeal that always presages a violent eruption of consumed knowledge – whether as vomiting (in the sickening *Cena*) or as an outpouring of verse ('pouring out words from the heart' *defundes pectore verba* (5, line 22); 'this effusion'; *hic impetus* (118.6); 'when Eumolpus had poured out these lines with immense fluency'; *cum haec Eumolpus ingenti volubilitate verborum effudisset* (124.2)). Is the retainability (the permanence) of knowledge, on which all Greco-Roman theory of education depends, now under threat? The accumulation and discovery of knowledge in the *Satyricon* looks ugly, violent, even physically menacing. As Quintilian comments, during his advice on how to educate the young: 'nothing is so bad for the memory as being overburdened' (*Inst.* 1.2.27). Foucault, fearing that plebeian readers might misunderstand his latest bestseller, *Les Mots et Les Choses*, famously said, 'A little knowledge is a dangerous thing';[54] Petronius seems to warn precisely the opposite, at the same time.

Instead of assuming information slowly and gradually, as Quintilian advises (*Inst.* 1.2.27–8), the characters of the *Satyricon* enact or discuss the speeding-up and concentration of learning.[55] In Neronian Rome's refocusing of human experience 'around the entry into social knowledge rather than the seasoned administration of the order of culture',[56] prematurity is idolised (especially in Petronian paedophilic fantasies) at the expense of

[54] Strangely, this enormously difficult book became a huge hit in France when it was published in 1966: Foucault was none too pleased, however, eager to command the powers of exclusivity with the warning that his work was not for everyone.

[55] Cf. Juv. 14.189ff, [Plut.] *De lib. educ.* 13, Tac. *Dial.* 30, Quint. *Inst.* 1.10.

[56] I quote Henderson (1993) 128.

traditional pedagogy. Agamemnon complains that boys are pushed *cruda* (raw) into the forum by over-ambitious parents; at 75.10 Trimalchio says he wanted to grow up too quickly, so used to oil his chin to stimulate/fake beard growth, and winds up the dinner party with his premature funeral, despite the fact that he has '30 years, 4 months and 2 days' left to live (77.2); his friend Habinnas likes nothing better than to make 'two days out of one' (72.4); and Quartilla's protegee Pannychis is deflowered at the age of seven by Giton, a mere boy himself, who is the focus or catalyst of most of Encolpius' adventures.

The notion that fast-forwarding (educational) experience actively imperils the preservation of memories (and hence knowledge)[57] is exemplified by the figure of the freedman in Trimalchio's *Cena*: the freedman is obsessed with memorialisation (Trimalchio is very impressed by the astrologer Serapa, who tells him things he has forgotten: 76.11), yet at the same time needs to forget and distance himself from his (slave) past, to pack a lifetime's worth of fun and privilege into his remaining 'free' years. He needs to forget who he was in order to learn who he is (and forgetting can fuel creation of the new, as my reading of Trimalchio's jumbling of myths hints): the buzzing University of Life teaches living for the moment, yet the freedman veers between living solely in the novelistic present and being reminded of his own mortality, and hence of the need to reaffirm and commemorate his evolving identity (which must ironically always be contingent on his previous life in captivity). His psychology echoes the contradictions of Encolpius' narration: fundamentally, the fiction of the *Satyricon* itself depends on it having been written down from Encolpius' memory, yet the central joke or dilemma of reading it is that our narrator's memory is apparently fallible, subjective and inadequate.[58]

REMEMBERING AND FORGETTING

Crucially, this paradox often seems calculated to jolt and unsettle *reader* memory (and unlike Trimalchio, we don't have an expert on hand to remind us what we may have forgotten in a long and complex fiction). In this final section, I want to discuss three points in the *Satyricon* where I think we are urged to do a double-take, as it were, and to re-analyse our initial readings or memory of previous passages. The first example concerns what emerges,

[57] On the importance of memory in education see Quint. *Inst.* 11.2.17ff, 33ff, 45ff, and 11.3.25; *Rhet. Her.* 3.16.29ff and 20.33–4. Also see Morgan (1998) 250.

[58] E.g., see *Sat.* 56.10 ('we laughed for ages; there were, like, six hundred of these jokes, which have now escaped my memory'; *diu risimus: sexcenta huiusmodi fuerunt, quae iam exciderunt memoriae meae*).

even in our broken text, as a carefully patterned mirroring between the 'beginning' and 'end' of the *Cena*. At *Sat.* 72–3 Ascyltos and Encolpius plan to dodge Trimalchio's furnace-hot bath, but as they head for the door they are so startled by the appearance of a snarling dog on a chain that Ascyltos falls straight into the fish-pond, followed shortly by a sloshed Encolpius. Our narrator comments: 'I, who had been frightened even of a painted dog (*qui etiam pictum timueram canem*) was also drunk, and while I was trying to help the swimmer, I fell into the same abyss'(72.7). The doorman saves them by pacifying the dog, and Encolpius reports that heroic Giton had already got the beast on side by throwing it all the titbits he had saved from the dinner. Clearly, the distraction of the guard dog in this way recalls the doping of Cerberus in Virgil, *Aeneid* 6.417–23, while the fall into the Charybdis-like whirlpool resembles a botched crossing of the Acheron: this is after all the gateway to Trimalchio's hell-on-earth. As we slip into hellish fantasy worlds, is the guard dog a figment of our narrator's imagination, the same dog we found painted on Trimalchio's hallway wall when the guests entered back in chapter 29 ('beware of the dog'; *cave canem* (29.1)) which Encolpius mistook, apparently, for the real thing? It is emphatic in the scene at 72 that we have been here before: Encolpius mentions that Giton leads him (back) 'through the gallery' (*per porticum* (72.7)). This, together with the reminder of the earlier scare with the guard dog sends us back to *Sat.* 29 – a neat ring composition. Yet now, it seems, the dog is real, not painted. Did we understand the joke at *Sat.* 29, and are we now double-crossed? Did the other guests laugh because (as we were first led to presume) Encolpius was startled by a mere picture? Given the fact that Encolpius is, by his own confession, legless, by *Sat.* 72, is the joke that he's still hallucinating guard dogs, that he still thinks the painted dog is real, even as he remembers 'I mustn't be scared of painted dogs *this* time'? Or did he make the more ludicrous error at *Sat.* 29 of believing the real dog was part of the frieze, an object for his artistic critique?

Meanwhile the presence of another, apparently real dog, named Scylax, in Trimalchio's dining room, adds further confusion. For Scylax and the painted dog at *Sat.* 29 look like one and the same: the dog Encolpius reports seeing at 29.1 ('to the left as you went in, not far from the porter's office (*ostiarii cella*)') is *canis ingens, catena vinctus* ('a huge dog tied up on a chain') while the real dog at *Sat.* 64 is also *ingentis formae . . . canis catena vinctus* ('a dog of huge proportions tied up on a chain') and belongs to the *ostiarius* who kicks it to heel under Encolpius' table (64.7). Hence the division of real from fake is complicated further: readers are made to get a taste of the same dizziness and self-doubt that (apparently) plagues Encolpius himself, but only, paradoxically, if their memory for what has already happened in this anti-narrative is razor-sharp.

A similar muddle occurs at *Sat.* 54.3, and this is my second example: Trimalchio pretends to have been hurt by a clumsy slave, and a mini-drama ensues in which doctors rush to his aid, Fortunata is in tears and the slave is begging for his life. Encolpius, suspecting a impending joke, says: 'I was afraid that his petition was leading up to some comic turn. The cook who had forgotten to gut the pig had not faded from my recollection'. He is referring us back to the earlier scene at *Sat.* 49, where the cook pretends to have forgotten to gut the pig, a sham unveiled when the beast is sliced open and sausages and black puddings slop out in the guise of intestines. Is the joke at 54.3 that Encolpius has *forgotten* that Daedalus didn't forget to gut the pig, or do we read 'forgotten' in inverted commas? Remembering what has already happened in the *Satyricon* uncovers booby traps as well as artful narrative patternings.

Like all good comics (we might imagine), Encolpius knows when he's onto a good formula. The sketch at the end of *Sat.* 108 plays the memory game for a third time. In the midst of the civil-war theatre on board Lichas' ship, Giton and Encolpius both turn tragic and stage mock suicides; Encolpius comments:

> 'Then the gallant Giton took a razor to his manhood, threatening to end all our troubles by self-mutilation . . . I also lifted a barber's knife to my throat several times, no more meaning to kill myself than Giton meant to do what he threatened. Still, he filled the tragic part more recklessly, because he knew he was holding the same razor he had already used to cut his throat'. (108.10–12)

This episode sends us rewinding to events at the end of *Sat.* 94, which showcased Giton's previous attempt at suicide (by drawing a barber's trainee razor twice across his throat (94.12)). The melodrama ended happily, as the fake blade meant that 'Giton was not marked by even a trace of a wound' (94.14), something Encolpius now seems to have forgotten, or 'forgotten'. Are we seeing our narrator's appalling memory in action (the joke embellished by the fact that the razor is a *learning* tool, as well as a pantomime prop), or his overactive imagination hyping up reality for the occasion (Giton's 'tragic role' *needs* just such a cue, he hints)? At 108, in any case, there is potential confusion over whether the razor Giton wields is sharp or fake, as the razor, belonging to Eumolpus' hired slave-boy (*mercennarius*) (94.12 cf. 103.1) has just been used on the ship to cut the hair and eyebrows of Encolpius, Ascyltos and Giton (103.1) Is there one razor, or two (and which is sharp/blunted)? Are *all* wounds and scars in this text metaphorical, operating as narrative markers, memory-like traces of events that have happened before, in the *Satyricon* or elsewhere? The stowaways' branding marks (written in the poet's ink on shaved foreheads) are fake, just as Trimalchio's wounded arm at 54.2 is a put-on, and Encolpius' chest-wound at 91.6 is psychological and has left no scar; at 99.2, Encolpius declares that Giton will have to rid his mind of scars caused by previous episodes if they are to get back together, at 113.7, every kiss Tryphaena plants on Giton's face is experienced as a wound, and at 113.8, Encolpius is afraid of reopening a tender 'scar' just as the wound of love has begun to heal. In the *Odyssey*, scarring is associated with the memory of narrative (in *Od.* 19, with the story of Odysseus' rite of passage on a boar-hunt), and Petronius triggers the memory of this concept when he makes Lichas recognise Encolpius by his crotch, just as Ulysses' nurse identified her master by his scar (105.9–10).

Scars are associated with, or are symbolic of recollection, that much is clear, yet the *Satyricon*'s layering of real, metaphorical, visible and invisible scars seems to typify the way in which this fiction makes memory significant, only to suggest that remembering functions not as affirmation or clarification but as obfuscation, forcing us to reconsider perspectives on either the event that is being recalled, or on the scene that has triggered the recollection. Whereas compilatory writers like Valerius Maximus emphasise hierarchy and pattern in order to facilitate learning and recall, Petronius seems to play a perversely opposite game, whereby 'truths' in the text are found to be dynamic rather than static, and a good memory uncovers further tangents and uncertainties. As ever, there is always the possibility that we might be missing a crucial section of narrative that would clear up the confusion. However, given the run of 'memory tricks', supported by precise verbal pointers and by Encolpius' characterisation as forgetful narrator, it is tempting, at the very least, to think that such jokes were more rather than less emphasised, and certainly not non-existent, in the original text. Potentially, the game would be made more challenging by a novel too bulky to digest in one, or even three or four sittings.

In questioning, as we have seen, both the retainability of knowledge and the status of memory as retained knowledge, the *Satyricon* takes on the core components of Greco-Roman educational theory. I have argued that remembering in the *Satyricon* erodes knowledge, demoting it at best to belief or conjecture, spurring scrutiny of what (we think) we know, or knew. As Connors puts it, 'The *Satyricon* becomes part of Neronian discourses about the power of the distant past in the present';[59] yet going back to the past, in literary terms and in the narratives of the *Satyricon* itself, involves not just

[59] Connors (1998) 5.

nostalgic regression and recontextualisation, but experimental reordering and rewriting – and I have tried to show that this is also precisely what we are made to engage in as readers in recalling a series of previous episodes in the text, only to have to reweigh our first impressions. The past comes back to haunt us, just as it does for the freedmen, and for Encolpius himself, whose scrapes and afflictions are all apparently retributions for his initial error of offending Priapus.

To conclude, if we learn anything from the *Satyricon*, it is lessons we have been forced to teach ourselves. This is an encyclopedia conspicuously devoid of expertise, authority and prescription of any kind, disabling any attempt to resolve whether it reflects, enacts or opposes the decline, corruption and transformation in contemporary education to which it frequently refers. Instead, the brilliant prank of extended first-person narration has the effect of situating readers inside that fictional world, in the hull of Eumolpus' sinking Neronian ship – buffeted on one flank by the literary storms of old, on the other by the seditious poetics of civil war. Putting knowledge up for sale makes for a participatory kind of reading – the *Satyricon* lets us taste 'Neronian excess' up close. And in many ways, Amazon's recent sales-pitch for Schott's *Miscellany* ('every toilet should have one') also captures this text to the tee, as a liberating mish-mash aptly focused on bodily process, tailored to both bitty sampling and lengthier rumination. But when Petronius vandalises literary memory (and the pedagogic systems underpinning its construction), he ensures that his elite readership can never be immune to the limitations and snares built into hyper-creativity: those already occluded from positions of (creative) power are exposed to uncensored rearrangings of their world, while the tyrant/author swaps Callimachean exclusivity for anxieties of ownership. Trimalchio's clownish alibi (stupidity) cleaves plenty of space for escapist experimentation – but not without its risks.

CHAPTER 6

Diogenes Laërtius, biographer of philosophy

James Warren

'Tell them I've had a wonderful life.'
L. Wittgenstein

I

One way to 'organise knowledge' is to write a history of a discipline or area of study, tracing its development from its origins to its present state. By choosing what to include and what to exclude the author of such a history delimits that area of knowledge and makes clear those figures or groups thought to have contributed to its development, and arranges them in a way which presents their relationship to one another as practitioners or inquirers into this area of study.

Philosophy, a notoriously difficult practice to define, attracts organisers and history-writers. Philosophers themselves have always been interested in the history of philosophy. That is not to say, of course, that they always agree about how one ought to be interested in the history of philosophy. Characteristically, they ask questions about the history of philosophy. First, should philosophy essentially be concerned with its own history? Is an awareness of the history of philosophy essential to one's being a philosopher and being engaged in thinking philosophically?[1] (Some philosophers are interested in the history of philosophy only to the extent of actively dismissing it as relevant to what they do as philosophers). Secondly, what ought to count as the history of philosophy? Should the history of philosophy be narrowly conceived as the history of a certain agreed set of philosophical problems and how they have been addressed by a certain agreed set of philosophers?[2] Or should the history of philosophy be expanded to cover a

My thanks to Jason König, David Sedley, Tim Whitmarsh, and the readers for the Press for their comments on earlier versions of this essay.

[1] See Taylor (1984).

[2] Cf. Rée (1978) 21 who argues well against the approach exemplified by, for example, Russell (1961): 'The "historical approach" offered by the History of Philosophy simply means discussing positions

wider scope, embracing 'intellectual history' more broadly conceived, the various intellectual, social and cultural trends which affect and are affected by philosophical ideas? Thirdly, if one is engaged in studying the history of philosophy – whatever that is thought to be – must one be a philosopher oneself? Is the history of philosophy itself a philosophical discipline?[3] (One would not insist, I suppose, that an historian of medicine ought to be a doctor).

These questions are outlined here as a background against which I will consider the practice of one historian of ancient philosophy, indeed an ancient historian of philosophy: Diogenes Laërtius. Diogenes' importance as a source for those working on the history of ancient philosophy has never been in doubt, but his credentials as a philosophical historian have not been so universally accepted. Often, Diogenes is praised for the virtue of having collected and ordered information from other, mainly Hellenistic, sources, and is thanked for his compilation but excused for his lack of philosophical acumen.[4] Of course, such damning criticism of his approach is possible only once we have established some more concrete answers to the sorts of questions with which I began, questions about how the history of philosophy *ought* to be written. I make no effort to do that here. In any case, although I cannot attempt to articulate fully and defend the view here, I suspect that there is no single definitive or superior conception of how the history of philosophy ought to be written. Rather, I will ask why Diogenes wrote as he did. What does the organisation of the work tell us about his conception of philosophy and its history? My central contention will be that Diogenes' work is an example of one way of writing and conceiving the history of philosophy – in terms of biography. But he does not limit himself to telling the life-stories of philosophers; he also wishes to construct from these philosophers' lives the 'life-story' of philosophy itself.

in the chronological order in which they happen to have been occupied by Great Philosophers and adding a little human interest by sketching something of the lives of the Philosophers; it does not mean discussing the history of the positions themselves'. Rorty (1984) 61–7 is similarly critical of this approach, which he calls 'doxography' (a term which I will reserve for a particular ancient genre). He identifies a mistaken belief that 'philosophers' and 'philosophical questions' are two natural kinds. Compare Frede (1992) 323–5 on 'historical' and 'philosophical' historiographies of philosophy.

3 See Aubenque (1992) and Brunschwig (1992). Further thoughts are offered by Barnes (2002).

4 E.g., Mejer (1978) 1: 'Indeed, nobody would deny that it is more appalling to imagine what the history of Greek philosophy would look like if Diogenes was our primary source than to think of what we would know without Diogenes.' Cf. Rorty (1984) 62: 'Diogenes Laërtius gave doxography a bad name by insisting on answering the question "What did X think the good was?" for every X in an antecedently formulated canon.'

II

When thinking about composing his work Diogenes was faced with a considerable range of options. There were examples available of a number of different approaches to writing about philosophical history. Earlier philosophers had themselves demonstrated an interest in previous philosophies and some even set out to picture themselves and their place in relation to what had gone before. In particular, in addition to his lost work *On philosophy* (περὶ φιλοσοφίας), Aristotle is often keen at the beginning of his physical and metaphysical treatises to offer reviews of his predecessors' views in order – in the main – to show their failings and his innovations.[5] This is a critical engagement with one's predecessors, not a purely 'historical' enterprise, but it does view philosophy as having a past and also sees philosophy in terms of gradual development, if not progress.[6] Later there arose the practice we now, after Diels, call doxography, or the *Placita*-tradition: the collection and presentation of different philosophers' opinions on a particular range of topics. This offers a more static view. The philosophers' views are listed without criticism or discussion and, although they sometimes appear in chronological order, the intention is not to describe a development but to present a range of opinions on a given topic.[7]

Two more kinds of work should be added – the biography and the 'successional list' (*diadochê*, διαδοχή). These two enjoyed a vogue in the Hellenistic period, from which most of Diogenes' sources derive (and, it must be noted, our knowledge of them is to a large extent founded on what is preserved in Diogenes). Diogenes is by no means unique in writing philosophical biographies. (Indeed, philosophers have even been known to write autobiographies).[8] Even in his general interest in the lifestyle of these philosophers, in particular in their unorthodox and unusual behaviour, Diogenes is following a strong literary and artistic tradition stretching back

5 See especially the opening books of the *Physics*, *De anima*, and *Metaphysics*. Aristotle's own credentials as a historian of philosophy were famously challenged by Cherniss (1935), (1944). Guthrie (1957) offers a critical response.

6 See especially *Metaph.* 982b11ff. It would be misleading to think, of course, that Plato did not engage with his predecessors in a similarly critical fashion, but his choice of the dialogue form crucially affects how this engagement is presented. See, e.g., McCabe (2000).

7 See, of course, Diels (1879) and Mansfeld and Runia (1997).

8 For example, see the successful biographies of Russell and Wittgenstein by Ray Monk: Monk (1990), (1996), (2000). See Monk (2001), (2002) for his views on philosophical biography. For philosophers' autobiographies see, e.g., Russell (1967–9) and, more recently, Warnock (2000), Honderich (2001), McGinn (2003).

at least to the classical period. The genre of biography, of poets, statesmen and intellectuals of all kinds, can be traced back at least as far as the fifth century BCE, and persisted as a popular form of writing throughout classical antiquity. Diogenes himself had various models of such writing to consider, from Xenophon's *Agesilaos*, through the works of Antigonus of Carystus and Philodemus, through to Plutarch and Philostratus.[9] His models could therefore be found not only among Hellenistic writers contemporary with the sources for much of his philosophical information, but also among writers – such as Philostratus – closer to his own time, which we can roughly date to the early third century CE. Biography, therefore, is not a method of writing confined to Diogenes' own age, or even recent past. Nor is Diogenes the first to write biographies of philosophers. Nevertheless, however much his biographies sit squarely in the general ancient tradition of using anecdotal episodes to reflect and illustrate the particular literary or intellectual output of the individual in question,[10] Diogenes is nevertheless perhaps unusual in concentrating solely on philosophers and solely on the biographies of philosophers. For example, Antigonus of Carystus, the source of much of Diogenes' biographical information, did not solely write biographies of philosophers, producing in addition works on art history, an *On animals*, and an *On diction* (περὶ λέξεων).[11] In contrast, Diogenes, so far as we know, wrote only about philosophers, whether in his work on their lives or in his poetry. His interests are quite specific and he excludes other intellectual pursuits – poetry, oratory, mathematics, medicine and so on – from his history. In philosophy, Diogenes found something for which the biographical mode of writing fitted perfectly. It allows him to concentrate both on the peculiar nature of those who practise philosophy (since philosophers behave differently and look different from non-philosophers) but also allowed a way of explaining the history and transmission of that practice.[12]

The second kind of work, the successional list (*diadochê*), is another form of categorisation of philosophical traditions. Sometimes containing chronological or biographical information and anecdote, in their most concise form they appear as a simple line of philosophers which proceeds chronologically, each link in the chain being cast in terms of master–pupil relationships. The motivations for composing such lists are diverse. They are not only the product of Hellenistic scholarship trying to organise previous

9 See Momigliano (1993). 10 See the discussion in Fairweather (1974).

11 The best recent treatment of Antigonus is the introduction to Dorandi (1999).

12 See, e.g., Pl. *Tht.* 174a4–176a1. For depictions of philosophers in ancient art see Zanker (1995). Compare the photographic portraits of modern philosophers by Pyke (1993).

philosophical history but also owe a great deal to the Hellenistic schools' own desire to fashion for themselves an antique and venerable philosophical heritage.[13]

It is clear, therefore, that Diogenes was offered a series of paradigms for works of philosophical history and his own work is clearly indebted to each of these different forms to some degree. The imprint of a variety of previous methods of organising, understanding and presenting the philosophical past can be seen very clearly throughout the work.

The imprint of the successional literature is evident in Diogenes' organisation of the work into two distinct lineages, the 'Ionian' and the 'Italian' (1.13–15) and in the constant refrain at the beginning of each *Life* which gives some combination of the philosopher's toponym, patronym and philosophical master, the last introduced often by the verb ἀκούειν or διάκουειν, ('to listen to', 'be a student of') a favourite of the writers of *diadochai*.[14] He also includes some more narrowly defined doxographical sections, notably the long exposition of Stoic philosophy (7.39–160).[15]

Since it bears the unmistakable signs of the use of and dependence on all these different kinds of earlier works of philosophical history and categorisation, Diogenes' work is a treasury for those who like to indulge in *Quellenforschung*, encouraged no doubt by Diogenes' willingness to name his sources. Beyond this mining of the text, no one who cares to think about Diogenes himself would deny that he combines successional literature's interest in philosophical lineages with biography's interest in colourful anecdotes which supposedly illustrate the philosopher and his outlook. But this is about as far as many writers on Diogenes have been prepared to venture, leaving him as the culmination and combination of a number of other traditions of works which survive only in piecemeal fashion. If Diogenes is accorded any credit at all it is as someone who attempted an amalgamation of these forms and he is to be thanked for preserving information which would otherwise have been lost.[16]

13 On the *diadochai* see von Kienle (1961), Giannantoni (1981) and Giannattasio Andria (1989). There are discussions of Diogenes' use of such works and whether he had access to the original full version of, e.g., Sotion, in Mejer (1978) 62–81 and Aronadio (1990). Cf. Desbordes (1990) vol. 1, 7–46.

14 4.24: 'Crantor of Soli was admired in his own country but moved to Athens to study under Xenocrates as a fellow student of Polemon'. For other examples see: 6.20–21 (Diogenes the Cynic), 7.2 (Zeno of Citium), 9.21 (Parmenides).

15 Diog. Laert. 7.38: 'I decided to give a general account of all the Stoic doctrines in the *Life* of Zeno since he was the founder of this school'.

16 See, e.g., Mejer (1978) 1–2, who tries to defend Diogenes against the accusation of being a poor historian of Greek philosophy, not by asking what sort of history this is but by denying that it is a history of philosophy at all. Cf. Giannattasio Andria (1989) 28. The closest cousin to Diogenes' work is Philodemus' contribution to the history of philosophy, the *Index of philosophers*, σύνταξις

III

Let us try to begin from an alternative starting-point. Faced with this range of possibilities, Diogenes must have made an active decision to offer a work in the form that he did. Even if he is an amalgamator of some sort, his amalgamation was surely designed for some end and not merely the product of someone led by a peculiar curiosity for a peculiar group – philosophers – and their peculiar lives.[17] What might that end be?

There are major difficulties to be faced in answering this question. Above all, Diogenes is not an author who makes his own presence felt strongly within the text. He gives no indication of his own biography, where he is working, or any personal philosophical allegiance.[18] There is relatively little first-personal intervention within the work. Diogenes offers no programmatic explanation of why he has written this work, nor does he offer any detailed explanation of why the work has taken the form it has, instead launching immediately into a discussion of some other authors' mistaken view that philosophy first emerged among the non-Greeks.[19] Rather, his own personal intervention within the work is limited in the main to the short epigrams which he sometimes offers on particular philosophers, and only occasionally will he offer his own view on matters of philosophical history or comment on the plausibility or otherwise of the information he retails.[20]

τῶν φιλοσόφων, referred to by Diogenes himself at 10.3. This work, or group of works, contained the *Index academicorum* (*PHerc.* 1021 and 164; see Gaiser (1988), Dorandi (1991)), the *Index stoicorum* (*PHerc.* 1018; see Dorandi (1994)) and perhaps others. For an overview see Dorandi (1990) and Arrighetti (2003). It is clear from the less fragmentary works that they share with Diogenes an interest in biographical anecdote as well as philosophical tradition. Philodemus often uses the same source as Diogenes. See, e.g., their use of Antigonus of Carystus as a source for the story of Polemo's conversion, discussed in Warren (2002) 161–3. Cf. Gigante (1986) 25–34.

[17] So, Mejer (1992) 3561.

[18] The fact that Diogenes concludes with a book dedicated to Epicurus has been seen by some as an indication of an Epicurean allegiance, but this is not necessary. There is also the famous address to a certain 'lover of Plato' at 3.47, but this person cannot be securely identified. See Goulet (1999) 45–6 and Desbordes (1990) vol. 1, 316–32.

[19] Diog. Laert. 1.1: 'Some say that the business of philosophy originated among barbarians' (τὸ τῆς φιλοσοφίας ἔργον ἔνιοί φασιν ἀπὸ βαρβάρων ἄρξαι).

[20] E.g., Diog. Laert. 2.45: 'I think Socrates also talked about natural philosophy, since he talks about providence too, as even Xenophon says, who nevertheless also says Socrates talked only about ethics'. Mejer (1978) 55 collects seventeen occasions on which Diogenes comments directly on his sources. On Diogenes' poetry see Mejer (1978) 46–50, Gigante (1986) 34–44, Goulet-Cazé (1999) 16–17. His poetical work had at least two books (1.39, 63) of poems in various metres which may have been exclusively about philosophers (a more complimentary version, perhaps, of Timon of Phlius' *Silloi* – which Diogenes also finds room to cite). Critics have in general been unimpressed by Diogenes' poetic abilities, a reaction he perhaps anticipated: see his remarks at 4.15.

In constructing Diogenes' approach to writing the history of philosophy, therefore, and in the absence of strong authorial signposts within the text, we are for the most part left looking at the form, content and organisation of the work itself.

First, the title of the work can offer some clues.[21]

Diogenes Laërtius: Lives and Opinions of the eminent philosophers and a brief collection of the views of each school. (Index in *Parisinus graecus* 1759)[22]

The title promises a dual exposition of philosophers' lives and thoughts, *bioi* and *gnômai*. Diogenes is interested not merely in events and facts about the individuals' lives but also in what sets them apart as philosophers, namely the thoughts and theories they propounded. Of course, a strict separation between these two aspects is not possible: the anecdotes offered about the philosophers are intended to illustrate and reflect that person's view of the world, their values and outlook. Clearly, philosophers' lives are different because philosophers view the world differently. This is the premise behind Diogenes' biographical interest, and is what makes biography a viable method for the presentation of philosophical history.[23] To be a philosopher is to live one's life in a certain way. Philosophy is a kind of life, a βίος.[24]

Secondly, we should again notice that Diogenes was perfectly aware that he could have chosen other methods of arranging his material. First we have a brief presentation of the contents of the work (1.13–15): the seven (plus four)[25] sages, the Ionian school founded by Anaximander, pupil of Thales, which ends with Clitomachus (of the Academy), Chrysippus (of the Stoa) and Theophrastus (of the Lyceum), and the Italian school, founded by Pythagoras, which ends with Epicurus. But then Diogenes offers other classificatory modes (1.16–18). Philosophers could be sorted into dogmatists and sceptics,[26] or into those who left writings and those who did not (and

[21] I use Marcovich's text. The *inscriptio* to book 1 has similar wording. Cf. Dorandi (1992).

[22] Λαερτ(ίου) Διογένους βίοι καὶ γνῶ(μαι) τῶν ἐν φιλο[σοφίᾳ] εὐδοκιμησάντων και τῶν ἐν ἑκάστῃ αἱρέσει ἀρεσκόντων ἐν ἐπιτόμῳ συναγωγή.

[23] Gigante (1986) 17 terms this 'biodoxography'.

[24] Cf. LSJ *s.v.* II. A recent provocative discussion of the relationship between a modern professional philosopher's philosophy and his life is offered by Honderich (2001), particularly in the *coda* to his autobiography. Cf. Conant (2001).

[25] Diogenes also includes *Lives* of Anacharsis, Myson, Pherecydes and Epimenides. Peisistratus, included in the list by 'some', does not receive a separate *Life* but appears as a correspondent in the *Life* of Solon.

[26] Compare a similar classification in Sext. Emp. *Pyr.* 1.4. However, Sextus includes among the dogmatists those who hold the view that things are unknowable (ἀκατάληπτα), Diogenes' sceptics, contrasting these with Pyrrhonists who 'suspend judgement'.

those who did leave writings can be further organised in order of the number of their publications),[27] or into how the particular school is named, or even according to their interest in the distinct parts of philosophy: physics, ethics and dialectic.[28] This is not simply the sign of Diogenes' interest in classificatory schemes;[29] Diogenes' mention of these possibilities serves to emphasise his own particular choice of biographical and successional presentation. In contrast to these other methods, Diogenes' approach will be resolutely diachronic: the history of philosophy from its origins to its culmination.

This is what the reader might indeed expect given the opening of the work. Here, the concern is clearly about the origin of philosophy, its birthplace. 'Some say that the business of philosophy began with the barbarians' (1.1). In fact, Diogenes goes to some lengths to explicate this view, citing various authorities for the notion that philosophy arose in Persia, Assyria, Egypt and the like. He considers carefully the claims of Orpheus the Thracian to be a philosopher, and also the apparent positive result of this theory of barbarian origins, namely its ability to explain the various forms of 'philosophy' in different peoples since, presumably, its multiple origins among different cultures would produce a diversity of forms of intellectual life and inquiry which are nevertheless sufficiently related to one another to be recognisable as philosophy. Still, although he is prepared to outline this view, Diogenes sets out to correct the various authorities who have sought to defend it.

> But these people forget the excellent deeds[30] of the Greeks, from whom not only philosophy but also the human race began, assigning them instead to the barbarians. (1.3)[31]

Diogenes later, famously, offers as evidence of the Greek-ness of philosophy the purported fact that the Greek word for 'philosophy', φιλοσοφία, cannot

[27] Chrysippus wins, beating Epicurus, Aristotle and Democritus. The catalogue of Chrysippus' works is at Diog. Laert. 7.189–202. At 10.26–7, however, Epicurus is 'very prolific' (πολυγραφώτατος) and it is said that Chrysippus expressly set out to outdo him, resorting even to repeating himself in multiple works.

[28] Archelaus, Socrates' teacher, marks a chronological turning-point in this respect. Socrates inaugurated a primary interest in ethics rather than physics (1.18, 2.21). Diogenes is willing to point out that neat divisions are not so easy to make. At 2.16 he agrees that Archelaus did some ethical thinking too.

[29] *Pace* Mejer (1978) 52.

[30] κατορθώματα is a word borrowed from Stoic ethical terminology. Diogenes' choice of vocabulary reinforces his claim for the achievements of Greek philosophical thought.

[31] λανθάνουσι δ' αὑτοὺς τὰ τῶν Ἑλλήνων κατορθώματα, ἀφ' ὧν μὴ ὅτι γε φιλοσοφία, ἀλλὰ καὶ γένος ἀνθρώπων ἦρξε, βαρβάροις προσάπτοντες.

be translated into any other language (1.4).[32] To be sure, this can be seen as part of an ongoing concern with cultural possessions and more specifically with a restatement of a specifically Greek intellectual heritage. But it also can be seen as part of more specific concerns. First, the origin of philosophy had been a contested topic. Jewish and Christian writers, for example, had even before Diogenes' work been quick to claim a dependency of all Greek philosophy on Biblical knowledge.[33] However, Diogenes does not seem to be attacking non-Greek writers in particular. Rather, it is important that the sources he cites for the barbarian-origins view are themselves Greek and, indeed, are often philosophers who will appear elsewhere in Diogenes' work: Hecataeus and Aristotle, for example. So Diogenes is correcting the Greeks' own misconceptions and restoring to people like Aristotle their true, Greek, philosophical lineage.[34] Diogenes' more extravagant claim that the Greeks founded the whole human race, which is not pursued or justified in what follows, may reinforce the idea that his concerns are with the *genesis* of philosophy, its Hellenic legitimacy.

Diogenes' view is that philosophy is a wholly Greek product and cannot be found anywhere outside of the Greek world.[35] Its parentage is pure Greek, and can in fact be traced historically to two 'founding fathers', or original principles (ἀρχαί): Anaximander and Pythagoras (1.13).[36] Here in the preface Diogenes then goes on to give a whistle-stop tour through the two major traditions of philosophy, the Ionian and the Italian, demonstrating how it is possible to move from these two founding fathers throughout the remaining history of philosophy in an unbroken succession. This schematic listing of the members of these two schools which follows in 1.13–15 does not, however, correspond entirely accurately to the full contents of Diogenes' work. The Cyrenaics, the Megarians, Empedocles, Pyrrho, Timon, and the Pyrrhonists and those philosophers Diogenes himself labels 'scattered' or 'ungrouped' (οἱ σποράδην (8.91, 9.20)), Xenophanes and Heraclitus, do

[32] Latin, for example, often simply uses the transliteration *philosophia*. Cf. Cic. *Off.* 2.5: 'Philosophy is nothing but the desire for wisdom' (*nec quicquam aliud est philosophia . . . praeter studium sapientiae*).

[33] See Ridings (1995), Boys-Stones (2001) 176–202. Canfora (1994) claims that Diogenes' prologue is aimed at writers such as Clement. Also on Diogenes' prologue see Spoerri (1959) 53–69, and Gigon (1960).

[34] Dicaearchus the Peripatetic may have followed Aristotle on this question. He too figures as a source for both Diogenes and Philodemus and seems to have offered his own biography of philosophy. See White (2001).

[35] Cf., e.g., Philostr. *VA* 2.30 and its description of Indian philosophy. Diogenes explains, for example, that Anacharsis, though Scythian had a Greek mother and spoke Greek (1.101).

[36] ἀρχή is also a philosophically charged word, used especially of the 'principles' of 'Presocratic' cosmologies: see Diog. Laert. 1.27 (Thales), 2.1 (Anaximander).

not appear in these lists but all figure in the *Lives*. This has been taken as evidence that Diogenes has not revised his work, relying in the preface on simple schematic successional lists but elsewhere seeing a more complex picture.[37] But it must also be agreed that at this point of the text a simple successional list serves Diogenes' purpose well. He is trying to establish a direct line from the two Greek sources for philosophy through to the final exponents of philosophy in each of the two traditions. He can offer more complexities later in the meat of the volume once he has secured agreement with his general view that philosophy is of pure Greek descent and does, in fact, explain how he has decided to proceed with the 'Socratics' only after he has offered the *Life* of Socrates (2.47).[38] We can excuse, therefore, a little conscious simplification for the sake of clarity in the preface.

The genetic thesis is the first crucial element in Diogenes' overall conception of his work. Secondly, he explains what he takes philosophy and 'the philosopher' to be. Importantly, these are to be contrasted with 'the wise', σοφοί. The first clue comes at 1.5 where Diogenes is reluctant to grant Orpheus the title of 'philosopher', principally because he is not sure whether someone who offered such pictures of the gods as did Orpheus deserves that title. The theological spin returns in 1.12 where Diogenes claims that Pythagoras was the first to use the term philosophy and to call himself a philosopher, 'since no human is wise, but only god'. Diogenes seems attracted to this air of humility. At least, he is himself insistent upon a sharp distinction between the wise men and the philosophers. Thales, for example, who regularly appears as the first 'philosopher' in modern accounts of Presocratic philosophy, and was also hailed by Aristotle as 'the originator of this kind of philosophy' (ὁ τῆς τοιαύτης φιλοσοφίας ἀρχηγός (*Met.* 983b20–21)), is not a philosopher – at least not at this point of the work[39] – in Diogenes' eyes, but rather a wise man, a σοφός. Thales' pupil, Anaximander, is the first philosopher of the Ionian line. The reason for this, it seems, is that Diogenes is insistent on the strict understanding of 'philosopher', φιλόσοφος as one who loves wisdom, with the implication that this lover of wisdom is not yet wise. A philosopher is a student, one who strives after the understanding and perfection of the wise man. If Thales is a wise man, a σοφός, therefore, he cannot be a philosopher, φιλόσοφος.

37 Goulet (1992).

38 Compare the similar explanation of his procedure with different Cyrenaic factions at 2.85–6. Cf. Laks (1993).

39 Cf. Goulet (1999) 49–52. At 1.122 Thales is the founder of the Ionian *philosophy*, and at 8.1 the Ionian philosophy is said to derive ἀπὸ Θαλοῦ (cf. 1.13: ἀπο Ἀναξιμάνδρου). Goulet concludes that Diogenes is trying to reflect two traditions, one in which Thales is a sage and another in which he is the first Ionian philosopher.

Rather too quickly this [study] came to be called 'wisdom' and the one who declaimed it a 'wise man', as being someone who had attained perfection in his soul. A 'philosopher', then, was someone who loves wisdom. (1.12)[40]

The contest for the possession and correct application of the term 'philosopher' has a long and ancient heritage, which can be traced at least as far back as the Platonic dialogues.[41] Diogenes himself may be alluding to a continuing discussion of the term when he goes on in 1.12 to note that 'sophist', σοφιστής, was also used as an equivalent for σοφός, which itself had been adopted as a term for those Diogenes wants more properly to designate by the term φιλόσοφος. 'Sophist', σοφιστής, he explains, was also used of people such as poets – people whom he wishes to exclude from the present work.[42] Diogenes passes rapidly through these potential confusions, but the complaint in 1.12 about the all-too-easy confusion of the philosopher and the wise man seems to indicate that Diogenes wishes to maintain a clear and specific distinction: there are sages and there are philosophers. Diogenes' interest lies primarily in the latter. To be a philosopher is on this account to be in pursuit of something, perhaps even in pursuit of something which might deserve to be called superhuman. At least, philosophers seem to be in pursuit of some kind of perfection and understanding – the understanding which the wise have, in contrast, already attained.[43]

There are clear parallels and possible sources for the sort of distinction which Diogenes presses here and in his distinction between the wise men of book 1 and the 'philosophers proper' of the following nine. First, there is the Platonic vision of philosophy as a desire for the knowledge or goodness which one lacks, which plays a prominent role in Diotima's speech in the *Symposium*. Second, there is the Hellenistic schools' own concentration on the promotion of the ideal exemplar of their particular philosophy, the wise man – an ideal which the Stoics, for example, thought was rarely if ever attained.[44] But Diogenes' classification is not merely a relic of previous attempts to classify and extol the philosophical life. In the economy of Diogenes' work this distinction adds a further dimension to the emerging

40 Θᾶττον δὲ ἐκαλεῖτο σοφία, καὶ σοφὸς ὁ ταύτην ἐπαγγελλόμενος, ὃς εἴη ἂν κατ' ἀκρότητα ψυχῆς ἀπηκριβωμένος, φιλόσοφος δὲ ὁ σοφίαν ἀσπαζόμενος.

41 See Nightingale (1995) 13–59. Cf. Burkert (1960).

42 We might compare Philostr. *VS* 479, 484 on the distinction between 'philosopher' and 'sophist'. Philostratus will deal with only those philosophers who also had a reputation for sophistry in addition to those who are 'proper' κυρίως sophists.

43 Again, note that Diogenes does not offer entirely hagiographical biographies of philosophers. These *Lives* are unlike those later treatments of, say, Pythagoras and Plotinus by Porphyry, or Origen by Eusebius. See on these: Cox (1983), Clark (2000).

44 See Alex. Aphrod. *Fat.* 199.14–18. For Stoic discussions of the definition of philosophy and wisdom see Sen. *Ep.* 89.4–8. Cf. Brouwer (2002), esp. 186–99.

portrait of what it is to be the sort of person whose life and thoughts are worthy of record and consideration – the *pursuit* of understanding.

Perhaps this much is not so unusual. However, Diogenes adds another theme to this conception and by doing so enlarges his picture of the nature of philosophy and the nature of philosophers. To this interest in what it is to be a philosopher and in the ways in which philosophers differ from everyone else he adds an interest in the question of how one comes to be a philosopher. A persistent theme through the work is the mechanism by which someone comes to take up philosophy – indeed, it is perhaps reasonable to term this a 'conversion' to philosophy since it is often a life-changing event – and, following from this, the mechanism by which philosophy is transmitted from one generation to the next. Here the two genres to which his work is indebted function perfectly in tandem. From the successional lists Diogenes takes the basic notion that philosophers are not born but made. In particular, they are formed by interaction with and interest in other, older, philosophers. Each new philosopher in these successions is the pupil of the last. From the biographical genre he can take a series of more detailed and elaborate stories of the conversion of various different philosophers. Biography's interest in the peculiar character of the philosopher, its emphasis on the charismatic strangeness of the philosopher, offers tremendous scope for tales of conversion, infatuation and instruction to colour and nuance the bare 'master–pupil' relationships offered by the successional lists. There are many such anecdotes throughout the *Lives*. Here, by way of an example, I offer a brief look at one case, the story of Crantor and Arcesilaus (4.28–9).[45]

> This is how he took up with Crantor . . . First of all, before moving to Athens, he was a pupil of Autolycus his countryman, with whom he travelled to Sardis. Then he was a pupil of Xanthus the Athenian musician. Next he was a pupil of Theophrastus. Finally he moved across to the Academy and Crantor. For Moireas – the brother I have already mentioned – encouraged him into rhetoric, but Arcesilaus was a lover of philosophy. Crantor desired him also and quoted the line from Euripides' *Andromeda* 'O maiden, if I save you will you be grateful to me?' (Nauck fr. 129), and was answered by the next line: 'Take me, stranger, as a slave or a wife' (Nauck fr. 130). From then on they lived together.

[45] For a discussion of the story of Polemo's conversion (Diog. Laert. 4.16) and of Nausiphanes' interest in Pyrrho (9.64), see Warren (2002) 160–4. Like these, the story of Crantor and Arcesilaus is from Antigonus of Carystus, although Diogenes does not name the source here. The parallel text is Phld. *Acad. Ind.* XVII, which breaks off before the Euripidean citations. Cf. Dorandi (1999) lx, who ends his Antigonus fr. 17B at φιλοσοφίας ἤρα (which becomes ἐπὶ φιλοσοφίαν ὁρμῆσαι in Philodemus). Also see on this conversion story: Long (1986) 439–40. On conversion generally see Nock (1933) who deals cursorily with philosophical conversions at 164–86.

Arcesilaus' intellectual formation begins close to home but moves from one teacher to the next and from place to place until he finds his true home and true love in the Athenian Academy. In particular, this story demonstrates Arcesilaus' search for his true calling in the face of pressure from family members and offers of favours from other potential teachers. Arcesilaus has a desire for philosophy, cast in terms of erotic attraction (ἤρα), and finally finds his philosophical life-partner after a process of courtship enacted here through the exchange of lines from tragic verse in which Crantor becomes the hero, rescuing the young maiden Arcesilaus from the clutches of rhetoric in return, presumably, for the fulfilment of his own desire for the gifted young philosopher. Theophrastus, the spurned suitor, goes on in the passage which immediately follows (4.30) to curse his luck at losing this promising student. Indeed, the Euripidean tragedy is entirely appropriate. Crantor plays the part of the heroic Perseus, rescuing Arcesilaus/Andromeda from the clutches of a sea monster (Theophrastus or, perhaps, rhetoric) to which he has been sacrificed by his own family.[46] Like Andromeda, once he is saved Arcesilaus never returns to his family but travels instead to live with his new saviour and lover.

The Euripidean quotations also emphasise the eroticised nature of philosophical pupillage, which is plain on the surface of this and similar stories. Equally clear is the notion that philosophy is not a pursuit in which people engage half-heartedly or as a mere stepping-stone towards other more practical matters. Rather, philosophy is the ultimate goal and the defining pursuit of Arcesilaus' life, and his previous life, characterised by a series of teachers in different subjects and constant travelling from place to place, contrasts markedly with the stable and constant life of philosophy which he finds with Crantor, with whom he stays until Crantor's death (4.32).

Diogenes would have found stories such as this in the Hellenistic biographies which figured as his sources, many of whom seem to have cast the master–pupil relationship in the terms of pederastic courtship.[47] We might also speculate that this general principle of transmission, that in order to be a philosopher one must be inspired and converted to philosophy somehow by another – usually older – philosopher, offers a further reason for Diogenes to begin his work with the σοφοί, who do not – it seems – need

[46] For the plot of Euripides' play see Eratosth. *Cat.* 15, 17 and the fragments in Nauck.

[47] The pederastic mode of casting philosophical pupillage is already present in Platonic works, and is certainly working in the background of Alcibiades' speech in the *Symposium*. Zeno of Elea is Parmenides' young lover (παιδικά) at *Prm.* 127b5, which is presumably the source for Diog. Laert. 9.25. Dumont (1987), Gaiser (1988) and Dorandi (1991) all try to downplay any erotic overtones in such phrasing.

such (inter)personal inspiration and conversion. Without them, an infinite regress is threatened. Thales and Pythagoras therefore can stand as the first converters to philosophy without themselves needing to be similarly converted.[48]

This general means of explaining the continued presence of philosophy, through direct personal inspiration from other philosophers, sits neatly with Diogenes' genetic thesis. Not only can he claim, therefore, that philosophy is purely Greek in its origin, he can also demonstrate through the biographical details of the *Lives* that this lineage is thoroughly preserved without dilution through the subsequent course of philosophical history. Philosophers attract, inspire and produce other philosophers. And since the first philosophers were Greek and philosophy is a uniquely Greek practice, this mode of transmission and propagation ensures the legitimacy of Greek philosophy from generation to generation. The *Lives*, therefore, offer a genealogy of philosophy, positioning each philosopher in his particular place on the family tree.

At this stage we may also be in a position to understand Diogenes' somewhat fossilised view of philosophy. The date of the work is not certain, but Diogenes is certainly writing some time in the early third century CE (the terminus *post quem* is given by the dates of Saturninus, the pupil of Sextus Empiricus, mentioned at 9.116 and Theodosius, mentioned at 9.70). If so, then it might indeed be thought surprising that Diogenes offers no extended discussion of developments after the early Hellenistic period. He mentions no Christian philosophy, nor any 'Middle' or 'Neo-' platonism. This concentration cannot be assigned merely to Diogenes' primarily Hellenistic sources, since this provokes the further question why Diogenes was so wedded to such aged sources. Further, Diogenes does mention some philosophers of later periods. He is well aware of the existence of Pyrrhonist philosophers up to Saturninus (9.116) and of Epicureans such as Philodemus (10.3). Later Stoics such as Posidonius (e.g., 7.39, 10.4) and Panaetius (7.41) are also mentioned as sources, even if they did not merit their own *Lives*.[49]

These later philosophers all tend to be following one or other of the major traditions of philosophy, all of which were either established by the end of the Hellenistic period or, in the case of Pyrrhonism, for example, explicitly

[48] Cf. Frede (1992) 319: 'Diogène suggère ainsi que la philosophie s'enracine en quelque façon dans cette sagesse pré-philosophique.'

[49] The contents in *Parisinus graecus* 1759 and its copies lists under book 7 the names of many later Stoics up to Cornutus. It is generally agreed that book 7 is incomplete, but it is not so certain that Diogenes would have included *Lives* corresponding to each of these names. See Dorandi (1992).

sought to claim for themselves a Hellenistic foundation. These philosophers cast themselves as followers or rediscoverers of the 'true' nature of the philosophies of Socrates, Plato, Epicurus and the others.[50] In that case, perhaps Diogenes took them at their word. Their claims against personal innovation disqualify them from needing to appear in their own right in the *Lives* since they themselves would have to agree that the major philosophical positions had been offered, at least in outline, already by the end of the Hellenistic period. In addition to the founders of the Hellenistic schools, therefore, Diogenes includes *Lives* of those Cyrenaics and Stoics, for example, who either dissented significantly from the orthodoxy (e.g., Ariston) or elaborated and added importantly to the founder's view (e.g., Chrysippus). Diogenes is once again not the poor prisoner of an outdated set of sources; he is responding to and implicitly endorsing a particular view of philosophical history.[51]

IV

Diogenes left us very little assistance in placing him and his work in any particular intellectual or cultural context. While in many ways it is tempting, mainly on the basis of the approximate dating we have for the work, to link Diogenes and his project with other roughly contemporary biographical works by Plutarch, Philostratus and Eunapius, swift assimilation to these other Imperial works does not seem to me to be particularly illuminating of the full nature of his project.[52] True, Philostratus does show some concern with pupil–teacher relationships and in tracing the origin of his particular interest – sophistic rhetoric – to a venerable Greek source (*VS* 482–4). He also follows a generally chronological order in the organisation of his *Lives*. Diogenes' concern for mapping successions, however, is far more pervasive, to the extent that he marks pointedly the two 'unordered' philosophers whom he cannot omit but also cannot place within any successional line (8.91, 9.20). It is of enormous importance for him to be able to produce as full as possible a genealogy of philosophy, encompassing all the members of his cast list, and he sometimes includes long lists of names, the only purpose of which besides, perhaps, the impression of scholarship

50 Cf. Frede (1992) 319–21.

51 Sedley (2003) argues that Diogenes' history of philosophy ends at roughly the same time as Athens' decline as a philosophical centre. Neo-Pyrrhonism, which does continue in Diogenes' work into the Imperial period, was never an Athenocentric movement.

52 On these and similar biographical works of the period see, for example, Edwards and Swain (eds.) (1997). Eunapius, although he refers to both Plutarch and Philostratus, seems to have no knowledge of Diogenes' work (Eunap. *VS* 454).

and comprehensiveness, is to link together important and known philosophers (see, for example, the list at 9.115–16). The entire work is carefully and systematically ordered to follow the two lines of progression introduced at 1.13. The first branch of the family tree takes up books 1 to 7; the second half books 8 to 10 (see 8.1).

Diogenes' other constant concern is the frequent references to the sources from which he has gathered his information. He is extremely careful to demonstrate his use of and dependence on other historians and mines of information. His own presence in the text is downplayed. Further, the philosophers in Diogenes' work hardly appear to function as ethical *exempla*, positive or negative, as they do in, for example, Plutarch, except in so far as we might think Diogenes to be recommending and praising most generally a 'life of philosophy', following no matter which particular philosophical outlook.[53] In fact, Diogenes seems deliberately to cultivate an air of Hellenistic rather than Imperial times. Perhaps unsurprisingly, given his decision to write the history of philosophy only up to the end of the Hellenistic period, Rome and Roman contexts for Greek philosophers do not figure at all prominently in his work.[54] He surprisingly fails to mention, for example, Carneades' famous speeches for and against justice delivered during an embassy to Rome in 155 BCE, and which form much of – so far as we can tell – the third book of Cicero's *Republic*. All this makes Diogenes again unlike Philostratus, whose works can certainly be read as generally preoccupied with the relationship between sophists and sophistry, power, the emperor and the like.[55] It is conceivable, I suppose, that the lack of reference to Rome in Diogenes is meant to be a conspicuous absence, a deliberate erasure of Rome from philosophical history. After all, Diogenes' genetic hypothesis and his philosophical genealogy stress and ensure the pure Greek nature of the philosophical life so he might be taking a staunch and defiant Greek outlook on philosophy. But it equally possible that Diogenes was simply unconcerned with tackling Rome's relationship to philosophy, and should not therefore be placed alongside those other writers whose concerns were those of 'being Greek under Rome'. As I have outlined, there are few if any elements of Diogenes' work which could not plausibly be argued to derive in the main from his Hellenistic sources, perhaps even the

53 Cf. Cox Miller (2000) 217–9. On Plutarch's paradigmatic lives see Duff (1999).

54 The references to Rome are all 'in passing': 2.104, an Aristippus wrote on the Romans; 5.61, a Strato wrote about Philip and Persaeus' war with Rome; 7.35, a Zeno wrote a work on Rome; 8.14, Romans among those who came to hear Pythagoras; 8.72, a statue of Empedocles moved to Rome from Agrigentum; 9.84, Roman burial customs (in a Pyrrhonist 'mode'); 9.109, Tiberius the dedicatee of a commentary on Timon's *Silloi*.

55 See Whitmarsh (2001) 225–46.

interest in biography itself as a literary form. I suspect that had we more of Philodemus', Antigonus' and even Dicaearchus' works we would find many more similarities with those of Diogenes. There is no positive evidence for seeing cultural identity and self-definition as overarching concerns in Diogenes' work except on the question-begging premise that such concerns are a characteristic of all elite Greek writing of this period.[56] Nevertheless, to resist pushing Diogenes into a group with other Imperial biographers does not require that we return to viewing him as a mere compiler of antiquated works. Indeed, both these approaches could be faulted for neglecting to consider Diogenes' work and its own ambitions and innovations in full and on its own terms.

Diogenes is not a fine philosopher, nor is he a fine historian of philosophy in the sense that someone now might set out to write a history of philosophy.[57] Rather, Diogenes' work sets out to do something different, and Diogenes' method and organisation suits this particular purpose perfectly; it is not a mere haphazard joining of previously distinct doxographical and biographical methods. That purpose is the presentation of philosophy as a distinctive, and distinctively Greek, way of life. Philosophy arose only in Greece and was transmitted from philosopher to philosopher, each new generation attracted and converted to the pursuit of wisdom by the charisma and intellect of the previous generation. Philosophy is inherited. It is passed down from the very first philosophers through the different branching and forking family trees traced by the successional literature. Diogenes offers, therefore, a genealogy of philosophy, a story of the philosophical family, their lives, and interrelations. We should surely expect a genealogy of this sort to combine precisely the two elements of biography and *diadochê*, the dual focus on the lives of individuals and the succession between one philosopher and the next which, unless they are viewed in the light of this overarching conception of philosophy and its history, might otherwise be thought the respective relics of two earlier genres, imperfectly amalgamated.

[56] Cf. Whitmarsh (2001) 2.

[57] Compare Frede (1992) 312, 325 who also argues that Diogenes should not be criticised for his failings as a historian of philosophy, because Diogenes' project is not like that of a modern historian of philosophy.

CHAPTER 7

The creation of Isidore's Etymologies or Origins

John Henderson

I believe that almost everyone who uses the book finds it more convenient to have recourse to the Index first.

(John Roget, Introduction to *Roget's Thesaurus* (1879, 2nd edn), cited by Roget (2002), Introduction, p. xv)

Dear Alexander Valkner,

. . . it was a relief to come across your long, brilliant piece in a recent issue of *Comment*, namely: The History of Dictionaries.

. . . From intimacy you travelled to grandeur, then back and forth, like a marvellously controlled metronome. I admired the way your essay builds on itself so meticulously, and the way it is anecdotal, accessible, and, finally, shading toward the confessional. I recognized only too well the moment in which you were tempted to approach some of our great writers to see whether or not they 'indulge', keeping a thesaurus hidden in their desk drawer.

(Reta Winters, in *Shields* (2002), 163–4)

When it's ajar . . .[1]

Almost every publication on Latin literature today practises citation from Isidore. Through the twentieth century, this was a matter of itemic consultation through a modern Index. Until 1991, the closest that many, perhaps most, scholars ever came to *reading* Isidore's *magnum opus* was, for sure, the *Index verborum* of Wallace Lindsay's *OCT* (1911a), vol. II, 371–442: Latin, and 443: Greek (with *ibid.*, 444–50: *Loci citati*).[2] Then, at a stroke, the publication of Robert Maltby's invaluable *Lexicon of Ancient*

[1] This chapter (p)re-works Part I of Henderson (2007).

[2] Lindsay's work has incurred foreseen criticism for cavalier (classicising) orthography: in particular, Greek script is unwarranted (e.g., Marshall (ed.) (1983) 12). His editorial policies did indeed privilege and favour generations of classical users, and the resultant accessibility of his edition came eventually to equip and facilitate the characteristic logophily of contemporary Latin studies. Lindsay's preface (with his article, (1911b)) sketches the *impossibility* of a full edition, given the myriad manuscripts awaiting collation and the superhuman range of knowledges required of the editor. Modern scholarship is dominated by the three-volume synthesis by Fontaine (1959; 1983; supplemented by his collected papers: (1988)). A Paris-based series of editions with commentary on individual books now covers a haphazard third of the text, but we still depend on Grial's sparse notes in Arévallo's edition

Latin Etymologies (1991) finessed this reflex from the Latinist's apparatus of automatic procedures: '[M.] has assembled all the explicitly attested etymologies of Latin antiquity, from the predecessors of Varro to Isidore of Seville; he has covered glossaries and scholia as well as the standard ancient etymological source-works.'[3] So, why bother further with Isidore, neat?[4] For the twenty-first century, he is the officially designated patron saint of computer users and programmers and of the internet: recommended for prayer by whichever site-seeker, worldwide.[5]

The prolific and polymathic seventh-century Bishop of Seville was a vastly influential conduit for classical antiquity into the medieval world, but his encyclopedic storehouse of Latin commands attention from Latinists strictly as a putative witness to earlier etymological 'lore', otherwise lost to us. While it is possible to write an outline history of etymological scholarship (spanning from 'philosophical' theorising to ebullient bullshit), we have *only* Isidore extant as (anything like) a complete text. But because of his date, he is virtually absent from classical scholarship. The *Cambridge History of Classical Literature* is typical: Isidore appears in just the one (tralatician) sentence that begins with Varro, then runs through the list of names down to Isidore, in order to gesture towards his 'all-pervasive', if 'not always at first-hand', 'influence'.[6] Publications and reviews on Isidore scarcely figure at all in classical journals.[7]

Yet contemporary logophilia and, in general, the graphematic turn in criticism have emphasised the semiotic prevalence of etymologising mentalities in Roman culture,[8] so that this particular titan of taxonomic knowledge is coming under the sort of pressure to deliver bona fide goods to the marketplace that it has not known for a millennium. While (truth to tell) it has proved enough for most purposes to back up a proposed 'word-truth' with a bare reference culled from *any place* within the curtilage of 'the ancient world', it has been necessary to repress interest in the genesis (the

(1797–1803) for the Latin Patrology for large stretches (cf. Fontaine (1959) 20). Hillgarth (1983) remains the best survey (+ *id.* (1990)). The paper by Codoñer (1994) re-theorises Isidoran etymology; Reydellet (ed.) (1984) is a model commentary (on book 9). Amsler (1989) 133–72 introduces discourse and power to Isidore's encyclopedic order of knowledge.

3 Jacket blurb.

4 By my reckoning, something like two-fifths of the entries (excluding cross references) include Isidore's testimony; perhaps one in eight of the words etymologised, and around a quarter of the etymologies compiled, are given by Isidore alone.

5 www.scborromeo.org/saints/isidores.htm. Bulletin from the Vatican, 14.06.99. San Isidoro saved Spain by politely warning Mohammed off in a dream (for his full legend, see Migne *PL* 81).

6 Kenney and Clausen (eds.) (1982) vol. II: 286.

7 Concentrated in Iberian *Helmantica*. For the rich culture of Iberia from 470 (through 1500) CE, see Gerli (ed.) (2003).

8 Recent work by Ahl (1985), O'Hara (1996), Paschalis (1997).

origins) of Isidore's materials, and (in particular) recognition of Isidore's agenda. In fact, the shock that awaits anyone prepared to *read* the *Etymologies* is closely analogous to the shock that hits users of *Roget's Thesaurus* when it dawns that *that* monumental word-store represents the bastardisation of a determined attempt to systematise a forceful ideology. Mediated through successive revisions of Roget's original scheme, the teleology of the classification led (leads) from abstract concepts through the material universe, to humanity, and, for climax, the apex of significance: morality and religion: 'the imperfect forerunner of that *Universal Language* to which Roget and his fellow reformists aspired'.[9] The editions we use today ring-fence the *Thesaurus* with dense paratextual assurances that stylistics and crosswords dwell here, so as to tuck the distracting creationist programming just out of our eyeline. But *Roget* had a theoretical tendency – *was* an agenda.[10]

What of the *Etymologiae*? When editors such as Lindsay preface Isidore, to the contrary, with a run of his letters followed by an agglomerated index plus scheme of chapter-headings through the twenty books ahead, the sloganised value-system of the *Etymologiae* is loudly advertised up-front, and heavily underscored for attention. The present essay will follow these *praemissa* ('prefaces', 'premises') in waymarking the main principles, categories and hierarchies of Isidore's taxonomy, on the path that leads from *grammatica* and *rhetorica* through to the finale of *instrumenta hortorum* and *equorum* (*Equipment, for Garden* and *Horse*). One eye will be trained on the difference brought to his Varronian inheritance by Isidore's position as a Latinate Christian authority who had no Hebrew and little Greek,[11] but who posited revelation of the creator's design through these three sacred tongues; and the other eye will light upon the textuality and *writing* that shape this vast icon of conceptual order. As the great Curtius outlines in

[9] Roget (2002) xvii, cf. Roget (1852, 1st edn) Introduction (*cit. ibid.*, xxx–xxxi), 'a *Philosophical Language* . . . the establishment of a *Universal Language*'. John Roget (*filius*) extended the system of cross-references and recognised the importance of the Index – Peter (*pater*) 'Roget himself had thought of it only as a last resort – his original notebooks had not had one', as the 150th-anniversary edition rather dimly notes ((2002) xv).

[10] Bp John Wilkins is the key reference here (Roget (2002), xxx n. 2). If the logical ordering of all items nameable within the universe could be verified by systematic science and certified by the Royal Society, the future of mankind would be to live in a world where words tell true, cf. McArthur (1986) 119–23. For the semantic programming of *Roget* see Hüllen (2005).

[11] This is strenuously argued by Fontaine, esp. (1988) IV, 535 (*pace*, e.g., Courcelle (1959)). Besides operating at the most basic of levels (*perhaps* considerately), Isidore does make corking mistakes with Greek (cf. Lindsay (1911b) 44), but when did the question 'What is it to know ancient Greek?' ever sustain a simple answer? (Ditto Hebrew – and Latin).

his basic book on the entire Middle Ages, in *this* pangram, it's downhill all the way:[12]

> the great Isidore of Seville, who in his great compilation of all human knowledge chose the road from designation to essence, from *uerba* ['words'] to *res* ['things'], and accordingly named his work *Etymologiarum libri* . . . The importance of this work . . . can hardly be overestimated; it may be called the basic book of the entire Middle Ages (*Grundbuch des ganzen Mittelalters*). It not only established the canonical stock of knowledge for eight centuries but also moulded their thought categories.

Where Lindsay could only *disqualify* his text – 'this encyclopedia is not a literary work of art'[13] – Curtius invokes the requisite category of 'power-book'. For sure, this mighty spree of articulated knowledge plays across the world visible to etymology while enactively *creating* a world from etymology. In relays of feedback loops, what can and cannot be learned about the universe from its dividends for etymology feeds into and on what can and cannot be learned about etymology from its stocktake of Creation. All the while, the writing models the belief system it subtends, as the monumental text morphs through its taxonomers' paradise of totalisation through system, dramatising power *as* power over knowledge. Performative display of the power/knowledge nexus takes a scholastic turn to cosmogonic enumeration as interpretation builds its shrine to interpretation. Here mapping inculcates an unfolding world of values, as exegetic modality delivers its particular blend of protreptic regimes: pedagogic, devotional, ecumenical, revelatory – classical/post-classical, through pre-medieval, to pre-modern . . . ? The clerical authority valorises his schedule of the known, affixes limits to the knowable; but this clerk of words writes *with* appreciation and awe, warms or boils as he records and preaches. He discriminates for and against, as well as between . . . the items and grids he chooses to love or list; favours his favourites, scolds the demonised, recommends this tool or that technology, plots his narrative to beguile the reader no less than shape nature. One constant dimension is, happily-cum-necessarily, language consciousness, a textualising semiotic of book culture – a cult of the book that lines up a 'liberal arts' manual as flagship and bearer of a mission to educate. Isidore profiles civilisation as inherent within the structures he avers: fixing origins for a permanence glossed as eternity; placing, contextualizing, opening out traditional schooling as conduit to the one-world superpower ideology of universal Rome, Rome pegged to Christ as the continuous present realm of *Latinitas* wherein 'we' dwell. Language-centred, -obsessed, but

[12] Curtius (1953) 496–7, cf. Bischoff (1966), Hillgarth (1983) 883–93. [13] Lindsay (1911b) 50.

not -blinded or -bound, Isidore's lexicographical, indexical, sign-fixated world nevertheless serves a specific Iberian catholicising politics within a durable Mediterranean cultural habitus.[14]

Our preliminaries begin with brisk epistolary crossfire between Isidore and Braulio of Saragossa, followed by a one-liner dedication to King Sisebut ('I–VI'). In the 'Spanish' group of manuscripts,[15] two letters from Isidore establish paradigmatic terms of Christian 'friendship' with the archdeacon Braulio ('A–B').[16]

A

The first sends (1) a pair of etymologies to go with presents of a band and a gown:

the *band* on account of our *bonding*, and the gown for the *garment* of our *agreement*, which is where in days of old it got this name from

anulus *propter nostrum* animum *et pallium pro* amicitiarum *nostrarum* amictu, *unde antiquitus hoc traxit uocabulum*

and (2) a folder of regulations. Neither of these genial puns, take note, will recur in the *Etymologiae*, or indeed any other Latin text. But twinning and (so?) equating of (seal-)ring and cloak with enclosed prescriptions entwines mandate with pledged affect in a double-barrelled or two-pronged offer of prayerful humility and playful amiability that cannot be refused. The programmatic work effected here is substitution of word for thing that forms the core of etymological mentation. 'Embracing' the letter as stand-in presence for its writer sets up its inaugural play on words ahead *as* word play, while contriving to consecrate its symbolic grasp on the trinity 'friend-cloak-Isidore' in one and the same all-embracing gesture (*pro* amico amplectere . . . *pro eo* . . . *complexentur* . . . *pro* . . . *amictu* . . . *ora igitur pro me*; '*Grab tight* for being *agreeable* . . . for that . . . be they hugged tight . . . for *garment* . . . So pray for me'). And the arch-Christian etiquette is formally re-affirmed when the letter is sealed with the sign-off 'My darlingest master and dearest son' (*dilectissime mihi domine et karissime fili*), where affective rhetoric annuls hierarchic distance.

14 Cf. esp. Fontaine (1959) 733–888, 'La culture d'Isidore de Séville'.

15 [Hence Lindsay's enclosure of both letters within square brackets. See (1911b) 45–9 for diagnosis of double recension and supplementation by the Spaniards.]

16 For Braulio, see Lynch (1938).

B

The second letter redoubles the formula, still more pointedly. For this time the missive trades a parcel of Isidore's own work for a package of canonised scripture: Braulio has 'the Sixth Decade of St Augustine' that Isidore asked for;[17] he now gets Isidore's *Synonimarum libellus* that he wanted. Together with its bearer, a boy, who is Isidore's stand-in – as innocent, vulnerable and promising as the sender is afflicted, and languishing 'for infirm Flesh and sinful Psyche':

I entrust this boy; and I entrust myself also . . . plus both weaknesses of the flesh and guilt of mind. In both respects I claim your protection, since shares in me trade at zilch.

commendo autem hunc puerum; commendo et me ipsum . . . et infirmitatibus carnis et culpa mentis. in utroque tuum praesidium posco, quia per me nihil mereor.

This boy is fetching Isidore's most passionate and poetic work, *Synonyms: or On lamenting the sinning soul*, and it must be worth a sackful of sacrosanctity.[18] No etymology tasters here, or in the rest of this prefatory epistolarity, but the insistent lesson of humility from the master extending over the whole field of mutually supportive exchange lodges the work ahead where it belongs in the prelate's mission on earth. It will be Braulio's service to image and valorise this latest latter-day 'Church Father''s addition to scripture. Expect no vainglory straight from its author's pen, whether here or in the creation to come.

Next (in the 'Spanish recension'; for openers, in the 'French' and 'Italian' manuscripts), a chain exchange of five letters dominated by Braulio, now raised to the episcopate, ushers us toward the dedication.

I

Isidore starts the ball rolling with the plain unprompted wish to see Braulio 'before I die' – and, thereunto, to feature in his prayers. Graceful warmth derives from pressurising the formulaic 'so *God fulfil my* prayer some day before I die' (*aliquando* impleret Deus *uotum* meum, *antequam moriar*) to divulge the prayer:

To *God* . . . , that both in this life He *fulfil my* hope and in what will be grant me a share of your blessing.

Deo . . . *ut et in hac uita spem* meam impleret *et in futuro beatitudinis tuae consortium mihi concedat.*

[17] This probably means the *Explanatio in Psalmos LI–LX* (so Migne *PL* 80: 649, n. a).

[18] A fiercely ascetic self-dialogue, 'On the Lamentation of the Sinner Soul': Migne *PL* 82: 826–68.

A sentence or two, no more, and maestro's generous address rises to accord the pupil *beatissime domne et frater* ('most blessed master and brother').

II

Braulio in response pours a cascade of (would-be) super-suave supplication: such a long time, such a hard time, since they met: 'a thousand obligations, and a thousand cares'. Enough of a profusion to warrant another double-barrelled request: (1) first, a full-throttle demand for Isidore's *Book of etymologies*, at long last. He's heard it's completed, and reminds the Archbishop, full circle, who it was in the first place that helped get the big 'sweat' started (*sudasti*). Matched by (2), an equally urgent request for the Conference Papers of the Synod (*Gesta Synodi*) starring Isidore as scourge of the heretic.[19] This time twinning of the requests is thuddingly emphatic ('I suggest . . . to despatch –> be they despatched . . . , and our. . suggestion'; *suggero . . . . dirigere –> dirigantur . . . et nostra . . . suggestio*,). A clumsy conceit has it that Isidore should shell out 'first' to Braulio, who helped get him going, so he can star 'first', in the councils of the saints, and this alibis the hyperbolic slide into the spectre of the fire-wielding scourge Isidore in the Universal Synod ('for that reason, *first* prove generous, so that *in* the *congregations* of the Saints you may be counted fortunate and *first*. The Proceedings of the *Council*, too, *in which* . . .'; . . . *ideo in me* primum *existe munificus, sic* in *sanctorum* coetibus *et felix habearis et primus. gesta etiam* synodi in qua). This letter's huff and puff measure in advance the mighty epic on its way to us. It will be a pressing part of Isidore's crusade for the truth: 'a blazing lamp, never dimming', as the sign-off pictures him (*lucerna ardens et non marcescens*).

III

Isidore's reply briskly deflates Braulio's puffery, nicely dents our anticipation: he still longs to meet up, but at least knows his friend's alive – because his letter did *arrive*. As to the message, umm, he got called away by a royal page before he could read it through; on his return, the letter had got lost, plus anything else in the envelope. So, shucks – could you please give it another shot?

This charming little shuffle somewhere between cock-up / put-down works by making us see Isidore seeing the damn letter clear as clear – had it right here in my hand I did:

[19] Viz. King Svinthila (*Sintharius*, 621-?31?). For the proceedings, see *Collectio Canonum S. Isidoro Hispalensi Ascripta*, Migne *PL* 84: 364–90.

I wish while in this body I could see the sight of the person whose well-being I have ascertained. I shall make clear . . . I was not worthy of reading your pearls through . . . I took delivery of your card . . . I gave that card . . . in order to read through afterwards and *write back* . . . I couldn't find the words you wrote . . . I didn't read through your pearls . . . *Write* it me *back* over . . . I'll take delivery . . . again

utinam cuius cognoui salutem, in hoc corpore aspicerem et uisionem . . . manifestabo . . . non fui dignus tua perlegere *eloquia . . . accepi* pittacium *tuum . . . dedi . . . illum* pittacium . . . *ut postea* perlegerem *et* rescriberem . . . *scripta tua non inueni . . . non* perlegi *eloquium tuum . . .* rescribe *mihi . . . iterum . . . recipiam.*

The engaging simplicity of this epistolarity polarises tellingly against the unctuous courtesies that sandwich it. In '*write back . . . Write* it me *back* over' here, Isidore parades the embarrassment of having to write *back* 'please *re*-write'. He doesn't *mean* to demand a re-think by 'return to sender', by 'return post'. But the door is open for some self-reflection. What would a Christian snub look like? If a saint could chide . . .

IV

Invitation accepted. Braulio does what he's told. And how, in a mock formal (re-)petition, full-blown as any oration and out to tell us as much. Announced with: 'Now without more ado, I shall start to set out the case' (*sed iam causam exordiar*). Out pour out-pourings, learned quibbling on 'tearful calumnies, calumniating tears' (*et lacrimabiles calumniae et calumniabiles lacrimae*). He gives it the works, then tells us so:

I have spared wordiness not a whit and, it may be, gone full-tilt for the foolhardy . . . How about all that for bravura . . . ?

nec uerbositatem carui et temeritatem fortassis incurri. . . . ecce quantum audaciae . . .

You see: the *Etymologiae* deserve all the stops pulled out, they will be worth the wait, they call to be imaged by, and weighed up against, nothing less than a litany of (eight) citations from Holy Scripture: Matthew 7, 2 Corinthians 11, Luke 14, Galatians 5, 1 Peter 4, Romans 12, 1 Corinthians 12.11 and 13, 21–23, 1 John 4. Such an apostolic salvo, dominated by Pauline epistolarity. All power to this bishop's elbow.

It is now over six years since Braulio first tried to winkle the *Origins* out of Isidore. It has been one story of prevarication after another story of no reply: 'Not yet finished; not yet written up; yr. letter never arrived; etc etc'. It's enough to try a saint: nothing for it but to blast the book out of him with a hail of querulous clamorous flagitation – a sonorous tumultuous incantation:

All this so you'll at least bestow what you refused through {your} humility, through {my} riotous skulduggery

tantum ut, quod noluisti per humilitatem, saltim tribuas per tumultuantem improbitatem

To quell this storm, Isidore must just post to Braulio his *Etymologies* – not as already in many a reader's hands, 'a headless chicken, chewed to bits' (*detruncatos corrososque*), but specially for him, 'copied out whole, checked, good and coherent' (*transcriptos, integros, emendatos et bene coaptatos*). If this ultimate plea has pushed polite rhetoric to (– beyond (!) –) the limit and risked pumping up the volume to the point of challenging gospel, it has melodramatically set before us the vital importance to the fulfilment of the *Etymologies* of their – *its* – 'accurate, full-length, perfectly transmitted, thoroughly articulated' state. The 'sacredness' of this precious consignment of text is signalled here at the gate. *Fortissimo.*

One more twist to come, before this uproarious screed is done. A mutual friend: Braulio once again links the fate of 'their' book with their ministry, by adding the recommendation that Isidore make a particular recommendation to fill a newly vacant see. Here, administrative patter between colleagues performs the return to ordinary businesslike lobbying now the big book has received its meed of turmoil:

I suggest that . . . you suggest . . . Same for the suggestions I've made here as for these complaints I made above

suggero ut . . . suggeras . . . tam de his quae hic suggessimus quam etiam de his quae supra questi fuimus

Rude, importunate, hybristic, 'complaints' they may style themselves, but these scripture-invoking urgencies have couched as characterful a trailer in anticipation of the long-awaited blockbuster as any urbanity could wish. Publication is needed – *urgently*: 'Seek and you will find + Knock and open sesame – so I am *yelling* to you "OPEN UP!" I'm an oaf, you're the word emperor, so do what the good book says: Suffer a fool gladly.'

The book of words is a public service, benefaction to the community: 'stop hoarding the talents entrusted to you to share out. Feed your 5,000, with self-renewing fuel that belongs to everyone'. This is no academic study: 'this gift is yours to give, ours to receive'. The governing metaphor is of 'stewardship appointed of God': 'to administer the *treasury* departments of wealth, salvation, wisdom and knowledge' (*oeconomiam thesauri sui et diuitiarum, salutis, sapientiae et scientiae*). The special focus is on encyclopedic 'multi-tasking': on the one hand 'this grace of God is multiform', and on the other 'the faith measures as limbs making up one organism'. The operation at the heart of everything in creation and in the book is, exactly, comprehension through compart-mentalisation, or 'division':

One and the same Spirit provides this comprehensive service, delivering individual portions by *division*, according to plan.

haec omnia operatur unus atque idem Spiritus, diuidens *singulis, prout uolt.*

At the same time as risking assimilation of Isidore's work to 'divine creation and administration' (*sic itaque creator noster et dispensator cuncta moderatur*), this outspoken pretend rant from Braulio next huffs and jests its mock-vulgar way toward installing his Archbishop atop the power-structure of their church: there's 'a foot, anyhow one of the less glorious body-parts, there's a belly grumbling, ingesting, digesting, and . . . and at the top there is the head running the outfit' (*sitientibus . . . fame*; *pes*; *aluo*; *membra*; *inhonestiora membra*; *egerere*; *principatui capitis*).

All this brash hype and rudery selflessly unblocks the proper path of saintly reluctance. The vast tome must be prised from our author's clutches without compromising his exemplary humility. Where it is OK for the lowly servant to push and puff, we shall find, Isidore can only deliver once he has no choice, when the Synod that is the purpose of his ministry is here, and *only if* he is meant to arrive, supposing he gets there in this life. But along the way, these meta- as well as para-textual preliminaries will have brokered a deal with publicity: Braulio's picture of the *Etymologies* as already in a parlous state of neglect, damaged samizdats bootlegging abroad, even before they are finished and ready to satisfy an eagerly waiting world, busts through Christian reserve, supplying terms under which the steward can trade. This confessedly sorry half-baked project is part of a transaction in process, aetiologised as rooted in accessing God's truth to the faithful, and inaugurated as an enterprise designed to require for its realisation participation on the part of editor, publisher and readership. The whole, 'multiform', *Team Isidore.*

V

Turns out, you see, next and only just in the nick of time, that Braulio's letter found Isidore in his element, and in fact at the apex of his spell on earth, at the (Fourth, and supremely magnificent) Council of Toledo, which in truth he utterly dominated. He received Braulio's deacon messenger plus message at court, and, buoyed up by his own New Testament motto

(Romans 5.5: Hope does not fuddle through charity, which floods in our hearts, *con*fundit . . . *dif* fusa), still hopes to see him again: there is good news and there is bad. The bad news is (1) that Isidore has been 'too ill to correct *Etymologies*' (*codicem . . . inemendatum prae ualitudine*) and (2) the King is not minded to prefer Braulio's nominee. The good news is (2) that the King hasn't yet made up his mind for sure, and (1) that Isidore *is* finally 'coughing up the manuscript, in a clutch of manuscripts, posted *en route*' (*codicem Etymologiarum cum aliis codicibus de itinere transmisi*). Always provided, that *if* in the middle of life, that providence means for him to 'reach the fateful arena set for the Synod' (si *ad destinatum concilii locum peruenissem*).

Since said Council took place in 633, and Isidore was to die in 636 aged seventy-six, we may take it that this equable covernote, short and sweet as ever but at last allowed to be informative, definitively places the *Etymologiae* where they belong, among the Bishop's entire oeuvre, blessed as the final culmination of his writing for Christendom. This is important to register. Authorship of dictionaries and encyclopedias, as of other monumental compositions, regularly construes as a special category of author-ity. It requires a long life of privileged prestige. Finishes off its servant and scribe. Embodies the sum of scholarly devotion and privilege.

And as first reader, Braulio has been deliberately, even ponderously, conscripted to join in with clinching this crowning achievement, for, as we've seen, Isidore's bargain with himself had been to 'offer it' to Braulio 'for proofreading' (*ad emendandum . . . offerre*), *if Isidore made it to Toledo for the Council* (to end all Councils: *si ad destinatum concilii locum peruenissem*). There, the assembled bishops of Spain were indeed imperiously instructed to establish seminaries in their cathedral cities on the model of the school at Seville where the boy Isidore had himself long ago worked through the liberal arts programme under his brother Leandro's supervision.[20] The *Etymologiae* address themselves, not just to Braulio's Saragossan see, and to every other site of learning in the land, but to the promulgation of a national policy promoting classical education; and these introductory letters find their own caring idiom to intimate as much.[21]

Imprimatur.

VI

After this momentous prefatory correspondence, the one short sentence (mightily confusing many of the scribes before they even got started) of

[20] See Canon 24 of the fourth Council of Toledo (Migne *PL* 84: 374). On Leandro: Navarra (1987).

[21] See Aherne (1966).

Isidore's (earlier: original?) dedication to King Sisebut, the Christianised Vizigoth dynast whose claim was to have unified Spain in partnership with its Church Fathers,[22] need not, and does not, *insist* on its bearing on the book it heads. The *point* here must be to herald intrication of mundane politics and pastoral business with scholarly synthesis.[23] As promised, we are told, find (enc.) one work of vast effort to realise the present regime within a seamless continuity with the world of classical antiquity. You don't have to be a Goth to get it, but if you were it would plug you in, to the power of the written wor(l)ds of Latin:

Here you are! Just the way I promised, I've launched a work on the origin of one thing and another, compiled from reading remembered from way back when, and at one point and another provided with commentary just the way it survives as penned in the classical style of Antiquity.

en tibi, sicut pollicitus sum, misi opus de origine quarundam rerum ex ueteris lectionis recordatione collectum atque ita in quibusdam locis adnotatum, sicut extat conscriptum stilo maiorum.

Bp Braulio has already displaced for any reader, beyond any possibility of misrecognition, the modesty of Christian catachresis incumbent on his feted author. Through epistolarity, he serves to take readers of the scholarly tome in prospect behind the scenes, and to lead us back from the scene of writing out to the impact of the writing upon its world of readers. Intimacy and humility conscript us to live, re-live, ourselves the vocational devotion that powers this author's authorisation of faith in his topic of topics: the sacred bond between text and church, word and world, man and God. Besides positioning the *Etymologies* at the apex and as the apogee of Isidore's ministry, Braulio models, too, as the privileged first reader in whose wake we stumble and trail. For Braulio models an emphatically interventionist reader, one that we *un*ordained (unappointed of Isidore) must, but can only, aspire to emulate. Reading the book is to be participatory, for no dictionary is ever done (God's work goes on). Yet, *after Braulio*, these pages have been canonised (emend, and be damned). Samples of unfinishedness that we shall encounter in the text will compete with features of supplementation in the transmission: both sets of traces propose unfinishability in the

[22] On Sisebut: Domínguez del Val (1986) 1: 327–30.

[23] Sisebut was dead well before Braulio got his hands on the *Etymologiae* (Lindsay (1911b) 51). But he was very much Isidore's king, in a dynastic partnership that stretched back to the third Council of Toledo in 587, where King Reccared (accession in 586) ceremonially converted, with his queen, Baddo, from Arianism to Catholicism, midwifed by Leandro, bishop of Seville from 578, where he was succeeded by his brother Isidore in 599/600: Migne *PL* 84: 342–63. On the political harnessing of *continuidad romana*: Diesner (1978) 84–107.

consciousness of the project; and they bind into its reception the imperative of to-be-finished fulfilment. Braulio embroils us all – Braulio gets you started.

What is more (though Lindsay did not see fit to pass this on), Braulio himself hails Isidore's entire polymathic library[24] with the all-important if miniscule bio-bibliography of 637 CE,[25] where he informs us that he, *Braulio*, was the one responsible, as part of his apostolic duty to perfect the chef d'oeuvre, for its division into '20 books':[26]

The book of the *Etymologiae* is really enormous. It was given diacritical subheadings but not book divisions by Isidore. Because he created it at my request, for all that he himself left it short of perfection, I have split it into 20 volumes. This work suits all varieties of philosophy: *anyone who reads it through in cumulative reflection*, will deservedly be not undistinguished for factual knowledge of divinity and humanity both.

Etymologiarum codicem nimia magnitudine, distinctum ab eo titulis, non libris; quem quia rogatu meo fecit, quamuis imperfectum ipse reliquerit, ego in uiginti libros diuisi. quod opus omnimodo/ae philosophiae conueniens quisquis crebra meditatione perlegerit, *non ignotus diuinarum humanarumque rerum scientia merito erit.*

After the dedication, Lindsay presents two tabular guides which he dubs *INDEX LIBRORUM* and *CAPITULA LIBRORUM* ('Index of books' and 'The chapters of books'). The first of these lists of contents avows (as well as betrays) that it is not Isidore's work (it presupposes and may refer to, or be, Braulio's). Readers are given summary notice, book by book from *One* through *Twenty*, that 'this page enables speedy answers to queries by pointing to the topics discussed by the book's architect in each book' (*de quibus rebus in libris singulis conditor huius codicis disputauit*):

I. De Grammatica et Partibus eius.
. . .
XX. De Mensis et Escis et Potibus et Vasculis eorum, de Vasis Vinariis, Aquariis et Oleariis, Cocorum, Pistorum, et Luminariorum, de Lectis, Sellis et Vehiculis, Rusticis et Hortorum, sive De Instrumentis Equorum.

The second index is (neither Isidore's nor Braulio's, but Braulio's successor *Lindsay*'s) confection – but re-made from the ancient (post-Braulio) paradosis:

[24] Braulio's encomium re-cycles Cicero's puff for Varro (*Ac.* 1.9): Fontaine (1988) essay III: 89 and n. 2.

[25] Migne *PL* 82:65–68 (cf. 81: 15–16). This hagiography, the *Renotatio* (originally appended to Isidore's *De uiris illustribus*), was traditionally known as the *Praenotatio librorum D. Isidori*, cf. Fontaine (1959) 866.

[26] On Braulio's editorial intervention: Díaz y Díaz (ed.) (2000) 177–80. On Braulio in Isidore's circle: Domínguez del Val (1986) I: 331–7.

These book-chapters that I have compiled here appear in the codices, each at the outset of its book or part of book.

Haec capitula librorum quae hic congessi apparent in codd. in initio sui quodque libri uel partis libri.

In this (relocated) articulation, the first and last books present thus:

i. De disciplina et arte.
ii. De septem liberalibus disciplinis.
iii. De grammatica.
iv. De partibus orationis.
v. De uoce et litteris.
vi. De syllabis.
vii. De pedibus.
viii. De accentibus.
ix. De posituris.
x. De notis sententiarum.
[De notis uulgaribus et aliarum rerum.]
xi. De orthographia.
xii. De analogia.
xiii. De etymologia.
xiv. De glossis.
xv. De differentiis.
xvi. De barbarismo.
xvii. De soloecismo.
xviii. De ceteris uitiis.
xix. De metaplasmis.
xx. De schematibus.
xxi. De tropis.
xxii. De prosa.
xxiii. De metris.
xxiv. De fabula.
xxv. De historia.
and thus:
i. De mensis et escis.
ii. De potu.
iii. De uasis escariis.
iv. De uasis potatoriis.
v. De uasis uinariis et aquariis.
vi. De uasis oleariis.
vii. De uasis coquinariis et pistoriis.
viii. De uasis repositoriis.
ix. De uasis luminariorum.
x. De lectis et sellis.
xi. De uehiculis.
xii. De reliquis quae in usu habentur.
xiii. De instrumentis rusticis.
xiv. De instrumentis hortorum.
xv. De instrumentis equorum.

On the one hand, Isidore's headings seem 'to have rather designed 22 (or 24) books rather than [Braulio's] 20'.[27] But on the other, Lindsay emphasises 'a great deal of confusion' and 'wide divergence' in the placing of title-headings and large-scale 'discrepancies in arrangement (as well as mixture of text)', while 'division of the whole work into 'pars I' (= books 1–10) and 'pars II' (= books 11–20) was a mere matter of convenience' (for both the manuscripts and his own *Oxford Classical Text*).[28] Yet closer scrutiny shows how the three-tier system of nested rubrics could function

[27] Lindsay (1911b) 50. Did alphabetic reckoning feature in Braulio's scheme? Fontaine ((1988) IV: 531 n. 25) cites Nonius and Gellius for compilations in *twenty* books.
Did *Isidore* not begin his account DE MVNDO ET PARTIBVS with the words *In hoc uero libello* . . . ('Well in this *book* . . .', 13.1.1)?
The rhyme of I, *De Grammatica et Partibus eius* with XI, *De Homine et Partibus eius* does flag a rewarding division between 'the two halves'.

[28] Lindsay (1911b) 50–1: 'More obscure are the traces of a division by triads'.

more serviceably than may be apparent, and how robust it could prove in the teeth of turbulent transmission.

To take the brass tacks of book 20 first, the initial book contents helpfully signals three subdivisions: the first groups 'containers for eating and drinking' together with 'tables, food and drink'; these topics will correspond with headings i through iv at the start of book 20, and resolve easily into the five sub-headings that open paragraphs 1 through 5 of the text (separating 'On vessels for eating from', *DE VASIS ESCARIIS*, from 'On vessels for drinking from', *DE VASIS POTATORIIS*; our editions' '20.iv–v'). The middle cluster then wraps up the other categories of specialised containers – six contents items, that shade into the two triads of: wine, water and oil; cooks', bakers' and luminaries'; five topic headings at the incipit of 20 and five matching paragraph sub-headings in the text, in both cases bracketing wine-and-water together, and then spoiling the heralded second trio by interposing 'On vessels for serving food' (*DE VASIS REPOSITORIIS*) (i.e. 'XX.viii.' and 'XX.ix', respectively) between their undifferentiated run of *-ariis*, *-ariis*, *-ariis*, *-ariis*, *-oriis*, and now *-oriis*, and their final odd-man-out (odd men out), *lumin-ariorum*. In both of book 20's sub-systems, that is to say, the salient breaks of topic come after i–ii, and after ix/x, with a solid run of 'On vessels' (*DE VASIS*) from iii-ix/iv–x. Finally, the contents align the triplet 'beds, chairs, carriages', marking two sub-divisions of the last '[carriages for] countryside and for gardens', before uncertainly appending 'Horse equipment' with a bizarre *siue* as well as its own would-be major paratactic *de*, as if to promise a fourth segment-in-one-bite. The actual text of book 20, however, knows better, offering two triads of its own, so that these final 'horses' complete instead three lots of 'On implements' (*DE INSTRVMENTIS*; viz. 'countryside; gardens' and horses''), and 'beds-and-chairs' snuggle up, off-set by 'carriages', and a catch-all round-up of miscellanea: 'Everything else that serves a use' (*DE RELIQVIS QVAE IN VSV HABENTVR*).[29]

Now, whether the team who produced this triple search engine saw this no-nonsense book of 'kit' as a single diagrammatic suite structured more or less seductively by Isidore is (I admit) not *entirely* clear. But this is how that story will go: first the dining table traces its ware back *out to the kitchen*, bakehouse, cupboards and stores, and suppliers of light. Next the remaining articles of the household furniture line up – *out onto the street*, where wheels turn them into conveyances, before walking-sticks become

[29] Confine reading Isidore to the list of contents, and all you can see will be: 'The last books are curious because of the odd pairings of their contents' (Conte (1994) 721).

clubs, clubs crowbars, and out tumble scissors, razors, combs and curlers, with key, chain, and clock to scrape the barrel. Barbers, locks on doors, sundials belong back there, but we shall travel on, *out of town*, to plough and dig and prune and harrow, once on the farm, twice in the (market-) garden; harness, rein, and spur our horse *off into the sunset* . . .

So it is that, as a reference machine, *Etymologies* can lead its users into the right neck of the woods, so long as they can pick the right rubric from the Contents; finding the right page by coordinating start-of-book heading with sub-head paragraphing within the text leaves us to make what we can of the more-or-less solid wall of items crammed into the space allotted to the topic. And yet, all the while, consultation this way virtually debars the user from perceiving any sense of logic or meaningful composition to the particular string of items consulted. Indeed a satisfied customer need never even notice whether a cluster of material chosen for thematic treatment has been exhaustively or sparsely, ludicrously, robotically, or respectfully treated: the quality and modality of the Thesaurus escape attention – and even notice. Which user will pause to reckon whether the patch of the moment is the product of negligence or intelligence? Will any frustrated user play lucky dip, get burned off by impatience with the Contents, blame the manual for missing the right hookline – but never *prove* the wordhoard's blindspot? Does Isidore exude perspective or pilot automatism?

Book 20 prompts just such questions: it invites us to play 'spot the missing item'. As if the completeness of an *inventory* must inhere in the *Etymologies* project. As if that is the driving teleology that *constitutes* the scheme. The presumption soon fades once you (re)turn to book 1. Here, the Contents list gives a bare title: 'On Grammar', plus the promise of complexity, '& its Parts' (*DE GRAMMATICA ET PARTIBVS EIVS*). Twenty-five headings mostly spit out one word subjects, while the *ad loc.* sub-headings tot up to forty-four entries. Contents' minimalesque cues skip those master categories of the whole enterprise, 'discipline' and 'art', which dominate the entrée paragraph (1.1), and entirely suppresses the grand parade of the 'Seven Liberal Disciplines [or Arts]' that launches the first three or four books ahead.[30] True, in appropriating the lemmata for I.iii–iv/I.v–vi, 'On Grammar . . . On the parts of rhetoric' (*DE GRAMMATICA . . . DE PARTIBVS ORATIONIS*), Contents does home on the organising categories of the book, viz. the count of thirty topics of grammar (1.5.4).[31] By contrast,

[30] After the holy programme of Cassiod. *Inst.* 2, preface: see O'Donnell (1979).

[31] Lindsay's Oxford Classical Text muddles its punctuation here: *id est, partes orationis octo*: must be corrected to *id est: partes orationis octo*, for the line-up of topics to reach that total of thirty (8 + 22).

however, headings and sub-headings will diverge – in order to cue location of each paragraph as they arrive. Thus the heading I.iv subsumes the various 'Parts of Speech' it trumpets, where sub-headings flag them up in bold as I.vi + vii–xv. Similarly, the manuscripts are content *either* to gather a run of entries on symbols used in various cultural arenas, *or* to run a double entry plus '*et cetera*' formula (I.x), where(as) the paragraphs themselves feature a suite of six specified *loci* for them (I.xxi–xxvi). Finally, four paragraphs on aspects of 'history' close the book, where the book headings in most manuscripts leave just the one entry: 'On History' (I.xli–xliv *vs* I.xxv). Otherwise, neither system makes the slightest attempt to interlink, subjoin, or relate any of the categories employed, with a single exception where 'On other vices' (*DE CETERIS VITIIS*, I.xviii) makes an exegetic incision missing from 'On vices' (*DE VITIIS*, I.xxxiv).[32] Yet Grammar makes a quintessentially pre-fabricated opening tableau, told and re-told down the centuries without the slightest concern to administer *etymology* as anything more than one incidental (sub-?)topic tucked away for its half-page meed of glory in its place. Grammar takes Isidore fifty-eight *OCT* pages, where book 20 musters just twenty-six. If we juxtapose beginning and ending this way, the question will obtrude: does *Etymologiae* earn its perfunctory wind-down *by* having revved up at the outset? If Isidore isn't plain stuck for stirringly materialist material in 20,[33] have his earlier heroics from 20 on eliminated the need – does he coast on through, once readers are trained to work with, *even for*, him? Or is the contrast down to textual morphing in tandem with the contours of existence, and our apprehension of it – a matter of *epistemontics*, as words will twine worlds as the world (un)ravels through words? Only a reading open to telling narrative self-transmutation can even contemplate compositional strategy in the trek from 1 to 20, front to back. Yet finding Creation in horse trappings *must*

The decision to expound the alphabet (*litterae communes, litterae Latinae*) before starting on Grammar spoiled at once the count-up and the first topic(s) after the Eight Parts of Speech, which appear(s) as *De uoce et litteris* in the heading, as DE LITTERIS APVD GRAMMATICOS in the sub-heading, but as *uox articulata, littera* in the *diuisio* at 1.5.4: a chunk of the paradosis drops the paragraph(s), one manuscript orphans the bare title, another remarks *iam in principio huius operis disputatum est*; otherwise, 1.15 appears as the skeletal jotting: *quot sint articulatae uoces. et dicta littera quasi legitera, eo quod legentibus iter praebeat uel in legendo iteretur* (which resumes 1.3.4, *litterae autem dictae quasi legiterae, quod iter legentibus praestent, uel quod in legendo iterentur.* This is a price paid for disrupting Donatus' grammar (flagged at 1.6.1), partly in line with the hint in Cassiod. *Inst.* 2.1.2: *de uoce articulata–de littera–de syllaba–de pedibus–de posituris siue distinctionibus*–et iterum *de partibus orationis VIII–de schematibus–de etymologiis–de orthographia*.

[32] DE SOLOECISMO (I.xvii) *vs* DE SOLOECISMIS (I.xxxiii) helps set up this divergence.

[33] Cf. esp. Fontaine (1959) 96 n. 29, on the Varronian antiquarianism of 'material culture' in book 20.

literally and logically depend on learning to learn through 'grammatical' competence how to find out *anything*.

As Curtius plainly saw, etymology is harnessed to different work as Isidore construes the grammar *of the universe*. And in the process, etymology unfolds to disclose eventually all-embracing semiotic power: obiter, sequencing and starring of topics process a multi-track adventure of progressively re-programmed induction into a multi-verse design where totalised wholeness is set as the product of models of partitioning, aggregation and multiplication. Learning (to understanding and appreciate) how heterogeneous systems (can, and in this vision, do) interact to attain holism for their product is the prize promised by reading – reading Isidore. My proposal is that we should track the *Etymologiae* in its main outline, resisting the peremptory intercession of the apparatus of headings, as so many obstacles and deterrents to *reading*, and instead paying them respect only where they point up exegetic continuity, proportion, or direction. If intruded titling (red ink and capital letters) aims to boss the text, it is by the same token easy to finesse from reading. There are, for sure, neither etymologies nor reference to etymology to be found *anywhere* in the paratext.[34]

Need it be said, such a project practises contempt for facile diagnosis of authorial patchwork, autopilot somnambulation, or mindless compilation.

First things first (that trauma of The Encyclopedia), digest Isidore's opening gambit. *Etymologiae* starts as it will continue, prefacing a block of material from the principal proximate precursor with a tag from one of the great Church Fathers: here, at 1.1.1, Cassiodorus primed with Augustine. (Jerome will be the other staple supplier of inspirational mottoes).[35] We are here to learn, to learn *learning*, plenary source of knowledge: as we discover, 'discipline' (*disciplina*) derives from *discere* ('to learn'; which in turn derives from *scire*, 'to know) + *plena* ('complete'). There is, on the other hand, also an 'art' to this, encapsulated, so Isidore at once explains, in a bilingual set of word-truths which knot together 'art' with its constitutive rules and regulations, and with knowledge as a 'virtue' ('ard-and-fast rules': in Latin artis *praeceptis regulisque*, and *aretē*, a Greek term, he claims, for 'knowledge').[36] We are now under starter's orders for knowledge: first

[34] Excluding, as noted, the first of Isidore's letters [in the augmented 'Spanish' paradosis].

[35] At a superstructural level, Augustine and Jerome (esp. *De doctrina Christiana* 2; e.g., *Epistulae* 33) gave Isidore the licence to build his generous encyclopedia of mundane knowledge derived from the order of pagan civilisation (cf. Fontaine esp. (1959) 32, (1988) v: 79).

[36] On 1.1: Fontaine (1959) 51–2; on 1.2.1: *ibid.* esp. 52 n. 4.

differentiated, as 'things that cannot, vs. things that can, be otherwise than they are' – matters of truth, and of likely-true opinion;[37] and then calibrated, as our educational programme is at once set to multiply sevenfold *disciplines*: the Liberal *Arts*. This syllabus will lay down hard-and-fast rules within which to exercise free conjecture: these standing orders for learning the art of learning the arts are set to run the show, and they will go the distance. Rigorously regulated formatting controls and patrols the course set for our craft to navigate best we can. Isidore's own art is to discipline his representation of each of the domains of human existence: Etymological Creation will customise the 'lettering' that writes each territory differently. From Grammar to Garden Tools, all the way, from A to Z.

As ever with works of reference, users must begin as *advanced* users. And the work of reference that uses language – script, text, index systematics – to peddle linguistic theory is bound to presuppose meta-level interpretative engagement with the interpretative analyses offered over the shoulders of the hypothetical classmates envisaged as here and now learning their lessons in school – 'right away', and 'later' (*iam . . . post . . .*).[38] *Etymologies* would get *them* to learn in an orderly fashion, but does get *us* to appreciate the methodology that produces order, knowledge *of* knowledge. We must see the letter as the *origin* that originates everything – everything that can be spelled out in the *Origins*[39] – and look within language for truth-production through language. Specifically, the significance of Isidore's introductory tour of the alphabet is to instate indelibly the grand historical narrative of the Christian West as the field of operations for his hermeneutics.

Take Cassiodorus, the sixth-century Italian intellectual and founding father. *He* had been intent in his guide to monastic education, the *Institutiones*, on subordinating the Liberal Arts to masterminding an ascesis for religious devotion. Spent much of book 1 warning the brethren not to over-rate book 2 at the expense of the spirit, however faith may depend on reverence through (and so, necessarily, for) books. His book 2 started by consecrating book 1 as thirty-three chapters set to stand for the Lord's *years* spent on earth, whereas the cursory dash through the seven Arts will mimic the *weekly* cycle rolling round till world's end. Isidore, by contrast,

37 This initiatory *differentia*, sheltered by ascription to 'Plato and Aristotle', does *not* figure in Isidore's work *On Differences*; it transcribes Cassiod., *Inst.* 2.20.

38 Hadot (1997), esp. 28–32, insists that texts on the Liberal Arts were 'philosophical' in nature, not for classroom use. Arguably, 'educational theory' even *grounds* 'Philosophy' within the ministry of an Isidore.

39 Cf. Amsler (1989) 147–8, 'The grammatical model'.

memorably implants the indelible lesson that letters were invented as aides-mémoire. Existence covers too wide a range for any human memory to hold all there is to learn. Called *letters*, for short, mind – really, they are *legit-ers* because readers must '*leg it*, or *let's iterate*' (*quasi* leg-iterae, *quod* iter legentibus *praestent, uel quod in* legendo iterentur, 1.3.3).[40]

Isidore's journey of reading re-traces the legendary itinerary of writing from Latin back through Greek to Hebrew, '*mother* of all tongues and letters', in the age-old phylogenetic metaphor for primitive classification (1.3.4). Thence the sequence Aleph -> alpha -> A launches alphabet sets of 22, 24, 23 signs, and these represent a first lesson in what lessons are: it is put before us that letters *are*, precisely, primary building-blocks, fit to found the elemental power that inheres in elementary education. Here is the grounding of Western Civilisation as an entirety.

A diffusionist story triple-tracks colonisation through Moses and Abraham; then Isis, Cadmus and supplementation by Palamedes, Simonides, Pythagoras; so to Latin, and the nymph Carmentis . . . as Isidore steers us toward his most fundamental proposition concerning his chosen field for disciplined performances of the etymologist's art. As his net of categories tries to catch the essential properties of the alphabetic, we find: 'nature dealt the sonority; volition accounts for positionality' (1.4.17). Which is to say, that there is nothing to *interpret* about the constitutive nature of vowel vs. consonant, etc. etc., whereas there is always a story there for us to figure out from the names given to letters by acts of will, such as the motivated 'names and shapes' of letters. Our first lesson makes history emerge as the efficient logonomy for the organisation and delimitation of linguistic knowledge. *Etymologies* thus promises to be a book about book-culture as medium for considering, tracking, discussing, *civilisation*, in so far as its genesis is open to its own inspection. As such it constitutes (a) theory – about theorisation.

Indeed Grammar claims its place at the head of the educational programme that heads *Etymologies* by *imposing* its theory. The text soon crawls with gerundives and gerunds, ensuring that writing is hedged around with traps and pitfalls, in a world where 'we' (*nos*) have departed very slightly from 'people of old' (*ueteres*) in a battle for validation that engaged even the 'Greats of Antiquity' (*antiqui*).[41] Eight modes of Analogy come to our rescue with rules for generating *correctness* – but they 'don't always apply'

[40] This traditional grammarians' lore was culled by Isidore from Diom. *Ars grammatica* (Maltby (1991) 343, *s.v. littera*).

[41] For Isidore and 'Late Latin' cf. Maltby (1999), Banniard (1992) 182–25; for Cassiodorus' *De orthographia*, see Keil (1857–80) VII: 126–210.

(1.28.4, *sed hoc non semper*). Rule-bound as it may be, *grammatical orthodoxy* doesn't come so easy. And, note, *this* is the context where Etymology nests within *Etymologies*.[42] That is to say, in the 'narrative' of *Bildung* that the book unrolls, here is a teletechnology to get us, wherever we may find ourselves, into *writing right*.[43]

Drill takes us on page by page through the traditional seven-branched curriculum, marooning us in interstellar overdrive across the universe as Astronomy leaves us with the sting in its trail: school, we learn on graduation, can warp as well as weave, for these heavens open into a fearful starburst of drubbing for the pagan civilisation that taught the teaching of all this integrated progressive curriculum (3.71.22–36): it was superstitious vanity, not mathematical science, that added monstrous figments to the *number* of stars. *Astrology*! Ye gods!

So to Medicine, positioned as the supplementary subject. At the death here we learn that a doctor must have been through the full heptadic syllabus, and we need telling why, taking each in turn from Grammar through in last place of all to Astronomy, too (4.13.4). 'Courses on the mind; course on the body': check (4.13.5). But we are not yet done with reading up on reading, wherever it takes readers and reading. Our first port of call, now that we have mastered college education, is the institution of Law. Which appears in the form of a combinatoire of religious-cum-social history, where the story of human civilisation is drummed in. *Chronology* ensues, not just an educational worry, but a writing project, climaxing with a fresh genre of 'diagrammatic' text that showcases the power of tabular writing to order knowledge: the laws of Chronography (*De descriptione temporum*). This manifests as a spectacular excursus in note-form, capturing the six Ages of the World, from the Creation to our Sixth Age (5.39).[44]

Readers here run through the blessings of the trinity of Hebrew-Greek-Roman culture: they alone have the gift of *letters* – of the grounding in Grammar that world culture has taught Isidore to teach us. Which is where

[42] This chapter has, very likely, been read more intently than the whole of the rest of *Etymologiae* put together: see esp. Codoñer (1994), rigorously excavating the 'theory' discoverable in 1.29; cf. Fontaine (1988) essays IV, V, X, Magallón García (1996) 277–87. For Isidoran etymology within ancient etymology, cf. Opelt (1965), Fontaine (1988) essay XI, Fresina (1991), and best of all Magallón García (1996).

[43] In the Donatan list of topics posted in Cassiodorus' summary, *schemata* precede the climactic pairing of etymology and orthography (*Inst.* 2.1.2).

[44] The insert bodily re-cycles Isidore's *Chronica mundi* (Migne *PL* 83: 1017–58), dated for Sisebut in 615/6, or 626 for King Svinthila.

we came in (5.39.9–11 ~ 1.3.4–6, 1.4.1, with the Latin word *litterae* embracing 'literacy, literature, and letter(s)):

Hebrews began literacy . . . Cadmus gave Greece literacy . . . Carmentis discovered Latin letters.

Hebraei litteras habere coeperunt . . . Cadmus litteras Graecis dedit . . . Carmentis Latinas litteras repperit.

The vista of History lifts our schooling to a still higher level, and we abide with the painstakingly revealed truth of the eternal Bible and the universal Church until . . . we subside into gushing pagan nonsense in the course of book 8. By then we shall be educated ready to attune correctly to maps of mundanity to come. We get there through contextualising the Bible as itself a library, an archive, a history; and the history of its writing is another, sacred, way to write the history of the chain between Moses to Christendom. Church history then enshrines every facet of the institution, not least the Council decisions that inform and stabilise every word of the *Etymologies* – and all the doctrines that council covenants will ever authorise, from the delegates at Toledo onwards (6.16.5, 10):[45]

Among the rest of the councils, there are 4 worshipful synods that enfold the whole of primary Faith, *just like the 4 gospels, or the same count of rivers of Paradise.*

. . . These are the 4 prime Councils, most fully preaching the orthodox Faith; but any Councils that are held as sanctioned by the Holy Fathers filled with the spirit of God, after this quartet's authorization, they abide fortified by all strength, and their Proceedings are comprehended *in the storehouse of this very Thesaurus.*

inter cetera autem concilia quattuor esse uenerabiles synodos, quae totam principaliter fidem complectunt, quasi quattuor euangelia, uel totidem paradisi flumina.

. . . *haec sunt quattuor synodi principales, fidei doctrinam plenissime praedicantes; sed et si qua sunt concilia quae sancti Patres spiritu Dei pleni sanxerunt, post istorum quattuor auctoritatem omni manent stabilita uigore,* quorum gesta in hoc opere condita continentur.

Isidore's hymn of praise to God, book 7, begins from the role of Hebrew language, from *Amen* and *Hosanna*, in inaugurating all subsequent acts of worship. To catalogue 'God, the Angels, the Saints' is to narrate, describe/prescribe, *and do homage.* And to know God is to read another *book.* More Jerome, most erudite multilingual and first translator of Hebrew

[45] A complete account of all the general councils of Christendom, area by area across the Mediterranean from Greece to ultimate Spain, is given as part of the *Collectio Canonum S. Isidoro Hispalensi Ascripta* (at Migne *PL* 84).

names to Latin.[46] The Catholic Church operates in a hermeneutic world that straddles both Greek interpretation of Hebrew foundations and Christian re-orientation and Latin interpretation of both prior traditions, beliefs and languages. It is one unified faith, that has spread worldwide (8.1.1). Once our education in the faith is far advanced, we are risked to exposure to the fully inspected version of the horror show of human folly, as preserved for memory in the books of every school, library and curriculum throughout Isidore's empire of Latin. Ready or not, time to read rotten Rome. Via the covens of pagan philosophy to the world of idolatry, demons, devilry, Satan, Antichrist, Baal, Beelzebub, Belial, Behemoth, Leviathan . . . Where, too, poets fake allegorised Nature – but only to doll the spooks up with metaphor, though their stories admit that the gods involved are notoriously bankrupt and disgraced (8.11.29):[47]

Empty space for *fictional* folly opens up in the absence of Truth.

omnino enim fingendi *locus uacat, ubi ueritas abest.*

Take a deep breath and plunge into Olympus: the flood of absurdities whoops it up to climax the book. There's Rome and then there's Rome, and catholic schools must learn the difference.

Protected by religion, we can go track the (hi)story of humanity. The founding postulate will be that the trio of Sacred Tongues (Hebrew-Greek-Latin) give intelligible access to the story (9.1.3 <- 1.3.4). 'Languages were prior to tribes', and 'tribes originated from languages, not vice versa', so we move 'from propositions about languages to propositions about tribes' (9.1.1, 14).[48] The founding moment of *cultural* diversification in this tale was, therefore, that world event, the Tower of Babel (9.1.1 <-> 5.39.6). In the predestined narrative to come, Isidore's Spain is, re-iteratively, where *language* completes its generative journey through the history of world culture.

(1) At point of departure, the pure naming system of Hebrew dynastics, etymologising semantics were hors de combat: the proper names functioned as patronymic markers, not signifiers, they were buried by later distortion, displacement and oblivion, and they are beyond the competence of all but the most learned of scholars in bilingual Rome's empire of Greek and Latin. (2) Greek found its geopolitical way through the epochs into every

[46] The Holy Ghost will soon chime in accord with this theme (7.3.24): *propterea autem diuersarum linguarum gratiam apostolis dedit, ut idonei efficerentur fidelium eruditioni populorum* ('The reason why He gave the apostles the grace of different languages was to render them suitable for the education of the peoples of the faithful').

[47] This section of scorn for poetic fiction is probably Isidore's own creation: Macfarlane (1980) 17.

[48] Reydellet (1984) 5.

corner of the globe, invading and invaded, always likely to surprise us with enlightenment on any kind of linguistic challenge, anytime, anywhere. (3) Latin *is* a layered history of multicultural waves of invasion by imperial incorporation of tongues-and-tribes. The mission to hold the three-ply linguistic line open between God's Creation and Vizigoth-Catholic orthodoxy means a never-ending, always mounting, battle for truth against the obliviation of the record of the past inscribed in the words of the present. In the *Etymologiae*'s key to Correct Speech, the ruin of the Plain stews mainly in Spain.

That books book what we know, teach, learn is by now established for the duration (of *this* book). We have begun to get used to looking through the words to the world without explicit waymarking of the archive, as the peoples of the earth divide into social-political groups, into categories of kin, into genealogies of self, when we run smack into the distinctive centrepiece of the whole work, where the ordering of knowledge is founded on a completely different, non-disciplinary, non-canonic and non-thematic, taxonomy: *the alphabetised dictionary.* Isidore does not come out and *say* so. Rather, the modest prefatory paragraph announces a gesture towards accessing an illustrative subset of data for the ordinary reader on the bus (10.1):

ON EPITHETS. The origin of some names, i.e. where they come from, *is not apparent to everybody.* I have therefore inserted some into this work in order to make them known.

. . .

As examples, I have put some of them in this work.

DE VOCABVLIS. Origo quorundam nominum, id est unde ueniant, non paene omnibus patet. *proinde quaedam noscendi gratia huic operi interiecimus.*

. . .

ex quibus exempli gratia quaedam in hoc opere posuimus.

But once we have starred the self (person, ego) with its range of descriptors, we return to macroscopic range, and explore the species: 'Mankind and Monsters', presented as panorama. Indeed the opening to Book 11 ushers *Etymologies* into a bigger work – a *maius opus* – that will stretch over four (of Braulio's) books.[49] The very best way to re-start is (it must be) from The Beginning. Hence the over-arching topic being born will be *Nature* (11.1.1):

Nature: named from creating *nascence* for things. Nature is mistress of *generation* and creation. Some have said She is God, by whom all things were created and exist.

[49] On books (11–)12 and the setting of 17: André (ed.) (1981) and (ed.) (1986).

Natura *dicta ab eo quod* nasci *aliquid faciat.* gignendi *enim et faciendi potens est. hanc quidam Deum esse dixerunt, a quo omnia creata sunt et existunt.*

Reviews of Man and of Beast, of cosmic elements air-fire-water-earth, of natural-historical earth and water, of geographical land and sea, contrive to tread and flow from Orient to Far East, all the way from God's creation of Man in the cradle of Mesopotamian Paradise to Spain's creation of Isidore for finale. 'Nature' unpacks still further, from dust and sand through rock and stone to metal and money,[50] which provides the stimulus to invent, and now record, all measurement and measure. Curtain-raisers on town and country have now preluded proper treatment of agriculture and botany (according to nature) and its twin, the war and games of urban man (given the full treatment of satire and sarcasm). For the final pairing of (Braulio's 20) books, we race through the history of culture, to run into civilisation as the product of art-and-craft, delivered by the invention of tools and equipment that permit the creation of our world, of boats, buildings, clothes and personal adornment, and ultimately our domestic and teletechnological mode of inhabitance therein, from the provision of food and drink, through packaging and transport, to tools for making tools, and equipage for accelerating culture: those gardens and horses' harness. Names, objects, usage, specialisation, industry, life-form, habitus, culture . . . and we have learned how to think about how to think about them all. How to think through the nexus of the 'word-thing'. And where such thinking puts us, end-product of God's creation. These words ring true, they wind up scarcely needing and rarely getting 'etymologisation', for they speak our language. These words bespeak 'us'.

In truth, I maintain, etymologisation has *throughout* played second fiddle to a project in cultural mnemonics according to which Isidore's reader is installed in the undeniably God-given position of destination for the entire thrust of geopolitical history and its authorised theorisation. There at his world's end – Toledo or Seville – to receive the accumulated store of technologies of knowledge, together with the circuits of reasoning, aligning, differentiating which operationalise them. If Isidore impresses one comprehensive lesson upon us, it must be the cosmogonic instrumentality of wordpower. From first to last, the letters brand (and cauterise) cultural memory. OK.

[50] See Díaz y Díaz (ed.) (1970).

PART III

Knowledge and social order

CHAPTER 8

Knowledge and power in Frontinus' On aqueducts

Alice König

INTRODUCTION

During his lifetime, Sextus Julius Frontinus was 'one of the most important figures of the Flavio-Trajanic era'.[1] Little is known for certain about his early career, but by the time of his death in 103/104 CE he had not only served as Governor of Britain, Proconsul of Asia, *Curator aquarum* (Guardian of the aqueducts) and *Augur*, but had also been made consul three times, twice with Trajan himself as his colleague.[2] Triple consulships were only rarely bestowed on men outside the imperial family in this period,[3] so scholars have suggested that these singular distinctions may be proof that Frontinus played a significant role in Trajan's succession as emperor, and perhaps even in Nerva's elevation to the 'post'.[4] indeed, it is tempting to wonder if Frontinus was himself ever a candidate for Nerva's job.[5] In spite of having been 'one of the most successful and influential senators of the latter

[1] Birley (1981) 69.

[2] For more details on his career, see, e.g., Rodgers (ed.) (2004); Eck (1982) 47–52; Birley (1981) 69–81. Frontinus and Trajan were consular colleagues in 98 and 100 CE (which marked the second and third consulships for both men); for a breakdown of all the consuls in 98–100, see Eck (2002) 217–19.

[3] Indeed, Eck (2002) 219 points out that it is not just the fact but also the speed of Frontinus' third consulship which makes it so remarkable: 'Such a rapid succession of second and third consulates, separated by only one year, is absolutely unique for a person not belonging to the imperial family.' On the 'supreme honour' of Frontinus' third consulship, see also Rodgers (ed.) (2004) 4; Grainger (2003) 14 and 124; Bennett (1997) 109; and Plin. *Pan.* 60.4–62.6.

[4] Eck (2002) 219–26 is in no doubt that Frontinus was one of the key players behind Nerva's adoption of Trajan as his heir; see also Rodgers (ed.) (2004) 4; Bennett (1997) 46–7; Birley (1981) 72. Dahm (1997) 174 argues that Frontinus helped to secure the succession not only of Trajan but also of Nerva, though this is more difficult to establish.

[5] Dio Cass. 67.15 makes it clear that Nerva was not the first man to be approached by the conspirators in 96 CE as they plotted Domitian's assassination – several were offered the chance of becoming Rome's new emperor. However, see Grainger (2003) 14–21, who speculates on Frontinus' involvement in the conspiracy, but does not consider him a potential candidate. Grainger later wonders if Frontinus could have been in the running to succeed Nerva instead of Trajan (pp. 96–7), but Eck (2002) 223 discounts this on account of his age and relatively humble origins (Frontinus was probably a *novus homo*).

part of the first century',[6] however, Frontinus is best remembered today as the author of a number of so-called 'technical' treatises and consequently languishes, largely forgotten, at the margins of classical scholarship.

Most of those who have had any contact with his surviving texts have argued, or assumed, that they are straightforward, practical manuals, which may be of interest to specialist historians but which have little to recommend themselves to a wider readership.[7] At first sight, they do perhaps give that impression. Frontinus' *Strategemata*, for instance, brings together hundreds of examples of successful military stratagems with the simple aim (so it claims) of aiding future generals in their endeavours in the field.[8] His fragmentary essay on Roman land surveying not only sketches a brief history of land measurement but also discusses some of the basic principles of the 'art', detailing (among other things) the different categories of land and types of dispute which a surveyor might encounter.[9] And his intricate account of Rome's aqueduct network – the *De aquis* (*On aqueducts*) – appears similarly narrow in its scope and purpose, having been written (Frontinus states in its preface) when he was appointed *Curator aquarum* to help him familiarise himself with the system's many 'ins' and 'outs'.[10] In spite of their appearance of straightforward practicality, however, it has begun to be recognised that all three are more complex than these first impressions suggest;[11] and this is in part (in my view) because all three experiment with the ordering of knowledge in a variety of ways.[12]

This article is about just one of these texts – the *On aqueducts* – and, indeed, it is about just one aspect of it, namely the relationship which it

6 Bruun (1991) 11. This is something which several of Pliny the Younger's letters underline: see, e.g., *Ep*. 5.1 (where Pliny describes Frontinus as one of the 'two most respected men of our era') and *Ep*. 4.8 and 9.19.

7 Frontinus' entry in the first edition of the *OCD* ((1949; repr. 1964) 371) is typical: 'Frontinus' writings are essentially practical, dealing with professional subjects in a straightforward style admirably suited to its purpose.' Indeed, Frontinus has not only been firmly confined to the 'Technical Writing' section of most modern companions to Latin literature, he has even been singled out as one of the writers who most clearly belongs in that category: 'If there was ever an author whose chief value was his technical content, it is surely Frontinus' (Hodge (1991) 169).

8 See *Str*. 1, pr., where Frontinus says that his book aims to 'surround' generals with 'examples of planning and foresight' in the hope that they might be both inspired and reassured when thinking up their own tactical moves.

9 No programmatic statement for this treatise survives, but it is possible that it also had a didactic purpose; on this point, see Campbell (ed.) (2000) xxx.

10 Thus it has been widely interpreted as 'a good concrete treatment of Rome's water supply system': Conte (1994) 503; McElwain (ed.) (1925) xxvii ('a simple and truthful narration of the facts').

11 There has been a revolution in particular in thinking on *On aqueducts* with a flurry of books and articles being published over the last decade which have challenged more straightforward readings of the treatise – see, e.g., DeLaine (1996); Cuomo (2000); Saastamoinen (2003); Peachin (2004).

12 I discuss the complexities of his *Strategemata* and gromatic treatise in a forthcoming longer study of all these works based on my Cambridge PhD thesis (Weeks (2004)).

constructs between Frontinus and his emperors, Nerva and Trajan. Clearly the representation of emperors and exploration of the nature and rhetoric of their reigns is an important feature of many different Imperial-period texts – but there are signs that it is a particular issue for many knowledge-ordering writers. Vitruvius' *On architecture*, for example, seems to spend as much time exploring and exploiting the authority of the Emperor Augustus as it does organising and elevating architectural knowledge. The explanation which Vitruvius gives for writing his treatise is not dissimilar to some of Frontinus' programmatic statements: as we will see below, he argues that his treatise brings together every aspect of the architectural 'discipline' in the hope that he might aid Augustus in his own building work.[13] At the beginning of his treatise, he portrays Augustus as omnipotent, and himself as his humble servant. As his text progresses, however, he elevates architects and architecture into all-conquering powers, simultaneously borrowing the rhetoric of and posing a potential challenge to Augustus' *imperium*.[14] Trevor Murphy has pointed out that Pliny the Elder claims imperial power and authority for his *Natural history* by dedicating it to the emperor Titus, but also notes that he has something of a tightrope to walk, for Pliny's knowledge-gathering activities could be seen not only to borrow but also to encroach upon the emperor's prerogative as ultimate arbiter and controller of knowledge.[15] And if one is happy to read Petronius' Trimalchio as a reflection – at least in part – of Nero and his reign, it is possible to see the *Satyricon*'s meditations on knowledge and learning as meditations also on the emperor's image and authority.[16]

In some ways, then, my study of Frontinus' engagement with the imperial authorities in his *On aqueducts* might serve as a case study for a much wider phenomenon. At the same time, however, I want to underline the very specific context of this text's composition – for the *On aqueducts* seems to have been written during the period of transition between Nerva and Trajan, a time of tension and uncertainty at Rome and in her empire.[17] Tacitus gives no hint of any tension or uncertainty when he lauds the principates of Nerva and Trajan at the beginning of his *Agricola* and *Histories*; indeed,

[13] See Vitr. *De arch.* 1.pr.; cf. below, pp. 184–5.

[14] I plan to discuss this more fully in a future article. On this topic, see McEwen (2003), who also explores some of the parallels drawn by Vitruvius between architectural knowledge and imperial power.

[15] See esp. Murphy (2004) 204–9.

[16] On Petronius, see Rimell's chapter in this volume.

[17] Frontinus' preface suggests that he began his *Aq.* as soon as he was made *Curator Aquarum*, which was in 97 CE while Nerva was still alive. References to Nerva as 'deified' at *Aq.* 102 and 118, however, suggest that the text was finished and published after his death in January 98. The composition of the treatise thus seems to span the end of Nerva's principate and the beginning of Trajan's (see Rodgers (ed.) (2004) 8).

he gives the impression that all anxiety and trouble melted away as soon as Nerva became emperor, and implies that Trajan's reign was then a natural continuation of his predecessor's glorious rule.[18] Pliny the Younger also attempts to present Trajan's succession as natural and undisputed, but his account in the *Panegyricus* also offers glimpses of the tension and even violence that preceded and accompanied it: on the one hand, he claims that Nerva's adoption of Trajan was written in the stars;[19] on the other, he suggests that prior to it Rome was on the brink of civil war,[20] and also implies that Trajan was forced upon Nerva in something of a coup:

> But there was a great dishonour done in our age, a great wound inflicted on the state: the emperor and father of the human race was besieged, captured and locked up, and from this most gentle of old men was taken his power to save mankind, and from this prince was removed that most blessed principle of his imperial power, namely that a prince cannot be *forced* into doing anything. However, if this was the only method which could move you [Trajan] to become captain of the state, I can bring myself to say that it was worth it.[21]

Though the prevailing rhetoric may have been about fresh starts, new-found freedom, peace and stability,[22] it is clear that the months between Nerva's accession and his death were full of uncertainty, disorder and back-stage plotting (reminiscent even of the atmosphere of 69 CE);[23] and, as I have already indicated, Frontinus was probably heavily involved in all of this. Indeed, his influence in Rome at this time may have gone beyond securing Nerva's adoption of Trajan – for, as Robert Rodgers points out, Trajan disappeared back to the provinces after he was made co-emperor and remained absent from Rome for at least a year after Nerva's death, leaving

[18] At *Ag.* 3 Tacitus characterises Nerva's reign as the beginning of a blessed new era, and then tells us that Trajan is building on the start made by Nerva by 'daily increasing the happiness of the times'. At *Hist.* 1.1 he reinforces the sense that Trajan's reign was a natural continuation of Nerva's, characterising them together as 'a time of rare good fortune'.

[19] See, for example, Plin. *Pan.* 1.5, where he claims that Trajan was divinely chosen; 5.3–4, where Rome's citizens predicted his succession as emperor; and 8.5, where Pliny tells us that as soon as Trajan was adopted 'every disturbance died away at once'.

[20] See, e.g., *Pan.* 5.6, where Pliny says that Trajan was only persuaded to assume imperial power because he saw that Rome was 'in danger'; and 6.4, where he describes the empire as 'stricken' and 'tottering'.

[21] Plin. *Pan.* 6.1–2. Eck (2002) supports this suggestion that Nerva was forced to adopt Trajan by Trajan's supporters; see also Bennett (1997) 46.

[22] Consider, for example, the coins issued by Nerva at the start of his reign 'to emphasise the dawn of a new age in which the peace-loving citizen could be at ease again' (Bennett (1997) 37).

[23] For further analysis of this, see esp. Berriman and Todd (2001), who characterise Nerva's reign as 'a period of great political instability at Rome and abroad' and describe Trajan as 'a usurper, who conspired against an aged and insecure emperor, welding an alliance with his peers at the head of the armies and provoking disquiet and disorder at Rome to wrest from Nerva, in all but name, the succession to Domitian's throne'.

a small circle of senators (which included Frontinus) '[in] control of the state's constitutional helm'.[24]

In view of this one might expect Frontinus' *On aqueducts* to reflect some of the change and uncertainty that marked the period of its composition, and, indeed, it is my argument that it does, more fully than scholars have so far appreciated. For while there have been several readings of the treatise which have explored its engagement with the political developments of 97/98, on the whole they have tended to overlook the tension of the times just as Tacitus did at the beginning of his *Agricola*. A. Trevor Hodge, for example, reads the *On aqueducts* as a text which supports and in fact embodies some of the positive reforms typically associated with Nerva and Trajan:

> Rome was at the time facing a dilemma . . . Is management essentially a profession in itself, so that a competent manager can run any enterprise, irrespective of its product, or, especially in technical fields, is first-hand experience of the trade essential to run the business properly? Hitherto, under the Julio-Claudian and Flavian emperors, Rome had preferred the second answer. Senior technical positions went to imperial freedmen . . . But now Nerva and Trajan changed the policy, appointing distinguished public servants from the senatorial ranks . . . Frontinus was one of the first representatives of this new order of things . . . no doubt there were many at Rome who looked forward with both pleasure and confidence to watching Frontinus, out of his depth in this highly technical quagmire, foundering ignominiously and dragging down with him into obloquy the whole new official policy. But Frontinus did master his hydraulics, did reform the Water Office . . . and what better way to prove the point – and so defend Trajan, the emperor who appointed him – than to publish a detailed and technical treatise on the Roman waterworks?[25]

In this analysis, Nerva and Trajan are at one, busy restoring power to the senatorial classes as well as reforming the general administration of the state, and Frontinus is firmly behind them supporting their efforts with the publication of his treatise. This is also the view taken by Harry Evans, who argues that Frontinus 'presents . . . himself as a loyal lieutenant of an enlightened *princeps* determined to correct the abuses of the past'[26] and sums up the *On aqueducts* as 'a document presented to celebrate its author

[24] Rodgers (ed.) (2004) 8.

[25] Hodge (2002) 16; for a similar argument, see also Blackman and Hodge (2001) 141. Hodge's analysis develops arguments put forward by Ashby (1935) 26–7 and Grimal (1944) xv, who also took the view that Nerva and Trajan were keen to restore some privileges and positions to the senatorial ranks.

[26] Evans (1994) 61.

and the policies of the emperor who appointed him'.[27] And Michael Peachin also argues that the treatise was designed to support positive reforms pushed through by Nerva and Trajan:

> Frontinus was composing in the midst of, and precisely for, a very specific political situation . . . Nerva, followed by Trajan, had decided that private use of water from the aqueducts, properly a *beneficium Caesaris*, would be monitored, as would the well-being generally of these structures; and Frontinus, the newly appointed *Curator aquarum*, was the man charged to oversee this action. However, since many of the private fish sure to be caught illicitly enjoying publicly owned waters would be strong, and potentially vicious, finesse was an urgent necessity for the anglers. I will argue, then, that the *On aqueducts* was meant primarily to inform the wealthy and powerful about the new water policy, and that the book aimed, in several ways, to mollify them with regard to it. In short, I want to suggest that Frontinus hoped to persuade his elite peers that this was an administrative action unlike many undertaken during a previous reign: Domitian, and his way of doing things, was dead. The book was designed not to praise, but to justify and explain a particular moment in Roman imperial administrative history.[28]

Readings of the text which have examined its political interests have tended, in other words, to buy into certain ideals about Nerva's and Trajan's principates, and to assume that Frontinus does too.

I will argue in what follows that in some ways this is right – that the *On aqueducts* does, at times, appear to enshrine and promote the kinds of policies and ideals identified by Hodge, Evans and Peachin. It is also my view, however, that it is more complicated than that, and I will be bringing some of the more 'technical' aspects of his treatise into dialogue with the political ones to explain how. In short, I will be suggesting that Frontinus organises and orders knowledge in his *On aqueducts* in ways which do not *straightforwardly* support Nerva or Trajan but which *reflect on* the new era which they claim to usher in, and in particular on the role of senators within it (clearly a subject close to Frontinus' heart).

INTRODUCING FRONTINUS' *ON AQUEDUCTS*

Looking first at his preface, we can see that Frontinus' opening paragraph might support the view that his treatise demonstrates wholehearted support

[27] Evans (1994) 63. See also Rodgers (1986) 353, who characterises it as 'a personal notebook which [Frontinus] compiled soon after he took office (subsequently revised and expanded for publication as a kind of political statement of the Trajanic ideal of senatorial administration).'

[28] Peachin (2004).

of his emperor(s).[29] As I have already mentioned, Frontinus begins his *On aqueducts* in a way very reminiscent of Vitruvius' introduction to his *De architectura*, though the effect in Frontinus' case is rather more intense. From his opening lines, Vitruvius insists on linking his literary endeavours with Augustus' political ones, arguing that his treatise will help his emperor to erect buildings which will reflect the magnitude of his achievements:

> But when I realised that you were concerned not only with the common life of all people and the constitution of the state but also with the provision of public buildings, so that the state might not only be increased by you with provinces but also that the majesty of the empire might be seen in the eminent authority of public buildings, I decided to wait no longer in publishing for you my ideas about these things . . . I have therefore written an exhaustive set of rules, which you may refer to when thinking both about buildings that are already complete and those planned for the future. (Vitr. *De arch.* 1.pr. 2–3)

He also reminds his imperial patron of their long-standing acquaintance,[30] and presents his treatise as a kind of (unexpected) return service to thank Augustus for all the favours he has received.[31]

Frontinus, however, is able to lay claim to a much greater level of both intimacy and importance, and does precisely that from the word go. His opening sentence echoes Vitruvius', but draws much closer connections between himself, his treatise and his emperor than Vitruvius manages: for while the latter's opening words initially open up some distance between his project and Augustus (instead of explaining why he published it, Vitruvius tells us that he actually hesitated to do so, for fear of distracting his addressee from his military operations),[32] Frontinus' opening lines establish his composition as a direct – and indeed the only proper – response to a specific commission from the Emperor Nerva:

> Since every assignment commissioned by the emperor demands particularly close attention, and since both my natural concern and my industrious loyalty prompt me to show not only diligence but also devotion in this matter, and since the office of *Curator aquarum* has been conferred on me by Nerva Augustus (it is hard to say

[29] The *Aq.* is addressed exclusively to Nerva, and Rodgers (ed.) (2004) in fact argues that no mention is made of Trajan anywhere in it (where the name *Traianus* occurs in the manuscripts he puts it down to copyists' errors). Given the period in which it was written, however, I think we must assume that Frontinus was partly addressing himself to and engaging in 'dialogue' with Trajan as well as Nerva.

[30] See Vitr. *De arch.* 1.pr.2, where he reminds Augustus that he has enjoyed the patronage both of Julius Caesar, his adoptive father, and his sister.

[31] Vitr. *De arch.* 1.pr.3: 'Since I am therefore obliged to your kindness, and need not fear poverty in my old age'.

[32] Vitr. *De arch.* 1.pr.1: 'When your divine mind and spirit, Sovereign Caesar, was taking control over the whole world . . . I did not dare, amid such great business, to publish my architectural writings . . . fearing that, by interrupting you at an inappropriate time, I might disturb your mind.'

whether he is more diligent or more devoted in his service to the state), an office which is concerned not only with the health but even with the safety of the city, and which has always been held by Rome's foremost men, I think that the first and most important thing to be done (which is what I have done in all previous posts) is to find out about what I have undertaken. (*Aq.* 1)

He makes it clear that this response is also the result of his peculiar conscientiousness, and as he explains this he takes care to talk his emperor's language: he not only uses words which echo some of the ideals of Nerva's reign,[33] he even ends up sharing the same vocabulary with him, for in the space of a few lines he matches his own 'diligence' and 'love' for the state ('not only diligence but also devotion', *non ad diligentiam modo verum ad amorem quoque*) exactly with Nerva's ('it is hard to say whether he is more diligent or more devoted in his service to the state', *nescio diligentiore an amantiore rei publicae imperatore* . . .). In addition, he manages to make his new job sound remarkably similar to his emperor's role: aside from the fact that the post of *Curator aquarum* was traditionally held by one of Rome's *chief* men (*per principes . . . viros*), it was concerned, he argues, with issues (like 'the very health and safety of the city') which were presumably of paramount importance to Rome's main leading man – the *princeps* ('emperor') himself. In several ways, then, Frontinus makes Vitruvius' attempts to ally himself closely with Octavian seem feeble by comparison; by the end of his first sentence we have been given the impression that his character, attitude, concerns and even responsibilities are entirely in tune with those of his emperor.

This aligning of himself so closely with Nerva supports Evans' view that Frontinus is keen to be seen as 'a loyal lieutenant' of his commander-in-chief. As his second paragraph develops, however, and Frontinus elaborates on his reasons for writing, a slightly more complicated picture begins to emerge:

I think that the first and most important thing to be done . . . is to find out about what I have undertaken. For I believe that there is no surer basis for any undertaking than this, nor is it possible in any other way to decide what must be done and what must be avoided; moreover, there is nothing so shameful for a capable man than to perform an office delegated to him on the advice of assistants, which is necessary when a man has to turn to their help because of his ignorance of the matter before him; for although these people are necessary cogs in the administration, they are, as it were, merely the hands and tools of the person in charge . . . Therefore, I have gathered together all the information which I could find that relates to this post,

[33] For example, at the start of his reign, Nerva issued coins with slogans like *salus publica* ('Public Health' – on this and other Nervan rhetoric, see Grainger (2003), 47 and 52–65).

and (just as I have done in previous jobs) I have set it in order and into a body of material, if you like, and set it down in this account, which I will be able to turn to as one might to an administrative handbook. (*Aq.* 2)

The sentiments expressed here are not ones we might expect to hear from a typical Roman senator,[34] though they are in keeping with Frontinus' insistence on his extraordinary diligence; but what is most interesting about them for the present purposes of this chapter is the connection which they make between knowledge and power. Frontinus does not simply suggest that knowledge *enables* managers to take control – he argues that it is absolutely key. Further, he appears to be worried about the knowledge – and, more particularly, the authority which might result from it – of administrative inferiors, whose proper place (as 'necessary cogs', but not the people in the driving seat) he is keen to underline.

This equation of knowledge and power may explain not only the existence of his *On aqueducts* but also its extraordinary thoroughness – as we will see, Frontinus goes into minute detail in his discussion of Rome's aqueduct network. At the same time, however, it raises questions about Frontinus' relationship to Nerva. On one level, as before, his arguments may seem supportive of his *princeps*. At a time when his emperor may have been replacing freedmen with senators in all sorts of administrative posts (as Hodge argued above), Frontinus' idealistic remarks set an impressive example for others of his rank: they show one influential senator recognising the responsibility which comes with his recent appointment and explaining how to maximise and protect the authority newly vested in him. The idea of knowing thoroughly about the work of one's administrative department is also very much in keeping with the ideals of transparency and accountability which were flagged as hallmarks of the new era.[35] However, as well as offering further evidence of their unanimity, this second paragraph threatens to throw relations between the pair slightly off-balance. It is striking, in the first place, that the words which Frontinus uses to describe the relationship between, say, a *Curator aquarum* (the manager) and the *aquarii* (the 'water-men') who work for him are ones which could as easily be applied to the relationship between the emperor and his officials. As well as putting

[34] As Peachin (2004) points out, Frontinus' peers do not seem to have shared his opinion: it was usual practice for high-ranking officials to rely on the support of much more knowledgeable subordinates.

[35] That is not to say that transparency and accountability *were* hallmarks of Nerva's reign, or Trajan's for that matter; but both emperors made intermittent efforts to suggest that this was so. Consider, for example, the commission set up by Nerva in 97 CE to investigate imperial spending, which Grainger (2003) 56, describes as 'a public relations exercise'. Frontinus himself was in fact one of the senators selected by Nerva to take part in this review of imperial finances.

himself on a par with his emperor, in other words, his discussion of the hierarchy that should exist between managers and their minions reminds us that Frontinus is himself a cog – one of the many 'hands' or 'tools' – in the imperial administration. And since this is so, the purpose of his treatise and quantity of knowledge which he goes on to gather might be sources of concern for his emperor: for if power lies in knowledge, where does Frontinus' *On aqueducts* leave Nerva (or Trajan)? Does the *princeps* need to apply the same principles as Frontinus to his own position? And if he does not, is his own authority diminished?

It may be in part to avoid such questions that Frontinus takes so much trouble in his first paragraph to underline parallels, not differences, between himself and his *princeps*. As his treatise progresses, however, Frontinus does expose important differences between them, which prompt reassessment both of the connection which he makes between knowledge and power and of his and his emperor's respective authority.

EMPOWERING *MINUTIAE*

Following on from his preface, Frontinus deluges his readers with a veritable flood of aqueduct-related data, a flood which sees narrative rapidly giving way to bare facts and hundreds of figures. First, he begins by sketching the history of the network (*On aqueducts* 4–16), and each of the accounts that he gives of the construction of Rome's nine different aqueducts moves relentlessly from historical story-lines into technicalities: his account of the building of the *Aqua Appia*, for instance, begins dramatically with the tale of Appius' out-manoeuvring of Gaius Plautus, a story about political intrigue and the lust for fame, but once that tale is told the narrative slides immediately from high drama into detailed statistics;[36] similarly, after telling the story of how the *Aqua Virgo* got its name, Frontinus once again follows it up with technical data,[37] as he has done for all the aqueducts in between and will do for all that follow. In moving repeatedly from historical

[36] *Aq.* 5: 'However, because Plautius resigned the censorship within a year and six months, having been deceived by his colleague (Appius) that he was about to do the same, the honour of naming the aqueduct was Appius' alone, who through many manoeuverings is said to have extended the length of his censorship until he could finish both the Appian way and this aqueduct. The intake of the Appia is in the Lucullan land, between the seventh and eighth milestones on the Via Praenestina, on a turn-off 780 paces to the left. It stretches in length from its intake to the Salinae at the Porta Trigemina, extending 11190 paces; 11130 of these paces run underground, while above ground it travels for 60 paces on substructures and (near the Porta Capena) on arches.'

[37] *Aq.* 10.

events to (what become formulaic) facts and figures, Frontinus effectively deconstructs each aqueduct as soon as it has been built, breaking it down into its particulars. And the effect of this is to drive home the point that, while the history of their induction is at times eventful, even momentous, and while the cumulative effect of their build-up is perhaps as awe-inspiring as Frontinus' exclamation in *On aqueducts* 16 suggests,[38] the story of Rome's aqueduct network is always going to involve – and will perhaps always boil down to – technical *minutiae.*

This is something that the ensuing chapters go on to confirm: for the second section of the treatise (*On aqueducts* 17–22) abandons anecdote entirely in favour of yet more detailed specifications, as Frontinus particularises each aqueduct's topographical co-ordinates;[39] and in the section which follows (*On aqueducts* 23–64) prose eventually breaks down into a list of numerical equations when we run into some systematised data of an even more specialised nature:

> The 35-pipe: diameter 6 97/144 digits; circumference 20 281/288 digits; capacity 28 49/96 *quinariae.* It is not in use.
> The 40-pipe: diameter 7 13/96 digits; circumference 22 61/144 digits; capacity 32 7/12 *quinariae.*
> The 45-pipe: diameter 7 41/72 digits; circumference 23 113/144 digits; capacity 36 47/72 *quinariae.* It is not in use. (*Aq.* 49–51)[40]

This list (which sets out the specifications for a total of twenty-five standard pipes) comes at the end of a digression on the origins and proportions of different types of gauges. The section begins more anecdotally at *On aqueducts* 24–5 with some speculation on etymology and possible inventors, but gradually numbers and fractions proliferate as Frontinus sets out different methods of measurement (*On aqueducts* 26–38) until the text is reduced to this series of formulaic calculations (*On aqueducts* 39–63). As the treatise proceeds, then, technical information not only abounds; it becomes

[38] *Aq.* 16 represents the climax of Frontinus' account of the network's construction, for here he suddenly exclaims: 'Compare, if you like, so many waters, being carried on so many indispensable structures, with the luxurious pyramids or the useless but well-renowned works of the Greeks!'

[39] At *Aq.* 18–22, Frontinus rearranges Rome's aqueduct network from the chronological order of the previous section into two new configurations, 'inducting' the waters for a second time in height order first (*Aq.* 18), and then rearranging them again by detailing the courses taken first by the six aqueducts which flow into 'catch-basins' (*Aq.* 19–21) and then by the three which do not (*Aq.* 22). These last two categories are subdivided in turn and the aqueducts reorganised once more, in different height orders again, as Frontinus specifies the route which each of them takes. Each conduit's journey into and through the city is thus increasingly particularised.

[40] Based on Robert Rodgers' online translation (http://www.uvm.edu/~rrodgers/frontinus.html).

increasingly detailed and increasingly prominent, as a barrage of *minutiae* transform the very look of the text, taking it over entirely.

By the time we reach the fourth section (*On aqueducts* 65–76), in which Frontinus lists the many discrepancies he has found between his own measurements of discharge figures and the system's official records, one reason for this level of detail has already been hinted at, namely fraud. Frontinus hints that Rome's water supply is being abused at the very end of his preface, when he explains that the final topic which his *On aqueducts* will cover is the law governing the building and maintenance of the network and the 'penalties' which have been established to protect it,[41] and as his account develops the fraudulent tapping of the water supply becomes a prominent leitmotif, as does the administrative incompetence of his predecessors.[42] When we see *both* at work early on in the treatise, Frontinus makes clear instantly what his and his emperor's attitude to both problems is, explaining (at *On aqueducts* 9) that, on his emperor's order, he immediately shut off the channel which was being diverted and 'restored' the water to its rightful recipients. While detailed knowledge of the system's ins and outs is not linked *explicitly* here to his tackling of incompetent and devious water-men, it is made much more explicit on the next occasion when error and fraud turn up together, at *On aqueducts* 31–4, where the reason behind Frontinus' intricate discussion of the methods of gauging pipe sizes suddenly becomes clear:

> The gauging of all the pipes from the 5-pipe to the 120-pipe is done just as I have shown . . . But, although the *aquarii* conform to the obvious measurements in most cases, they have altered them in four: in the 12-, the 20-, the 100- and the 120-pipes. (*Aq.* 31)[43]

The changes which water-men have made to the pipes are minute, but Frontinus' attention to detail has been able to detect them.

Both of these passages foreshadow what then follows at *On aqueducts* 65–76, where the technical nature of chapters 31–4 is combined with the impressive reaction described at *On aqueducts* 9. Here yet again Frontinus adopts a systematic approach, organising his material along the lines of the first 'map' of the aqueduct network presented at *On aqueducts* 4–15: he works his way one by one through each aqueduct in chronological order,

41 *Aq.* 3.

42 Fraud raises its ugly head again at *Aq.* 9 and 31–4; it becomes a constant undercurrent at 64–76; and is then alluded to again at 97, 103, 111–15 and 128–30. Administrative incompetence also becomes a recurrent theme, popping up in all sorts of places, including *Aq.* 7.1, 64–76, 91, 96–7 and 117.

43 Although Frontinus uses the word 'error' to describe one of these discrepancies at the beginning of *Aq.* 32, it soon becomes clear that the water-men have altered these pipe sizes deliberately.

tackling the oldest first, and where possible he also traces their course from intake to settling tanks or points of delivery, cataloguing at every step of the way all of the errors which he has found in earlier accounts.[44] The systematic and repetitive nature of this catalogue not only gives the impression that the incompetence and corruption which he is tackling is widespread but also underlines the thoroughness of Frontinus' knowledge of the system, and his framing of this section makes an even greater claim for his industry and diligence. Back at *On aqueducts* 9, we discover that Frontinus' swift action results in an amazing 'abundance' of water: the Tusculan people, we are told, 'wonder' at the unexpected quantity which reaches them after Frontinus has put a stop to illegal tapping.[45] On a much greater level, this is what Frontinus suggests he has done for the whole of Rome: for at *On aqueducts* 64, after reminding us of his scrupulous attention to detail,[46] he announces the grand total of missing *quinariae* (units of water) which his research has uncovered, a whopping 10,000 in total. Then, at *On aqueducts* 77, as he sums up this part of his account, he suggests that his findings almost amount to the discovery of a whole new supply of water;[47] and several chapters further on he repeats this claim, arguing that the volume of water now flowing through Rome's pipes has been almost doubled:

> Now . . . whatever was diverted through the fraud of water-men or lost through negligence has been added, as if through the discovery of new sources. And the supply has almost been doubled. (*Aq.* 87)

By the time we reach the end of Frontinus' fifth set of data (*On aqueducts* 78–86, where he sets out the amounts of water distributed to different parts of Rome), then, we have discovered that detailed knowledge of the water supply system enables Frontinus not only to *detect* fraud (which is

[44] See, e.g., *Aq.* 66: '*Anio Vetus* is credited in the records with 1541 *quinariae*. At its intake, I found 4398 *quinariae* . . . 2857 more than is stated in the records. A total of 262 *quinariae* were distributed before the aqueduct gets to its settling tank. The total at the settling tank . . . was 2362, which means that 1774 *quinariae* went missing between the intake and the settling tank. After the settling tank, 3408 *quinariae* were delivered, 69 *quinariae* more than than we have said was stated in the records, but 1014 fewer than the amount which we showed was received into the conduit after the settling tank. In total, the volume lost between the intake and the settling tank and afterwards came to 2788 *quinariae*, which I might have thought was due simply to errors of measurement if I had not found out where it was being diverted.'

[45] *Aq.* 9: 'On the order of the emperor, I returned the whole amount to the people of Tusculum, who now, no doubt with some astonishment, are not sure who to thank for this unusual abundance.'

[46] Frontinus not only tells us that the calculations which are about to follow were worked out 'through scrupulous research' (*scrupulosa inquisitione*), he also reminds us that he was working on behalf of 'the best and most diligent Nerva' (*optimi diligentissimi Nervae*).

[47] He begins this chapter: 'Enough has now been said about the volume of each aqueduct and a new method, if you like, of acquiring water . . .'

all that happens at *On aqueducts* 31–4) but also to combat it so effectively that life at Rome is transformed. The abundance of technical details in his treatise translates directly, it would seem, into a new abundance of water; and the link which Frontinus made at the beginning of his treatise between knowledge and control appears to have been strengthened, for we have seen him using his newly acquired learning not only to control Rome's aqueduct network but also to restore it, to put it back in order.

One of the things which Frontinus does as he celebrates the results of his detailed research is to weave himself and his emperor into close partnership once again. Indeed, immediately after he has finished his account of the network's distribution figures, he hands the credit for his work over to Nerva, claiming that it is ultimately thanks to him that Rome's water supply has been enlarged:

> Now, thanks to the foresight of our most industrious emperor all that was intercepted by fraud or lost by laziness has been added. (*Aq.* 87)

In so doing, he again employs vocabulary which he has also used for himself, highlighting the emperor's impressive 'diligence' (*diligentissimi principis*), and he does this several more times in the following chapters as he continues to wax lyrical about the 'care' which his 'pious' *princeps* has taken of Rome.[48] Some of this praise touches upon the claim which Frontinus made towards the beginning of his preface, that the office of *Curator aquarum* was concerned with 'the health and even the safety of the city'. At *On aqueducts* 88, for example, when he salutes Nerva's 'care' for Rome, he argues that one of its most significant effects has been to improve the general *salubritas* of the city, for thanks to the building of new fountains and so on, and an increase in the number of private grants given by the emperor, the city has apparently taken on a much more wholesome character:

> There is a new appearance of cleanness, the air is purer, and the reasons for the rather heavy atmosphere, for which the city was infamous in antiquity, have been removed. (*Aq.* 88)[49]

As in his preface, then, we see Frontinus and his emperor apparently sharing the same attitude and striving for (and ultimately achieving) the same things. In addition, the rhetoric of purity which becomes central in this section ties in nicely with ideals of transparency and purging, with the idea

[48] See, e.g., *Aq.* 88 ('[Rome] feels this care of the most pious Nerva each day'); and *Aq.* 89 ('What of the fact that even these improvements do not satisfy the emperor's diligence . . . ?').

[49] See also *Aq.* 89–93, where we see more evidence that Nerva's *cura* has resulted in purer waters as well as greater abundance.

of cleaning up the whole city and getting rid of corrupting factors; and the image which Frontinus gives of a Rome restored to its former glory – indeed, improved almost beyond recognition – helps to give a sense that the city has truly embarked upon a bright new era. Furthermore, the order which Frontinus imposes on the material in his text systematically dispels images of chaos and lack of rule and gives the impression (however wrong it may be) that under Nerva order and proper control were imposed on Rome itself.[50] As well as insisting on their common approach to aqueduct-related matters, in other words, Frontinus' text also appears once again to be more widely supportive of Nerva's principate in general, for here, as elsewhere, it reflects some of the pivotal aspirations and rhetoric of his reign.

All in all, then, what follows his preface appears to support Frontinus' initial representation of himself as a devoted public servant whose interests and activities are entirely in tune with, and thoroughly supportive of, his emperor. As we shall see, however, the knowledge which he amasses in the course of his *On aqueducts* also empowers Frontinus (as the second chapter of his preface hinted it would) in ways which potentially shatter the sense of collaborative partnership which he has taken such care to build up.

FRONTINUS EMPOWERED

In the first place, as well as amassing knowledge Frontinus appears to master it impressively. This is something which his preface prepares us for, for as well as explaining that he wants to gather together *all* data relating to the water supply system to help him succeed in his new post[51] he also indicates that it is his plan to 'order' it, to 'bring it into line'.[52] We have already seen him doing this most impressively with his list of pipe sizes at *On aqueducts* 39–63, where he responds to the fraudulent altering of pipes by water-men by putting everything back in order again, reducing his prose to a systematic list of equations which sets out each of the twenty-five pipes formulaically, in ascending size, not missing a single one out. Similarly, his various textual 'maps' of the aqueduct network organise and re-organise his newly acquired knowledge repeatedly, testifying to his command of his material. Moreover, as well as piling up an increasingly impressive body of facts and figures,

[50] This contrasts with the instability and disorder which, as I mentioned above, were probably more characteristic of Nerva's principate.

[51] At the beginning of *Aq.* 3, Frontinus makes clear his intention not to miss out anything 'that relates to an understanding of the entire subject'.

[52] *Aq.* 2: 'I have gathered together into this account everything that I could find that relates to the whole topic, and have arranged it in order and turned it into one body of material (*in ordinem et velut corpus diducta*)'.

each of these successive sections brings the city's water nearer and nearer to its points of delivery. As I have already explained, Frontinus' first 'map' (*On aqueducts* 4–16) begins in the distant past not long after Rome's founding, and as it traces the construction of each aqueduct it charts the distance which it travels from outside to inside the city. The second (*On aqueducts* 18–22) then provides more details on the various courses taken *within* the city; and, as we have already seen, Frontinus organises the flow of his list of accounting discrepancies likewise from outside the city inwards, in the direction of reservoirs and distribution centres (*On aqueducts* 65–76). Finally, the data presented at *On aqueducts* 78–86 eventually ends up at public buildings, private residences, street fountains and the like. The momentum of Frontinus' entire text (as well as of individual sections), in other words, is organised in such a way as to mirror the flow of water into and through Rome to its individual inhabitants, and this narrative current suggests a certain amount of mastery not just within but also beyond the confines of his text.

For just as the textual control which he demonstrates in sections like *On aqueducts* 39–63 is suggestive of Frontinus' actual ability to control and address real errors and abuses, so his on-page 'induction' of hydraulic data from outside Rome to its heart places Frontinus himself at the city's centre. Some time after Frontinus has set out the network's distribution figures he provides us with yet another carefully organised list, this time of people, not facts and figures: at *On aqueducts* 98, he explains that Marcus Agrippa was Rome's first ever *Curator aquarum*, and then (a few chapters on) recounts the names of all successive incumbents of the post in chronological order, ending the list at the present day with himself.[53] The history of Rome's aqueduct network therefore does not end at *On aqueducts* 15, when Frontinus has finished telling the story of its build-up, but continues into the rest of the text and eventually climaxes with our author. Moreover, in the section in which the history of their construction is narrated, each aqueduct's water reaches the city, as we have seen, but not its people; it is only after Frontinus has spent time tackling fraud and incompetence that he allows us to see water basins filled and fountains suddenly gushing forth, and this leaves us with the impression not simply that he is responsible for increasing the volume of water supplied but even that it is only really under his curatorship that *any* of Rome's water arrives at its destination. Frontinus' textual organisation, in other words, goes beyond the mere assertion of his ability to manage and mend the aqueduct network; for topographical and

[53] *Aq.* 102. This list is not, in fact, exhaustive – several names seem to be missing.

chronological undercurrents combine to turn him into the hero everybody has been waiting for.[54] The pupil of his preface has become an undisputed expert, and in the process he has claimed an enormous role for himself in both the running and the history of the city of Rome.

Furthermore, Frontinus not only masters and orders hydraulic knowledge; he withholds some from his readers. Right from the start of his treatise, he confuses his readers about whether or not they should actually be reading. On the one hand, his preface promises us a story about Rome, its *princeps* and its *principes viri*, a story, indeed, that is potentially of interest to all Roman citizens; on the other, and in the next breath, he insists that *he* is its only (or at least primary) intended audience,[55] and then proceeds to underline its narrowly technical focus with a 'contents page' of sorts which promises fact after fact and figure after figure.[56] In his opening chapters, in other words, he at once invites readers in and shuts them out, making his text seem momentous and all-embracing one moment and dry and exclusive the next; and this bewildering double message continues into the rest of the treatise. He begins the section following his preface, for example, with the words *Ab urbe condita* . . . , a phrase which recalls Livy's monumental history of Rome and thus prepares us for an equally weighty and inspiring work;[57] as we have seen, however, his account then slides relentlessly into technical statistics, and these statistics become increasingly frequent and specialised as he moves from history to topography and then into his lengthy digression on gauges and pipe sizes. I pointed out earlier that we learn of the value of that digression several chapters into it, when Frontinus shows how water-men have altered several pipes' sizes, but the danger is that we might not get as far as finding that out, for even before he embarks upon his list of complex equations at *On aqueducts* 39–63 the material which he is presenting is already of such a technical and specialised nature that it may

[54] As Evans (1994) 58–61 points out, Frontinus goes out of his way on occasions to draw connections between himself and Agrippa. As we have seen, he also insists that in between their curatorships the management of Rome's aqueduct network slipped into chaos, so in many ways he presents himself as a second Agrippa, Nerva's equivalent to Augustus' right-hand man, who restores the water supply to its Augustan glory. In addition, however, he is critical of several aqueduct builders at *Aq.* 4–16 (not least Augustus himself at *Aq.* 11), and this sets him up in competition not only with earlier administrators but also with those who were responsible for bringing water to Rome in the first place. The claims which he makes about overcoming fraud and discovering what almost amounts to a whole new supply of water thus counter problems which his treatise suggests date back to the earliest days of the network's existence, not just the post-Agrippan era, and set Frontinus up as the first man in its history to tap its full potential.

[55] *Aq.* 2: 'In the case of my other books, which I have written after gaining first-hand experience and practice, I looked to the needs of those who will follow me. This treatise, too, may be useful for my successor, but it will be most useful for my own instruction and guidance.'

[56] See the whole of *Aq.* 3. [57] *Aq.* 4; on this point, see esp. DeLaine (1996) 122–3.

well put many off.[58] His persistent progression from prose into numbers of an increasingly complex nature, in other words, relentlessly underlines the narrow, technical focus of the *On aqueducts* at the expense potentially of its wider appeal, and thus has the power to bewilder readers just as his preface does, not only because they may not be able to follow all of his calculations but also because they may end up wondering why they are trying.

Frontinus in fact acknowledges the possibility that some of his material may be 'dry' and 'difficult' in a passage which most commentators pass over without really discussing. This is one of several transitional passages where he concludes the preceding section and prepares for the next, and in this case he is about to present us with the aqueduct network's distribution figures (*On aqueducts* 78–86). Just before he does, however, he offers his readers a short cut through his text:

> It remains to explain the distribution [of water] according to each aqueduct and in each of the city's districts. I realise that this account may seem not only dry but also complex, but I will set it out in spite of that (though as briefly as I can) so that my 'administrative handbook' will not have any gaps. But for those who are satisfied with knowing the sum totals (*summa*), you may skip the details (*leviora*). (*Aq.* 77)[59]

The irony here is that, although his readers may well be tired by now of dry and complicated 'details', this is the very point in the text when they may really be of interest – for this is the point, as I have explained, when water finally reaches Rome's inhabitants. Moreover, that water only does so when 'lump sums' are broken down into minutiae, something which the text itself underlines. For if one follows Frontinus' advice and skips straight to his *summa*, one learns nothing at all: the only *summa* in this section occur at or towards the beginning of each chapter, *before* any water is delivered to its destinations.[60] On the other hand, if one does press on and trawl through his *leviora* – as readers must if they want to learn *anything* from this section – the narrative flow of each chapter once again works in harmony with the message being put across (that here 'details' are far more important than 'totals'), for it relentlessly works its way from broad brush strokes ('outside the city', 'inside the city') to smaller sub-categories ('in the

[58] In spite of his assertions in his preface, it is wrong to assume that Frontinus was writing just – if at all – for a specialist audience. His opening – 'from the foundation of the city' – at *Aq.* 4, for example, suggests a more generally educated readership.

[59] Rodgers (ed.) (2004) prints *summam*, not *summa* here. On this, and the translation of both *summa(m)* and *leviora*, see below.

[60] See the opening sentences of *Aq.* 78 and 79. In the previous section, the 'sum total' of *quinariae* which Frontinus has recovered is the climax of his account; here, however, the 'sum total' (14018 *quinariae*) is only the starting point of the discussion that follows.

name of Caesar', 'for private use', 'for public use'), and then on again to subdivisions of the last of those subcategories:

> So, of the 14018 *quinariae* which we set down as being the total number discharged from all the aqueducts together, only five are distributed by the Appia outside the city . . . The remaining 699 were distributed inside the city, through the second, eighth, ninth, eleventh, twelfth, thirteenth, and fourteenth regions into twenty reservoirs. Of these, 151 *quinariae* are distributed in the name of Caesar, 194 to private parties, and 354 for public use. Of the last category, four *quinariae* went to one camp, 123 *quinariae* went to fourteen public works, two *quinariae* were delivered to one fountain, and 226 *quinariae* went to 92 water basins. (*Aq.* 79)

The text itself, in other words, is busy fragmenting totals and subtotals into increasingly smaller fractions, underlining the point that in this section above all it is the 'details' which are the goal.[61]

Even though Frontinus explicitly invites us to overlook them, then, his treatise shows vividly what we would miss if we were to take his short cut: in theory the most vital information which the *On aqueducts* contains. And yet there is another, even greater, irony in this, for if we do struggle through and get to the end of this section we discover that the very figures which we have just decided were worth tracing to their destinations are already out of date. Frontinus hints at this when he sums up *On aqueducts* 78–86 with the phrase 'this is the volume of water which was calculated up to the time of Nerva, and this is how it was distributed',[62] but it is at the end of the next chapter where he really comes clean, when he explains:

> It does not escape me that an explanation of the new distribution figures has not yet been provided; but I will add this when I have done the sums; for it must be understood that these cannot be published until they have been added up. (*Aq.* 88)

What he initially implies are the magnificent results of his anti-corruption efforts of *On aqueducts* 65–76, then, turn out to be nothing of the sort; for instead of presenting us at *On aqueducts* 78–86 with the 'new supply' whose 'discovery' he has just been celebrating, he distributes only the old, incomplete quantities based on the faulty records. As we have seen, he appears all the way through the *On aqueducts* to be sharing *all* of his knowledge with us, often with mind-numbing thoroughness; and he goes out of his way in this particular section also to share the glory of his

61 Indeed, *Aq.* 79–86 are themselves subdivisions of *Aq.* 78, which sets out the discharge figures 'outside the city', 'inside the city' and so on for the aqueduct network as a whole before breaking them down for each aqueduct individually.

62 *Aq.* 87.

findings with his 'colleague' and emperor. However, it is *he* in the end who ultimately 'skips over' details – the very details which his entire treatise has been working up to; we, the readers, do not even find out the revised 'sum totals', for he withholds even them from us. Having seen him fighting error and fraud right from start, we could in fact be forgiven for feeling cheated ourselves at the culmination of his account.

So, at the end of his fifth and climactic set of data, Frontinus' *On aqueducts* suddenly becomes more exclusive than ever, for what we learn above all is that Frontinus is still the only person fully in the know; indeed, although he claims that Nerva is ultimately responsible for the great improvements in Rome's atmosphere and waters, in his treatise it is *he* who holds the key, it is Frontinus alone who has the information at his fingertips which will make the city clean again and bring health and safety to its people, information which he refuses to share. Thus far, I have talked about the way in which Frontinus models himself on his emperor, but in many ways his *On aqueducts* models Nerva on himself, for his descriptions of his emperor always come second, after he has set out his own qualities, and so it is Nerva who appears to be copying him, not only in his characteristics but even in his concern for Rome's *salubritas* and *securitas*.[63] As well as watching Frontinus make the post of *Curator aquarum* seem remarkably similar to the role of emperor, therefore, we almost see Nerva striving to keep up with his *Curator*, and at this point in his text Frontinus leaves him behind. For in spite of the sense of partnership, and even leadership, which Frontinus concedes at *On aqueducts* 87–93, his withholding of crucial information at the last minute gives the lie to both, turning his emperor into a figurehead who may take the credit but not the real authority behind these improvements. This takes us back to the second paragraph of his preface, where Frontinus worried about the loss of authority that might result from a manager knowing less about his business than his assistants; indeed, what we perhaps see here is the relationship between knowledge and control suggested in his introduction being played out very much to Frontinus' advantage. Frontinus' discovering and presentation of knowledge not only empowers himself but potentially disempowers both his readers and his *princeps*.[64] We are faced, then, with a text which seems at once supportive and challenging; and it

[63] It is long after Frontinus mentions that he, as *Curator Aquarum*, is concerned with 'the health and safety of the city' that we see Nerva taking an active interest in both (at *Aq*. 87–93).

[64] DeLaine and Evans both acknowledge Frontinus' tendency towards self-promotion (DeLaine particularly so), but they both suggest that this is entirely compatible with a supportive attitude towards his emperor (see esp. DeLaine (1996) 129 and Evans (1994) 57–63). I am arguing otherwise – that he empowers himself so much through his organisation of knowledge within the text that he diminishes the authority of his emperor.

becomes more complex still as cracks begin to appear in the control which Frontinus' knowledge enables him to claim for himself.

CALIBRATING KNOWLEDGE AND WEIGHING UP POWER

When Frontinus suggests at *On aqueducts* 77 that those who are struggling to keep up or stay awake might consider skipping his 'leviora', he uses vocabulary which carries intriguing value connotations. Even if Rodgers is right to print 'summam' (meaning numerical 'sum total') at *On aqueducts* 77 instead of the more suggestive 'summa' (meaning 'the most important things'),[65] Frontinus' choice of the word 'leviora' here encourages us to compare it with the adjective 'summus' as well as the noun which derives from it, for its primary meaning – 'lighter' or 'less consequential' – has more in common with the value-laden superlative than with its mathematical counterpart. Though 'summam' might prompt us to translate 'leviora' as numerical 'details', in other words, we must be aware that 'leviora' itself pushes us towards more subjective interpretations of both words, interpretations which are further complicated if we recall another author's use of the same word in a near-contemporary text.

Just a quick glance at Pliny the Elder's *Natural history* is enough to impress the reader with its enormous scale. At times in his preface, however, he plays its dimensions down: he begins by quoting Catullus' own jest on the size and significance of his poetry ('For you always used to think my trifles were worth something'),[66] playfully suggesting that his huge compendium is itself a mere trifle, and then goes on to describe it more explicitly as 'rather light-weight' (*levior*):

> My audacity has reached such a height that I have dedicated these slim volumes of rather inconsequential work (*levioris operae*) to you. (Plin. *HN* pr.12.)

His reasons for so summing it up are that his compilation 'does not accommodate any talent': it is not entertaining, it contains no digressions, no speeches, no 'interesting happenings'; in fact, he states outright (with tongue very much in cheek), his subject matter is 'sterile'.[67] He uses 'levioris' ironically, in other words, contrasting his work with publications which are 'pleasant' (*iucundus*) and 'charming' (*blandus*), and setting the *Natural history* up as 'lighter', 'more inconsequential', because it is not the type of

[65] McElwain (ed.) (1925) and Kunderewicz (ed.) (1973) both print 'summa'.

[66] Catull. 1.3–4: 'namque tu solebas / nugas esse aliquid meas putare'. See Plin. *HN* pr.1 for the quotation.

[67] Plin. *HN* pr.13.

text which tends to appeal to Roman readers. For having thus apparently belittled his work, he soon makes it clear that *he* values 'useful' texts over 'entertaining' ones ('I myself feel that there is a special place in learning for those who have opted for utility, in spite of the difficulties, instead of popularity through giving pleasure'),[68] and so 'levioris' morphs from being a negative apology into a positive claim. Pliny thus teases us with the possibility that his encyclopedia may be 'trifling', but goes on to hint at precisely the opposite.

Writing several years later than both Pliny and Frontinus, Tacitus also exploits the language of 'trifles' to weigh up the import of his *Annals*. In fact he does so at a crucial moment in his text, for he embarks upon a discussion of the potential 'lightness' (*levitas*) of his work at *Annals* IV, 32, a passage which is widely recognised as his 'second preface':

> 'I am well aware that much of what I have narrated and am about to narrate may seem rather trifling (*parva*) and not worth recording (*levia memoratu*); but no one should compare my history with the writings of those who have written about the Roman people's ancient past. Their subject included huge (*ingentia*) wars, the overthrow of cities, kings driven out and captured . . . However, it will not be entirely pointless (*sine usu*) to look closely at matters which may seem trifling (*levia*) at first sight, for out of them often come the beginnings of momentous happenings (*magnarum rerum*).' (Tac. *Ann.* 4.32.)

His emphasis here is more on the nature of his subject matter than his writing: his work is 'inglorious', he explains, because it reflects the less than momentous times which it describes. That does not stop it from being 'useful', however, for as he goes on to indicate 'those things which seem trifling at first view' (and therefore also his own writing) are sometimes more consequential than they might seem. Tacitus thus contrasts 'levis' first with *ingens* ('huge') and then with *magnus* ('great') to two different ends, for he succeeds both in denigrating the period (and particularly Tiberius' principate, the subject of *Annals* IV) and also in playing up the potential significance of his work. In different ways, in other words, both he and Pliny tease out the language of *levitas* in order to prompt us to weigh up the value of their writing, and in order to insist, in the end, on its importance.

Frontinus could not have read Tacitus' *Annals*, but he may have read – or at least known about – the preface to Pliny's *Natural History*, and he too toys with the comparative in a way which must prompt his readers to meditate on – and perhaps be persuaded of – the value of his own material, though in his case, in my view, questions remain about some

[68] Plin. *HN* pr.16.

of its relevance. As we have seen, for much of his text he insists on the vital *importance* (the *summus*-ness) of what he provocatively describes at *On aqueducts* 77 as 'less consequential' facts, on the necessity of having an intimate knowledge of every pipe and gauge of the water-supply system, and this is very much in keeping with the ironic way in which both Pliny and Tacitus play with *levis*/*levior*. We have also seen him manage and order all of that data impressively; but there is also a sense on occasions that the volume has got a little out of hand, that we have an overflow of information, even that some of it is superfluous. I noted above that one impressive feature of the end of his digression on pipe gauges is that the list which he presents at *On aqueducts* 39–63 does not miss a single *fistula* out; this in itself might end up worrying us, however, for ten of the twenty-five pipes whose dimensions he gives are 'no longer in use', a point which Frontinus drives home by repeating that very phrase (*in usu non est*) until it becomes almost a refrain.[69] In the process of demonstrating his extraordinary thoroughness, in other words, Frontinus includes material which he himself labels as being 'of no practical value', and thus prompts us to wonder if some of his details *are* in fact 'trifling' and 'inconsequential' and not nearly as valuable as his double bluff with the word 'leviora' has perhaps suggested.[70]

At the same time as flagging up the importance of detailed knowledge, then, Frontinus' inclusion of some apparently redundant material also questions it; and questions are raised too in his *On aqueducts* about the link which he draws between exhaustive knowledge and his ability to combat fraud. The final section of the *On aqueducts* (chapters 94–130) moves away from technical facts and figures and combines a potted history of the management of Rome's aqueduct network with the recitation of a number of laws connected with it. Along the way, Frontinus discusses the problems posed by fraud and decay more fully than he has done anywhere else: from *On aqueducts* 112 in particular, the illegal tapping of the system becomes an especially frequent topic.[71] Some of this fraud is talked of as being in the past, but most appears to be a problem which Frontinus and his men have to guard against in the present, and indeed the final chapter of his treatise

[69] See *Aq*. 41; 44; 47; 49; 51; 53; 55; 57; 59; 61, where 'in usu non est' concludes each chapter.

[70] In addition to hinting at the potential redundancy of some of the data at the end of his digression, Frontinus teases us with some superfluous information also at the beginning: the discussion which takes up *Aq*. 25), for example (where Frontinus debates the inventor and etymology of the *quinaria*), might strike us as not entirely *necessary*.

[71] See *Aq*. 112, 113, 114, 115, 120, 126, 128 and 130.

anticipates that it will continue into the future.[72] Following on from his glowing praise of Nerva for the wonderful transformation which Rome has undergone thanks to his care, in other words, we get a picture of the city's water supply system that is more in tune with the first image which his treatise conjures up: that of a network which is impressive on the surface but which has all sorts of cracks and defects if one begins to look closely.[73] In spite of his corrective and controlling rhetoric throughout the whole treatise, in other words, not even Frontinus' extraordinary *cura* (let alone Nerva's) or the vast body of knowledge which he has accumulated by the end are able to bung all the leaks and make the system watertight.

It is telling, too, that as he nears the close of his treatise Frontinus begins to lean on another type of authority as well as his own knowledge. In some ways his listing of the various laws which had been passed in connection with the maintenance of the aqueducts amounts to yet another display of his thorough command of the whole subject, yet another area which he has researched diligently and another body of data to add to the material which has been presented thus far. Legal knowledge, however, is slightly different from some of the other types of knowledge which Frontinus has been dealing with in his *On aqueducts*, for legal knowledge tends to borrow authority from other people; that is to say, while the authority which Frontinus derives from the calculations which he makes at *On aqueducts* 65–76 is entirely his own, since he alone did the research which underpins it, the authority which he acquires by quoting, for example, the resolution passed in the consulship of Quintus Aelius Tubero and Paulus Fabius Maximus about access to conduits which lie in private land[74] is on loan from the Senate, who enacted the law. Towards the end of his treatise, in other words, Frontinus begins to rely not just on his own actions but also on other authorities in his fight against physical rot and moral corruption.

Moreover, it is not just earlier legislators on whom we see him relying, for we also see him turning to contemporaries for support. In spite of his occasional reliance on laws enacted in the past, Frontinus does spend much

[72] Frontinus' final sentence at *Aq.* 130 begins *In reliquum* . . . ('As for the future . . .'), and though he 'hopes' that he will not need to prosecute any more people for water fraud, he indicates that he is perfectly prepared to should that become necessary.

[73] Although Frontinus ends *Aq.* 4–16 by comparing Rome's aqueduct network favourably with the 'luxurious pyramids' and the 'useless but well-renowned works of the Greeks' (*Aq.* 16), in the chapters immediately preceding this he has made it clear that several of the aqueducts are plagued by leaks and muddy waters (see esp. *Aq.* 11–15). Problems like those listed at *Aq.* 119, for example, therefore feel horribly familiar.

[74] *Aq.* 125.

of this final section, as ever, insisting on both his independence and his unmatched excellence. After he has finished describing some of the new powers and privileges conferred by the senate on *curatores aquarum* at *On aqueducts* 99–101, for example, he asserts his own superiority, claiming not to need any attendants himself, for when *he* goes out on his rounds he can trust simply to himself and to the authority given him by his emperor.[75] At *On aqueducts* 119, however, a slightly different picture emerges:

> For that reason, the *Curator Aquarum* ought to be furnished not only with the wisdom of experienced people but also with his own experience, and he ought to make use not only of architects of his rank but also call upon the trustworthiness and skill of many others, so that he might decide what must be done (*repraesentanda*) immediately, what must be postponed (*differenda*), and also what ought to be carried out by contractors and what by his own workmen.

The gerundives (*repraesentanda* and *differenda*) recall those at *On aqueducts* 2 (*facienda* and *vitanda*), and therefore invite comparison, but the message of this later passage is entirely at odds with Frontinus' earlier statement, for here Frontinus is positively encouraging a certain amount of reliance on the expertise of assistants. By the end of his treatise, in other words, the direct link between his own personal research and his ensuing power, which so much of his rhetoric has relied on, has not just been complicated; it has been revised and potentially overturned. With his ambiguous use of the word *leviora*, Frontinus has already eroded our confidence in the absolute necessity and value of some of his data; additionally, the continuing problems of fraud and dilapidation prompt us to question the idea that thorough knowledge equates to full control of Rome's waters; and now, finally, we begin to see Frontinus turning away from his newly acquired expertise to the wisdom and experience of others, and indeed promoting the latter.

CONCLUSIONS

This chapter has taken only a brief look at just one aspect of Frontinus' *On aqueducts*, but I hope that it is clear even from this that his account of Rome's aqueduct network is far more than a 'simple and truthful narration of the facts'. Further, and more importantly, I hope it has also become clear that it is neither straightforwardly supportive of Nerva, as several scholars argue, nor a simple self-promotion exercise. Rather, in my view, it is

[75] *Aq.* 101.4: 'As for me, when I do my rounds, my trustworthiness (*fides*) and the authority (*auctoritas*) given to me by my emperor will serve instead of attendants'.

fundamentally an exploratory text, which puts several things to the test without fully resolving the questions which it raises about any of them. Of course, it raises questions about many more issues than I have had the space to discuss here,[76] but in particular I have been keen to show how Frontinus interweaves his own authority as a knowledge-orderer with images of imperial authority and virtue in ways which do not finally determine but which rather interrogate the power of knowledge itself, the nature of Nerva's principate, and Frontinus' own position within it.

In some ways, as we have seen, Frontinus comes across as a model representative, perhaps even a mouthpiece, of the post-Domitianic era. At the same time, however, his treatise raises some doubts about it: for, although sections of the *On aqueducts* reflect the hope that under Nerva the corruption characteristic of Domitian's principate has become a thing of the past, we also learn what is risked when transparency and accountability become government aspirations. We see what it is like to lay open one of Rome's most important institutions to scrutiny: on the one hand, we may marvel at its size and complexity (as Christer Bruun suggests we will);[77] on the other, we may be a little discouraged by its ongoing problems with fraud and decay, and perhaps even begin to wonder how much really has changed. In addition, we have to contend with the problem that Frontinus ultimately withholds the most vital data from us. In other words, in spite of his very public demonstration of transparent accounting, we still have not got full transparency by the end of his treatise. This is not to say that Frontinus' account is necessarily subversive; the suggestion that it is *either* supportive *or* subversive is too crude a dichotomy. What we do see, however, is Frontinus not just reflecting but also reflecting *upon* several ideals central to the new administration.

This reflection is directed more introspectively too. As I noted earlier, Hodge believes that one specific reform which Frontinus is particularly interested in is the reappointment of senators to positions of authority. At times, we have seen Frontinus appearing to endorse this reform by taking his new responsibilities as *Curator aquarum* seriously and putting himself forward as a paradigm for others to follow. However, I think that we do not just see him prescribing but can also see him *wondering* how newly re-empowered senators should act. His play on the word *leviora*

[76] In particular, it raises important questions also about genre boundaries and ideals of Roman identity and *imperium*, as I explain in Weeks (2004).

[77] Bruun (1991) 18 argues that Frontinus' *Aq.* 'belongs, in part, to the same tradition as the *De architectura* of Vitruvius or even the *Natural history* of the Elder Pliny, in the sense that what these writers did was to present "Marvels of the World and of Roman Civilisation".'

and the questions which are subsequently raised about the relevance and currency of some of his knowledge, for example, prompt us to wonder if it *is* absolutely necessary to research one's field of work thoroughly, or if senators can operate much as they did in the past and rely on a circle of advisors to guide them in their post. Further, as well as considering the ideal conduct of such senators, we also see Frontinus exploring the extent of their reinstated powers. More specifically, his treatise prompts us to think about how senator-emperor relations in the new administration may be defined. Can prominent senators ever really be 'colleagues' of their emperor (as Frontinus' text occasionally suggests), or do they inevitably remain merely 'lieutenants'? And how far can one push the idea of partnership and collaboration before it begins to threaten the *princeps*' authority?[78] To put it another way, is the emperor still *summus* and Frontinus *levior*, and are those categories as straightforward as they seem?[79]

Ultimately, Frontinus leaves these questions unanswered, as his final chapter demonstrates. As I have mentioned, he ends his *On aqueducts* by explaining that in the past he sought the emperor's pardon for those who transgressed the laws but in future he will not be so lenient:

> As for the future, I hope that it will not be necessary for me to go to law, but I will guard the trustworthiness (*fides*) of my office even at the risk of giving offence. (*Aq.* 130)

In working Nerva (or possibly Trajan)[80] into his closing statements he implies that he is (and always has been) acting at the behest and on behalf of his emperor; at the same time, however, his mention of *fides* recalls his use of that word at the end of *On aqueducts* 101 (*fides nostra*) where (as we have just seen) he reasserted his relative independence. His final chapter thus draws attention both to the powers vested in him by his emperor and to his own (knowledge-based) authority, leaving us uncertain about their correspondence. On the one hand, much of his treatise (including this final chapter) teases us with the possibility that Frontinus' own authority outstrips that of his *princeps*, and even makes us wonder how close to being 'emperor' this *curator* might be. Although he hints that the answer is 'very',

78 It is worth recalling that Frontinus was Trajan's consular colleage at some point during the composition of his treatise.

79 In view not only of Nerva's elevation from the senate to the principate but also of Frontinus' and other important senators' probable roles both in Trajan's adoption and in the running of things at Rome after Nerva had died, the categories of emperor and senator, *summus* and *levior* must have seemed rather blurred at times.

80 The *imperator* mentioned at this point is not named, but since Frontinus is discussing the future here it might make more sense to understand him as referring to Trajan, for even if Nerva had not yet died, Trajan was probably already co-emperor when Frontinus wrote this.

he also positions himself ostentatiously from beginning to end as subordinate as well as similar and potentially superior, thereby underlining the gulf that lies between agent and emperor at the same time as interrogating and appearing to challenge it.

Frontinus' frequently ambitious self-positioning, then, explores the nature and extent of his authority without necessarily undermining his emperor's. He toys with the potential similarities between himself and his *princeps* in ways which prompt reflection not only on their collaboration and correspondence but also on what it is that sets them apart; and in so doing, he both draws on and invites reconsideration of the power and values that may be invested in certain types of knowledge. Among other things, we are left wondering what weight technical expertise really carries – what exactly its currency is, and what kind of authority it can confer (compared to authority from other sources). By showing others adding up incorrectly or deceitfully, and by failing to provide us with his own final calculations, Frontinus' account also threatens to undermine some of the qualities – specifically transparency and reliability – often associated with mathematics.[81] And though his organisation of facts and figures may be impressive on paper, we also discover (as we continue to encounter leaks and fraud) that highly systematised knowledge can sometimes be more rhetorically than actually effective.

A longer analysis would tease out some of these arguments further by looking at the different ways in which Frontinus' *Strategemata* and gromatic treatise explore, represent and order knowledge; for I have argued elsewhere that the exploration of knowledge is a guiding thread for his whole oeuvre. I have confined myself to his *On aqueducts*, however, to give just one illustration of how politicised as well as 'technical' Frontinus' writing can be, and how the ordering of knowledge within a text can be used to reflect on and take part in the ordering of self and society. Of course, I have only begun to peel back the layers of the complex relationship which Frontinus constructs in his *On aqueducts* between himself and Nerva, and also – if only implicitly – between himself and Trajan, who though not directly addressed hovers in the background.[82] I hope, though, that one important thing has become clear: that far from being an unambiguous

[81] On the association between mathematics and transparency and reliability in connection with Frontinus' *Aq.* specifically, see Cuomo (2000) 194, who argues that 'the emphasis put on measurement, and the overall mathematical outlook, provide Frontinus with a rhetoric of objectivity and accuracy in which to embed his presentation of himself as an honest and competent administrator'. My argument is that, while he exploits these associations at times, his treatise also undermines them.

[82] And I think that his relations with these two emperors need to be studied separately, not lumped together as they have been previously.

endorsement of the kinds of ideals we saw mentioned at the beginning of Tacitus' *Agricola*, Frontinus' *On aqueducts* is much more reflective than has been previously acknowledged of the uncertainty and political manoeuvring that characterised much of Nerva's brief principate. As I argued at the start, it was written not so much at the beginning of a 'glorious new era' as in a period of uncertain transition, when much *might* have changed, but quite what and how much was still being decided; as it tries out the new imperial rhetoric for size, and flexes Frontinus' muscles along the way, I think the *On aqueducts* is busy grappling with, and trying to shape, the answer to precisely those questions.

CHAPTER 9

Measures for an emperor: Volusius Maecianus' monetary pamphlet for Marcus Aurelius

Serafina Cuomo

> To the Greeks the Muse gave intellect and well-rounded speech; they are greedy only for praise.
> Roman children, with lengthy calculations, learn to divide the *as* into a hundred parts. (Horace, Ars Poetica 323–6)[1]

Like many clichés, Horace's sour depiction of Roman pragmatism has some truth to it – except that the Greeks were just as interested as the Romans in correctly dividing currency into parts. Metrology, the knowledge of measures of weight, length, volume and of money, was a major presence in ancient education and ancient life. The excavations of the Athenian *agora* have turned up some two hundred metrological objects.[2] A great many papyri from Hellenistic and Roman Egypt are accounts, bills, or contracts stipulating sizes or weights of things. Texts such as the so-called Athenian coinage decree[3] or the bilingual tax decree from Palmyra[4] are expressions of political and economic decisions variously translated into metrological policies.

The act of measuring creates a correspondence between things in the real world and symbols, be they numbers, units of measurement, or signs representing numbers or units of measurement. For instance, the Palmyra decree enforces a correspondence between, say, a sack of salt and two symbols: a number expressing its volume, expressed in terms of *modii* (units of measurement), and another number, tied by decree to the first one, expressing the tax tariff in terms of duty per *modius*. In the specific situation of the tax payment, the sack of salt 'becomes' its volume, that is, it exists in the

[1] This passage is mentioned in Dilke (1989) 50. Another *locus classicus* here is Cic. *Tusc*. 1.4. Translations are mine unless otherwise indicated.
[2] Lang and Crosby (1964).
[3] Meiggs and Lewis (eds.) (1969) 45 (450–446 BCE). See D. Lewis's and H. Mattingly's articles in Carradice (1987).
[4] *CISem*. 2.3.3913 (137 CE, in Greek and Palmyrene). See Matthews (1984).

form of a certain number – this translation is what makes the transaction possible.

Metrological documents are 'inscription devices', in a sense of the term which I adapt from Bruno Latour's work.[5] They are technologies that provide a representation of reality such that the representation becomes a necessary medium for interacting with that reality, because it (the translation, the 'something standing for something else') makes reality more manageable and more orderly. Examples of inscription devices include maps and graphs. Maps today can be said to have substituted landmarks as a means for orientation – in the case of the London Underground map, for instance, the coloured lines intersecting each other on paper are, for its users, by all accounts much more real than the real world of tunnels that 'are' London, underground. And graphs expressing, for instance, the outcome of a complex biochemical experiment substitute recourse to the mass of data produced by the various stages of the experiment. Once the graph is accepted by the scientific community, the reality it stands for, the mass of data, will in practice be obliterated. Inscription devices then write down (inscribe) reality in such a way that from now on we shall look at the inscription rather than at the reality it is supposed to represent. Indeed, according to Latour, the representation ends up 'being' the reality.

In some contexts, once the device is in place, the process that produced it disappears, like scaffolding, to the point that reality is now seen as *always* having contained within itself the inscription or the representation, independently from the human agents that produced it in the first place. What originates as basically a matter of useful convention, of approximation, of educated guess, is transformed into a law of nature. In a metrological context, a sack of salt may 'become' a certain number of *modii* to such a deep extent that one talks of a 'natural' or 'universal' system of measures, as if the number of *modii* had somehow been lurking within the apparently unordered mass of salt all along, waiting to be weighed and declared; just as Michelangelo's statues were said by the artist to be imprisoned in their block of marble, waiting to be freed from the superfluous matter in order to reveal their true shape.

Thus metrological documents can be a powerful tool to create a certain 'natural' order, and to make knowledge of certain taxonomies, relations and systems indispensable for correctly accessing reality. Both the Palmyra decree and the Athenian coinage decree are good examples of how metrologies – in particular metrologies having to do with money – serve as political

[5] Especially Latour (1987) 68–78.

instruments, and, given their extension of a certain order from a centre issuing the metrology over to a periphery, as instruments of empire.[6]

A thorough exploration of how these issues are articulated in the ancient world would warrant much more time and space than I have at present. Nevertheless, I shall attempt at least to whet the reader's appetite by focusing on one intriguing ancient metrological text, known as the *Division, as well as terms and signs of the parts in things which are reckoned by weight, number and measure* (*Distributio item vocabula ac notae partium in rebus quae constant pondere numero mensura*), which I shall henceforth refer to as the *Distributio*.[7] Its author, Lucius Volusius Maecianus (110?–166? CE) was a well-known and respected member of the Roman elite, widely appreciated for his expertise in the law. In addition to the *Distributio*, he wrote (as far as we know) sixteen books on legacies, fourteen books on criminal actions, and a treatise on the Rhodian law of the sea.[8] We have several inscriptions relating to him, including a *cursus honorum* or account of his political career; we know that he was a patron of the corporation of ferrymen, auxiliaries and record-keepers; and that he was appointed to the post *a libellis* (in charge of juridical petitions and co-ordinating responses to them) in 138 CE under Hadrian, and to two posts in the imperial bureaucracy (*a studiis* and *a bibliothecis*) under Antoninus Pius in around 150.[9] He taught law to Marcus Aurelius.[10] Between 159 and 161 CE he was prefect of Egypt; in charge of the corn supply in Rome by 161; a consul in 166, the year in which, according to some interpretations, he died. According to others, however, he was killed in 175, during an uprising in Alexandria.[11] It is abundantly clear from this that Maecianus' vicinity to three emperors, culminating in what appears to have been a direct and close relationship with Marcus

[6] See again Latour (1987) 215–57.

[7] The main edition of the *Distributio* (*Distrib.*) is by Hultsch (ed.) (1866), vol. II, vii, 17–22 and 61–71, also available in Seckel and Kübler (eds.) (1908) 408–18, and based on Mommsen's text (1857) 281–5 (*non vidi*).

[8] Rohden and Dessau (1898) 3.481–2; Seckel and Kübler (eds.) (1908) 408; Casavola (1980) 328–32; Fanizza (1982) 105; Kunkel (2001) 174–6.

[9] *CIL* 14.250=*ILS* 6174, found in a church wall at Ostia. See also SHA *Ant. Pius* 12.1. For similar examples of patronage of professional associations, see Clemente (1972); van Nijf (1997) 100–20.

[10] SHA *M. Ant.* 3.6. Marcus Aurelius and Lucius Verus referred to Maecianus as 'our friend' in a rescript (161–9 CE), *Dig.* 37.14.17. Marcus Aurelius also mentions Maecianus as someone close to him in a letter to Fronto, dated to between 140–3 CE, now in Fronto *Ep.* 1.74–8 (Loeb edition (1919), corresponding to 2.4 Naber). Marc. Aur. *Med.* 1.6, mentions a philosopher Marcianus that he resolves to listen to – the manuscript tradition is problematic, and some scholars think that Marcianus could be emended to Maecianus.

[11] A Maecianus is mentioned as having been killed by the army after conspiring against Marcus Aurelius in SHA *M. Ant.* 25.6 and *Avid. Cass.* 7.4, but not all scholars agree that he is the same as the jurist. For a death around 166 CE, see Fanizza (1982) 114.

Aurelius, puts him in a privileged position to comment on issues of order and knowledge, as indeed he appears to do in his work. We shall return to these points presently. I shall begin by describing the contents of the *Distributio*, and then tackling some of the issues arising from it.

THE TREATISE

The *Distributio* was written for Marcus Aurelius, probably in 146 CE when he was not yet emperor. Maecianus starts:

> I have often noticed, Caesar, that you are upset because you regard the subdivision of the *as*, which is necessary for inheritances and for many other things, as unknown. Thus, so that such a small thing does not impede your mind in any way, I have assessed how to set out both those parts and their names and signs. You can grasp then on the one hand the infinite subdivision of parts, on the other their utterly small names and signs.[12]

The tone is familiar, and there is an assumption that Marcus Aurelius is aware of Maecianus' other field of expertise (inheritance law), which is here hinted at, as if to bolster his credentials.[13] One of the main themes of the treatise is introduced: the relationship between various parts of the *as* (the piece of money), and their number and name and sign, small things that, amazingly, pin down the infinite.

Maecianus starts with the subdivision of the coin known as the *solidus*, also called *libra* or *as*. The three alternative nomenclatures are the prelude to a systematic taxonomy, where each part of the *as* is introduced in turn as a numerical fraction, a name (the formula is 'it is called' (*vocatur*) or 'its name is' (*nomen est*))[14] and a sign ('its sign is' (*cuius nota*)).[15] The *as* is subdivided into halves (*semisses*), thirds (*trientes*), fourths (*quadrantes*), sixths (*sextantes*), eighths (*sescunciae*), ninths (*unciae duae sextulae*) and twelfths (*unciae*) – the 'elements, as it were' of the first division (*distributio*).

[12] *Distrib.* 61.12–19: *Saepenumero, Caesar, animadverti aegre ferentem te, quod assis distributionem et in heredum institutione et in aliis multis necessariam ignotam haberes. Quare ne tam exigua res ingenium tuum ullo modo moraretur, cum partes ipsas tum vocabula et notas proponendas existimavi; et deprehendes distributionem quidem partium infinitam, oppido autem quam exigua vocabula et notas.*

[13] Fanizza (1982) sees a link between the *Distrib.* and Maecianus' work on *fideicommissa* (15), and between the *Distrib.* and Maecianus' post as *praefectus annonae* (112). Indeed, *Dig.* 35.2.32.4 (Maecianus from book 9 of the *Fideicommissa*, commenting on the *lex Falcidia*) contemplates a proportional contribution on the part of heir and recipient of legacy, in case they owe money as a result of a case involving the deceased. According to *Vindius noster*, the contribution should be proportional to their respective inheritances. Maecianus finds this idea both fair and logical (*aequitatem et rationem . . . habet*).

[14] On the importance of names for measures, see Heilbron (1990) 207–42, especially 214–15.

[15] In Pliny the Elder's discussion of Roman coinage, *nota* is the design on the coin: see *HN* 33.44–6.

Maecianus perhaps alludes here to Euclid's *Elements*, and hence to the fundamental, seminal nature of his present work. These elements, he says, 'preserve equality',[16] unless they are added or subtracted to each other, in which case they sometimes produce equal, sometimes unequal parts. For example, if you add a *sextans* to a *quadrans*, you obtain a *quincunx*, equivalent to five *unciae*, i.e., 5/12; or, if you add a *semis* to a *sextans*, you obtain a *bes*, i.e., 8/12.[17] Those are unequal parts. In general, the subdivisions of the *as* can be equal – a certain multiple of each subpart produces a whole *as*; for instance, six *sextantes* make an *as*, and so do two *semisses*, three *trientes*, and so on – or unequal. No multiple of a *quincunx* can produce a whole *as* – two will fall short of an *as* by a *sextans*, three will exceed an *as* by a *triens*.

The two parallel subdivisions are distinguished also by the fact that equal parts can only be characterised in one way, whereas unequal parts have several alternative definitions. For instance, a *semis* is, simply, one half, 1/2, and is obtained by dividing the *as* into two. A *bes*, on the other hand, can be obtained by adding 1/12 to 7/12, or by adding 1/2 to 1/6, or 5/12 to 1/4, or 1/3 to 1/3, and can be defined as eight *unciae* or four *sextants* or two *trientes* or even an *as* minus a third.[18] Even though unequal parts have a non-univocal nomenclature, and are characterised in a multiplicity of ways, their distinctive 'signs' (*notae*) remain the same. A *bes*, no matter how defined in terms of addition or multiplication of parts, is denoted by S =. The 'signs' of the unequal parts are in fact loaned from those of the equal ones: the sign of the *bes* reveals, and possibly privileges, one of its possible origins as the sum of a *semis* (denoted by S) and a *sextans* (denoted by =).

After the *as*, Maecianus moves to a 'less known, but not totally obscure', monetary sphere: that of the *uncia* or twelfth of an *as*,[19] which is posited as the mid-point between a tree of subdivisions upwards, towards the *as*, and an open-ended subdivision tree downwards, into ever smaller parts. The first subdivision of the *uncia* is along similar lines to that of the *as*: into halves (*semunciae*), thirds (*binae sextulae*), fourths (*sicilici*), sixths (*sextulae*), twelfths (*dimidiae sextulae*) and twenty-fourths (*scriptula*, also called *scripula*). In each case, as with the parts of the *as*, we are told what part of the *uncia* the unit is, what it is called and what its denoting sign is.[20]

The apparently easy symmetry, however, is immediately shattered, as Maecianus points out the complete arbitrariness of his own systematisation. 'These parts', he says, 'can be further divided into however many parts you

[16] *Distrib.* 62.13–15: *Haec velut elementa primae de asse distributionis aequalitatem servant.*

[17] *Distrib.* 62.15–18 and 22–4, respectively. [18] *Distrib.* 62.22–4; 63.28–31.

[19] *Distrib.* 64.12–17. [20] *Distrib.* 64.18–28.

want, but below them you do not find signs or proper names apart from those'.[21] The reason why some subdivisions have signs and names and others do not, is simply not given: Maecianus observes that, for instance, the *as* could be divided into eleven equal parts, but it is not. There are no names and no signs for elevenths or tenths, the way there are for ninths or sixths. In other words, the relationship between thing and name and symbol, which had seemed rather straightforward in the first subdivision and had acquired multiplicity in the second subdivision, has now been exploded – there are things that, although at some level they exist for us, do not have a name or a distinctive sign, unless they serve specific purposes.[22] Thus 'some accountants call the half *scriptulum* a *simplium*', or again a 1 per cent interest rate is endowed with a specific name.[23]

From mentioning interest and capital, the account moves on to a sort of interlude. Maecianus comments: 'The nomenclature of the *as* has to do with concrete things and bequests taken as a whole, while its division has to do with a description of the parts; it can also be applied to numbered wealth (*pecunia numerata*), which used to be in bronze, later started to be struck in silver, so that each silver coin had value depending on the quantity of bronze [it amounted to].'[24] A historical dimension is thus introduced.[25] For instance, Maecianus says that the *libella*, i.e., a tenth of a *denarius*, used to have the same function as the *as*, but is now associated with the past and the ways of the ancients (*exemplo maiorum*).[26] The *victoriatus* was once a foreign coin, 'as tetradrachma and drachma are today': originally from Illyria and Thessaly, it started to be issued in Rome between the First and

[21] *Distrib.* 65.13–15: *Has quoque partes in quantum libet dividere possis; verum infra eas neque notas neque propria vocabula invenies praeter ea.*

[22] *Distrib.* 65.19–66.14. Similar questions arise in contemporary literature (Gell. *NA* 2.26, on how there are more colours than there are names for them in either Greek or Latin) and jurisprudence (Neratius in *Dig.* 22.6.2 posits a contrast between the determinacy of law and the indeterminacy of facts subsumed under it; see discussion in Scarano Ussani (1979), 5–77).

[23] *Distrib.* 65.15–16: *Dimidium scriptulum audio quosdam ratiocinatores simplium vocare*) and 66.14–15, respectively.

[24] *Distrib.* 66.21–6: *Sicut autem assis appellatio ad rerum solidarum hereditatisque totius, divisio autem eius ad partium demonstrationem pertinet, ita etiam ad pecuniam numeratam refertur, quae olim in aere erat, postea in argento feriri coepit ita, ut omnis nummus argenteus ex numero aeris potestatem haberet.* I have translated *aes* throughout as 'bronze' for convenience, but in fact it could indifferently denote bronze or copper.

[25] Temporal adverbs abound; the Greeks and the Twelve Tables are mentioned at *Distrib.* 67.5–9. Gell. *NA* 20.1 features a discussion of the laws of the Twelve Tables, revolving around the changing value of the *as*: when the Tables were written, it was a remarkable sum, but no longer so in the second century CE. This sparks off the debate whether the ancestral laws were excessively cruel and are now outmoded. Historical awareness was a fundamental component of jurisprudence; see, e.g., Casavola (1980) 9–12; Bretone (1982) 10.

[26] *Distrib.* 70.16–30.

the Second Punic Wars, and gradually assimilated into 'normal' currency.[27] The parallels between appropriation of the coin and gradual incorporation of Illyria within the Roman state (effective by the early first century CE), sanctioned by the appropriation of the image of Victory, which gives the coin its name, are obvious.

Having gestured towards the potential infinity of micro-units lurking beneath the *as* and *uncia*, a whirlpool of nameless bits of money, a 'here be monsters' on the map of currency that he is drawing, and after the historical interlude, Maecianus continues his deliberate ordering. He launches into subdivisions of larger denominations, the *denarius* and the *sestertius*; the accounts are punctuated by direct appeals to the reader to denote each part with its sign and name.[28] The author declines to go into the subdivisions of the *victoriatus* or the *quinarius* because he does not know the Roman way of proceeding; he says, however, that the reader can work it out by analogy with other monetary units.[29]

Finally, Maecianus turns to weight, liquid and grain measures, which are organised along lines similar to money. In fact, units for weight and for money often share names and values, because at least in origin each coin was denoted by its weight. This relation was loosened and partly broken down when financial circumstances required devaluation of the currency.

The conclusion is tantalisingly fragmentary: 'The natural cause of the parts and of number remains unchanged, however much they may differ in name with each nation. The size of weights and measures is unstable, because its weighing and measuring out . . .'[30] It would seem that Maecianus was commenting, perhaps as a conclusion to his survey, on the complex relationship between the thing and its stand-ins (number that expresses its measure, coin that expresses its value and is also represented by a number, sign that denotes the coin that expresses the value of a certain quantity of a thing), and on the permanence and variability of these various classifications and correspondences – issues that have already emerged at several points in the text.

Using a coin is ultimately an act of trust in the correspondence established between the piece of metal and the thing one wants to buy, a correspondence represented by an equivalence of value between the price of the merchandise and the amount the coin is worth.[31] Money is in fact the prime example

[27] *Distrib.* 66.29–67.2. Cf. Mattingly (1928) 13–17.

[28] *Distrib.* 67.12–68.24; e.g., 67.14–15, 17–18, 19–20, *passim.*

[29] *Distrib.* 69.1–6.

[30] *Distrib.* 71.23–6: *Partium et numeri naturalis causa durat, quamvis nominibus apud quasque gentes different. Ponderis et mensurarum modus incertus est; nam eius dispensio ac dimensio . . .*

[31] Cf. Arist. *Eth. Nic.* 1133a20–b15. For assessments of the cultural and political background to classical Greek monetary economies, see von Reden (1997), Kurke (1999), Seaford (2004).

of a metrological object which has turned from representation of reality into reality – an inscription device where the signs meant to depict a thing are now taken as the thing itself. The Latin word for 'money', *pecunia,* is semantically multi-layered: fundamentally, and originally, it denotes sheep; it also comes to signify wealth in general, because those who owned a large number of sheep were wealthy; and hence also money, the translation of a certain number of sheep into a quantity of metal which can travel and be stored and exchanged in a way that sheep cannot. Money is quantified value, sheep or other bodies that have become a number: *pecunia numerata,* or 'reckoned *pecunia*'.[32] Of course the sheep are no longer important – what used to stand in for the object of value is now the object itself. In fact, money now signifies on the basis of other money rather than of things outside: the value of coins is expressed in terms of other coins, their universe of reference is self-contained and independent of its original meaning.

The story of Rome could be narrated as that of the changing relations between things and the inscription devices which stand in for them: a metrological story.[33] It is in parallel with the development of the empire, the accumulation of riches and various devaluations, regulations and deregulations, that *pecunia,* the substance for which coins stand in, becomes 'reckoned' (*numerata*) – it becomes inscribed and entangled in an intricate network of correspondences. The network pictured in Maecianus' short treatise is revealing of wider webs and ramifications, and it gives rise to several questions. How does the *Distributio* relate to contemporary metrological literature? How does it relate to jurisprudence, which was Maecianus' main field of expertise? And, finally, why should the emperor know about the subdivisions of the *as*?

THE TREATISE IN A METROLOGICAL CONTEXT

In many respects, the *Distributio* is quite unique in ancient metrology. The closest comparable account is a passage from the fifth book of Varro's *On the Latin Language* (47–45 BCE), where the author performs a sort of naming ceremony for all aspects of reality, including public offices and elements of religious ritual, bestowing Latin names upon them.[34] One of the underlying

32 *Dig.* 50.16.178 (Ulpian): 'The word *pecunia* consists of not only counted money (*pecunia numerata*), but absolutely all money, that is, all bodies: for nobody will doubt that 'bodies' are included in the designation of money'.

33 See Nicolet (1991).

34 Varro *Ling.* 5.36 (169–74), excerpted in Hultsch (ed.) (1866) II, 49–51. Unsurprisingly, given the time gap between them, Varro's and Maecianus' subdivisions overlap but do not coincide.

issues is, on what basis is this naming operation performed: what makes a certain name go with a certain thing? The section on money ('stamped (*signata*) *pecunia*') starts rationally enough: *as* comes from *aes*, the metal it is made of; *dupondius* from its 'double weights' (*duo pondera*), and so on. But by the time we get to a hundred *asses*, the stable, almost natural, connection between thing and name breaks down: '*ducenti* (two hundred) and higher numbers which are made by analogy do not indicate *asses* any more than they do *denarii* or any other thing'.[35] What this brings home, once again, is the ambiguous nature of money and of its relationship with reality.

For Pliny the Elder, the creation of money is just another of the crimes committed in the name of greed. In the section of his *Natural History* (published in 77 CE) devoted to precious metals, he tells us that initially the Romans used raw metal, then King Servius introduced stamped bronze, and then 'stamped silver' (*argentum signatum*) came after the victory over King Pyrrhus. Pliny points out the original relation between coins and 'stuff', reminding the reader of some weight-linked etymology: 'expenditure' derives from *expensa*, sums weighed out, and *pecunia* from the design stamped on the metal, which was an ox or a sheep.[36] Whereas Varro's order is, one could say, linguistic, trying to show that the relation between name and thing has a rationale, and Maecianus' account is anchored to simple arithmetic, Pliny's monetary map is shaped like a historical and moral narrative, where the explanation of various currency values, signs and names, is to be found in the circumstances of Roman history, down to the present.[37] Take this passage: 'Next according to a law of Papirius *asses* of half an ounce were made. When Livius Drusus was tribune of the plebs he mixed the silver with an eighth of bronze. The coin now called *victoriatus* was struck under the Clodian law; but previously this coin imported from Illyria was used as an article of trade. It is in fact stamped with a Victory, hence the name'.[38] The value and composition of coins are often changed by deed of Roman officers – the stability of the original relation between 'real' thing and monetary value, in its turn signified by the stable relation between the name of the coin and its composition or weight, both grow weaker with time, and are more and more subject to the vicissitudes and even whims of power. 'The emperors gradually made the gold *denarius* smaller,

[35] Varro *Ling.* 5.170, Engl. tr. R.G. Kent with my modifications (Cambridge, MA: Harvard University Press 1951).

[36] Plin. *HN* 33.42–45. Crawford (1985) 19–20; Burnett (1987) 15, and Savio (2001) 109–10 think Pliny's testimony is not to be taken literally.

[37] Plin. *HN* 33.44–7. [38] Plin. *HN* 33.46.

and most recently Nero had forty-five *denarii* stamped from a pound of gold'.[39]

The third 'metrological' work I shall discuss is Columella's *On agriculture*. While he does not talk about currency specifically, Columella discusses measures of land; his work begs comparison with Varro's treatise on the same subject, both being repositories of useful knowledge and at the same time of ethical guidelines for the estate-owning members of the upper orders. Showing an attitude quite at odds with Maecianus, Columella states that metrological matters do not really pertain to him, but are rather the job of surveyors. He compares his role as a farmer to that of an architect, who plans the building project, but delegates measuring and cost-calculating to other people. Nevertheless, he proceeds to provide a discussion of measurements for the benefit of his reader and friend Silvinus – even architects, after all, have to be acquainted with the 'account of measurements' (*ratio mensurarum*).[40] Columella thus lists units of land measurement, drawing on Varro on a couple of occasions, and, like Varro, occasionally providing etymologies and local variations in nomenclature and subdivisions. In a manner analogous to that of Varro's piece on money, the temporal dimension sneaks in, as a factor that loosens the relation between thing and 'stand-in' for the thing:

> [F]ormerly the *centuria* was so called because it contained 100 *iugera* [approximately 2/3 of an acre], but afterwards when it was doubled it retained the same name, just as the tribes were so called because the people were divided into three parts but now, though many times more numerous, still keep their old name.[41]

After a further disclaimer, in which Columella says that the smaller fractions of the *iugerum* are superfluous because no transaction depends on them, he goes on to list subdivisions of the *iugerum* anyway, from its smallest fraction, the half-*scripulum*, to the *iugerum* itself, which is explicitly compared to an *as*.[42] The next section applies these measurements to pieces of land of different shape, used as *formulae*.[43] In other words, a further level is inserted between thing and measure: the geometrical representation of a field. A piece of land becomes a geometrical figure, becomes a certain quantity of *iugera*; and from there it can come to represent a certain quantity of any other units of measurement, even non-Roman ones, even those no longer in use, provided one can establish a relation between those

39 Plin. *HN* 33.47.

40 Columella *Rust.* 5.1.2–4. Engl. tr. E. Heffner (London/Cambridge, MA: Heinemann and Harvard University Press 1954).

41 Columella *Rust.* 5.1.7.

42 Columella *Rust.* 5.1.8–12.

43 Columella *Rust.* 5.1.13–2.10.

and the *iugera*. The well-established, solidified, inscription device allows connections and comparability across time and space.

The surveyors Columella refers to often feel it necessary to impart some metrological knowledge onto the readers of their treatises. The authors in the collection we now call the *Corpus agrimensorum romanorum* frequently and explicitly equate measure and order in a wider political sense, and their closeness to the centres of power can be argued from their biography, as in the case of Frontinus, or from their own statements, as in the case of Balbus. The job of the surveyor consisted to a great extent in negotiating the metrology of a territory: converting un-measured pieces of land into measured ones, converting non-Roman or pre-Roman measures into Roman ones, or juxtaposing them to Roman ones, and making sure that the relation between measure and thing remained stable through the use of boundary stones, indeed making it stable by producing maps.[44]

A Roman surveyor from probably the first century CE states that official measurements should be given both in Roman and in local units. But 'if there was a dispute whether a *versus* [a Dalmatian unit of measurement] had 8,640 feet, confidence (*fides*) could nevertheless be placed in the *iugera* [. . .] When the *iugera* have been recorded, even if something can be <done> using local terminology, a system involving *iugera* will be inherently reliable for us'.[45] The desire for stability and systematisation of measures unsurprisingly tends to privilege Roman standards, but is compounded with the recognition of local realities and local networks of consensus: 'each region follows its own practice so that a trustworthy method can be agreed upon'.[46] In general, 'we must watch out <for the practices of> different regions in case we seem to be doing something unusual. For our profession

44 E.g., Frontinus, *De limitibus* 10.16–25 (ed. and Engl. tr. Campbell (2000)); Hyginus 1, *De condicionibus agrorum* 88.22–90.9 (Campbell); Hyginus 2, *Constitutio limitum* 136.28–38 (Campbell); Balbus, *Expositio et ratio omnium formarum* 204.19 (Campbell); *Deformatio* 240.15–22 (Campbell); *De mensuris agrorum* 270.10–34 (Campbell); *De agris* 272.22–5 (Campbell); Marcus Nipsus, *Podismus* 296.4–26 (ed. F. Blume, K. Lachmann, A. Rudorff, *Gromatici veteres*, Berlin: Reimer 1848–52); *Mensurarum genera* 339.1–340.8 (Blume); *De mensuris* 371.1–376.13 (Blume); [Boethius], *Demonstratio artis geometricae* 407.1–408.2 (Blume). *De mensuris agrorum*, *Mensurarum genera*, *De mensuris* and the pseudo-Boethius would warrant further study, but, given their late date, not as part of this paper. All quotations from Campbell's edition reproduce his translation. For Frontinus, see Alice König's chapter in this volume.

45 Hyginus 1, *De condicionibus agrorum* 88.23–90.12, especially 88.25–32. Cf. also *ibid.* 96.23–24 (Campbell).

46 Hyginus 1, *De generibus controversiarum* 92.24–25. See also *Ordines finitionum. Latinus et Mysrontius togati Augustorum auctores. De locis suburbanis vel diversis itineribus pergentium in suas regiones* 254.13: 'In many lands trust (*fides*) is required in different markers' (Campbell).

will retain its integrity if we also conduct our investigations principally according to the practice of the region'.[47]

Sometimes the similarities with Maecianus' small treatise are striking, for example when Siculus Flaccus talks about subdivision of the main Roman unit of measurement for land areas: '*Centuriae* do not contain 200 *iugera* in all regions. For in some we find 210, in others 240. So this matter also will have to be carefully examined, since it follows that *limites* will not be of an equal length between the boundary stones if *centuriae* have more than 200 *iugera*. For example, if a *centuria* has 240 *iugera*, it follows that there will be 24 *actus* from stone to stone along one *limes*, . . . and 20 *actus* along the other . . . I have discovered that in some lands that had been divided, although the *centuriae* contained 200 *iugera*, they had not been given equal lengths of 20 *actus* between the marker stones, along the *limites*. In the territory of Beneventum there are 25 *actus* along the *decumani*, and 16 along the *kardines*. Nevertheless, 200 *iugera* are enclosed by this type of measurement, but square *centuriae* are not thereby produced'.[48] As in the case of the *as*, there can be various subdivisions, and they can be related to the passage of time or political events in certain regions: the surveyor, the administrator and, by extension, the emperor have to be aware of these fluctuations in the relations between things and measures.

Balbus' treatise *The description and account for all shapes* (*Expositio et ratio omnium formarum*) is again a foil to the *Distributio*. Its declared aim is to set out the basics of the surveying profession, starting from measurements, i.e., 'anything that is defined by weight, capacity or by judgement', although Balbus is thinking essentially of measures of length.[49] He proceeds to expound the twelve names of the measurements in use, and some of their subdivisions: for instance, a *sextans*, also called *dodrans*, encompasses three *palmi*, nine *unciae* and twelve *digiti*. The objects of Balbus' account start in a two-dimensional world, as it were, and expand into further dimensions: the 'concave square foot' (*pes quadratus concavus*), for instance, 'has the capacity of an amphora of three *modii*'.[50] In fact, it is when explaining this expansion that he invokes the real world behind the intricate web of names, equivalences and subdivisions: 'Measurements are taken in three ways, by length, by breadth, and by height. That is, a straight line, a plane figure,

[47] Hyginus 1, *De generibus controversiarum* 94.25–7. Cf. also Siculus Flaccus, *De condicionibus agrorum* 104.34–106.13, 17–18, 108.20–21, 26–27, 114.34; Agennius Urbicus, *De controversiis agrorum* 20.16–21, 30.31–33, 34.19–21, 36.11–12, 40.4–6, 42.10–13 (Campbell).

[48] Siculus Flaccus, *De condicionibus agrorum* 126.6–17 (Campbell).

[49] Balbus, *Expositio* 206.5–6 (Campbell).

[50] Balbus, *Expositio* 206.8–27, in particular 27 (Campbell).

and a solid figure. A straight (line) is where we measure the length without the breadth, for example, lines, porticos, running-tracks, length in miles, the length of rivers, and similar things. A plane (*planum*) is what the Greeks call *epipedon*; we refer to "level feet" (*pedes constrati*).'[51] A correspondence is established between a thing (a river), the geometrical representation of that thing (a straight line) and what we call that representation (the name of the measurement, in Latin or Greek). Whereas we cannot really manipulate the real thing at will, we can operate on its representations, especially on the measurement, which can be further ordered according to divisions and correspondences. This aspect becomes crucial in the 'taming' of wild territories, which are subsumed, if only in an imperfect and approximate way, under a geometrical representation – are inscribed in the various senses we have given this word – and thus domesticated and made part of the empire.

Finally, we have archaeological and epigraphic evidence on the regulation of weights and measures.[52] One of the duties of the official known as an aedile was to inspect weights and measures in use in a market to prevent frauds, and it is well known that officially approved weights and measures had to be used in cities across the Empire: this is testified by archaeological finds of measuring tables (*mensae ponderariae*) in many marketplaces,[53] by inscriptions[54] and by legal rescripts such as the following: 'If a seller or a

[51] Balbus, *Expositio* 206.34–7 (Campbell).

[52] A further category of metrological texts is papyri dealing with units of measurement, including monetary units. E.g., *PSI* 763 (first century BCE, provenance unknown); *PLond.* 2.265 (first century CE, ed. F. G. Kenyon, London: British Museum Publications 1898); *POxy.* 9 *verso*, 669, 3455–60 (ranging from the first to the fourth century CE); *PRyl.* 64, 538 (second to fourth century CE); *PVindob.* G 26012 (third to fourth century CE) in Sijpesteijn (1980). See also Boyaval (1971) and Pintaudi and Sijpestijn (1989) 114–15, relative to Ammonios' notebook, Louvre MNE 911, probably sixth century. The *Distributio* often shares with them a didactic approach, the familiar tone, the frequent direct appeals to the reader in the second person singular, the exhortations to 'say' or 'write', as a sort of exercise after the author has shown the reader how to do something. Interestingly, some of the techniques of subdivision found in the *Distributio* had been in use since ancient Egyptian times. As is well known, Egyptian arithmetic used parts (what we would today call fractions) of the type 1/2, 1/3, 1/4 and so on, the only exception being 2/3. This also meant that each part which was not 2/3 and did not have one as a denominator had to be expressed in terms of the sum of 'recognised' fractions. For instance, 3/5 was denoted by 1/2 plus 1/10, but could also be denoted by 1/5 plus 1/5 plus 1/5. As in the case of Maecianus' 'unequal' subdivision of the *as*, there were several alternative sequences for each fraction, some of which seem to have been preferred to others. Part of the calculator's skill, and thus of the training he received, consisted in doing these sums and in choosing from among the many alternatives the one best suited to the purpose, see Gillings (1972) 45–50; Harrauer and Sijpesteijn (1985) 151–64.

[53] See Frayn (1993) 108–14, 123; Corti (2001), both with further references. Particularly interesting is the *mensa ponderaria* from Pompeii (*CIL* 10.793), which is inscribed with Latin indications of measures and weights but still shows traces of the previous, Oscan, measures, which have been erased.

[54] E.g., *CIL* 9.2854 (from Histonium in Puglia, no date given); *CIL* 10.6017 (Minturno, ca. CE 40); *CIL* 11.6375 (Pesaro, no date given) – all three refer to the supervision of metrological standards in

buyer tampers with the publicly agreed measurements of wine, corn, or anything else, or deceives with malicious intention, he is sentenced to a fine double the value of the thing in question; and it was decreed by the deified Hadrian that those who had falsified weights or measures should be exiled to an island'.[55]

In sum, the *Distributio* can be seen against a wider background of metrological texts and indeed objects: it is part of a strong interest in standardisation, which I take to mean establishing a stable connection between thing and measure. Once a standard is set in place, the universe of inscription devices can be considered self-sufficient and self-referential, reality with its messiness and disorder can be black-boxed, information can be effectively stored, communicated and transported. The process is not simple, and is never completely successful: it always appears to be the fruit of negotiations between Rome's present and her eventful past, and between the different cultures present within the empire and the allegedly dominant one.

THE TREATISE IN A LEGAL CONTEXT

Another interesting context for the *Distributio* is offered by contemporary legal literature. Hadrian and Antoninus Pius gave great importance to overhauling the bureaucracy, and reorganising jurisprudence. Maecianus' experience both as administrator and as jurist puts him in a privileged position as observer and participant in this process. Unfortunately, his own contributions to the law are no longer extant in their original form, having been selected and collected in Justinian's *Digest*.[56] Some fragments are, however, rather revealing. In one of them, Maecianus refers to the rationale (*ratio*) underlying a decision: 'Slaves who are pre-adolescent are excepted . . . But the legate Trebius Germanus ordered even a pre-adolescent to be executed, and yet not without reason'. This has been seen as an appeal to the common 'principle' or even 'rationality' at the basis of law and administration, which is held to be more cogent than rules explicitly laid down.[57] Again, Maecianus wrote on the *lex Falcidia*, which granted free

terms of *aequitas*. For an example from the Greek world cf. *IG* V, 1.1156 (from Gythium in Laconia, second century CE).

[55] Modestinus (third century CE) in *Dig.*, 48.10.32. See also Paul (early third century CE) in *Dig.* 4.3.3 and Ulpian (early third century CE) in *Dig.* 19.1.32 on using false weights.

[56] See, e.g., *Dig.* 29.5.14; 32.9; 32.11.2; 32.11.15; 32.13; 32.15; 32.17; 35.1.86; 35.1.91; 35.2.28; 35.2.30; 35.2.32; 37.14.17; 40.5.42.

[57] *Dig.* 29.5.14. See Fanizza (1982) 115–17. Scarano Ussani (1987) 34–5, 114, sees a foreshadowing of Salvio Giuliano's teachings, in their turn based on the notion of common interest and aimed at maintaining social and political order.

power to dispose by bequest of up to three quarters of one's substance, thus: 'Suppose that Titius's share is reduced in a legacy of twenty through the Falcidian law, Titius himself being charged to give five to Seius; . . . a proportional reduction is to be made in Seius's five comparable to that in Titius's twenty. This decision is both more just and more reasonable.'[58] A slightly different approach is revealed in another fragment, on the Rhodian law of jettison: 'Volusius Maecianus, *From the Rhodian Law*: Petition of Eudaemon of Nicomedia to the Emperor Antoninus: "Antoninus, Ruler and Lord, we were shipwrecked in Icaria and robbed by the people of the Cyclades". Antoninus replied to Eudaemon: "I am master of the world, but the law of the sea must be judged according to the sea law of the Rhodians when our own law does not contradict it".'[59] Finally, on the topic of money, Maecianus, again commenting on the Falcidian law, deals with the complications of legacies and bequests in cases where a bequest has been specified in kind or in weight, number or measurement (as in 'three talents of silver', rather than 'the silver which I have in the warehouse'), and what happens when the goods become damaged before the heirs come into them. The question, indirectly, is again about the dialectic between *pecunia*, a valuable body, and counted ('numerata') *pecunia*: not just 'coined' money but (to stretch the sense) valuables that have been expressed 'by weight, number and measuring' (*pondere numero mensura*).[60]

Maecianus' fragments encapsulate a number of questions that were being debated in second-century law. One is the ontological status of money, and how that affects everyday transactions. For example, Gaius considers the case of whether, in a sale, the price agreed must be in counted money (*pecunia numerata*) or can be in other items, such as a slave, a piece of land or a toga. Gaius' teachers thought that it could, because they thought that since time immemorial (and Homer is quoted in Greek to this effect) an exchange (*permutatio*) is a sale. The authorities of Proculus say, however, that exchange and sale are different: 'In particular, they think it impossible in an exchange of goods to determine which thing has been sold and which given as price; they see it as absurd, again, that each thing should be regarded as both sold and paid as the price'.[61] The question, it seems to me, revolves around whether 'counted *pecunia*' is the only stable way to effect a transaction. According to the second opinion, the lack of a measure throws the whole process into confusion. *Pecunia* by itself is disorderly and

58 *Dig.* 35.2.32.

59 *Dig.* 14.2.9. The central passage can also be translated: 'I am master of the world, but the law is mistress of the sea'; cf. Manfredini (1983).

60 *Dig.* 35.2.30.3–5, from book 8 of Maecianus' *Fideicommissa.*

61 Gai., *Inst.* 3.141.

difficult to manage; its numerical stand-in has, in a sense, become more real than the real thing. But we should not forget that there are contrasting opinions here.

A remarkable passage by Paul (late second to early third century CE) states the terms of the question even more explicitly: 'Buying and selling started from exchange. Once in fact there was no coined money (*nummus*) and it did not happen that one thing was called "wares" and the other "price" . . . But since it did not always and easily happen that when you had something which I wanted, I, for my part, had something that you were willing to accept, a material was chosen, the official and permanent assessment of whose value would remedy the problems in exchanges thanks to the uniformity of quantity. That material, struck with an official figure, demonstrates its utility and dominion *not so much on the basis of its substance as of its quantity*, so that no longer are the things exchanged both called wares but one of them is termed the price' (emphasis added).[62] The fact that money now has a value that depends not on its substance, but on a convention, ratified by the official figure struck on it, is the result of what we have called an inscription process. *Pecunia numerata* has almost become *the* reality by this time, and the jurists, including Maecianus, are engaged in reconstructing the genealogy of their present situation.

In practice, a lot of the money that the *Distributio* discusses only existed in the form of signs and names. It has been observed that small units of currency would have been little used in antiquity because the *as* 'would have been adequate for many of the purchases of everyday life'.[63] A cursory look at what we know of actual prices from the Roman Empire reveals, in the East, figures of 1/24 of a *denarius* and 1/48 of a *denarius* for bread.[64] The graffiti in Pompeii mention *uncia* and *semiuncia*, even though the context seems jocular,[65] more frequently *semisses*[66] and most frequently of all *asses*. Often there are numbers, or even itemised bills, with no indication of what unit is being referred to. A couple of inscriptions[67] might have the symbol for *scrupulus*. There are also occurrences of what could be a *sicilicus*, and perhaps of *quadrans*. One could debate how representative these scattered testimonies are, and how tentative our reading of currency symbols, but overall there does seem to be a mismatch between the small bronze that may have been in circulation in antiquity and our finds of small bronze, a gap

62 *Dig.* 18.1.1 (Paul *Edict* 33). Paul continues with a discussion which is almost identical to the one in Gaius, *Inst.* 3.141, and ends up siding with Proculus' school.

63 Howgego (1992) 19. 64 Sperber (1974) 118–19. 65 *CIL* 4.4227.

66 E.g., *CIL* 4.8561, 4.8565, 4.8566, 4.8789, 4.8968 (in Greek with price in Latin).

67 *CIL* 4.2029, 4.2030.

wider than in the case of silver and gold coinage. This is hardly surprising, if we consider that smaller coins are found as isolated and casual finds rather than as part of hoards. Their lesser value means that they would not have been treasured, and not actively sought if lost.[68]

Even providing for these accidents of survival, if one examines the distribution of Roman bronze coins in the Western Empire from 81 to 192 CE, the presence of 'small bronze' (anything smaller than *asses*, mostly *quadrantes* and a few *semisses*) is negligible. The *quadrans* has been found rather sporadically, more on Italian sites than in the northern provinces. The only surviving examples of *semuncia*, *quartuncia*, *sextans*, *triens*, *quincunx*, and *bes* coins date from the third or second century BCE.[69] Overall, the production of *asses* declines and that of sesterces increases from the first to the second century CE. By Trajan's time, the smaller coins (nothing smaller than *quadrantes* in any case) may have disappeared because of inflation. In the Eastern coinage, there are more often smaller coins (obol and smaller, down to *chalkos*), but even then, at least in the case of Egypt, the frequency of the smaller bronze coins seems to decline from around the time of Hadrian.[70] The obol seems to have been the smallest unit actually used in tax receipts and private accounts in Egypt, but there is also second-century evidence from Karanis that a very small unit, the *dichalcon*, was in use in tax receipts and ledgers, probably as an accounting device.[71]

At least in the case of the subdivisions of the *uncia* Maecianus is therefore talking about 'symbolic', accountant money, used in calculations, not about 'real' money.[72]

MONEY, MEASURE AND THE EMPEROR

There is a practical aspect to measured wealth: if one agrees on standard weights or lengths, or at least on exchange systems, transactions and translations are made possible. Metrology allows control, a certain degree of order

[68] Savio (2001) 160, 186.

[69] Mattingly (1928); Hobley (1998), esp. 12–14. For money units smaller than the *as*, see Crawford (1985) 60–5; Burnett (1987) 95–7. On the problems of calculating coinage output, see Howgego (1992); Duncan-Jones (1994) 95–247; Savio (2001) 50, 303–8. Also useful are Strack (1937); Sear (2000).

[70] West and Johnson (1967) 18–20.

[71] West and Johnson (1967) 17–18, 20–1. Rathbone (1991) 318–30 describes a system (Egypt, third century CE) which is basically monetised without necessarily using actual coins.

[72] Mrozek (2001) 9, 94–101 argues that the 'abstractness' of money was evident since at least late Republican times, because people invested and made debts, sometimes debts so huge that they could not possibly be paid back. A potentially infinite debt cannot correspond to actual, material, amounts of money. There was the idea, thanks to debit, interest and profit (*faenus*) that money, even when expressed in the language of money units, does not necessarily exist in the form of coins.

and centralisation. On the other hand, the significance of measures lies in the fact that they are symbols. Because the relation between things and their representation is not immediate or univocal, any decision concerning that relation is invested with a special authority,[73] which can be religious and/or political. For instance, in the Middle Ages in parts of Europe measures of grain were established by the king, supported by his God-given power, and they acquired a sacred character; breaking them was akin to sacrilege.[74] Alternatively, decisions about measures can be based on science, and justified as reflecting nature itself: Hyginus, a probably first-century CE land-surveyor, argued that the *kardo* and *decumanus*, two perpendicular lines which were the main reference points when laying out a land-division grid, were grounded in nothing less than the heavens and the *ratio* of the universe.[75] Or again, expediency or utility can be invoked in the choice of one metrological network over another: this seems to have partly motivated Frontinus' decision to use the *quinaria* as standard over the many other possibilities, because it was the best known, and its subdivisions the most accurate.[76] The difference between recourse to utility and recourse to science is that the former tends to acknowledge the man-made, artificial or conventional aspect of the decision, which is presented as preferable given the circumstances, hence somewhat arbitrary, rather than as the most true or rational thing to do.

We can try to reconstruct what Maecianus may have thought on the issue. Perhaps his position was contained in the missing part of the treatise. In the extant text, he does not seem to take a stand on the question of whose authority is behind the money system he describes. He points out historical dimensions, the presence of economic interests, hints at local differences, but the fact that, for instance, the *as* is divided one way rather than another is not justified on the basis of nature or even of expediency: it is just given as a fact. Then again, Maecianus reveals the tentativeness of his arrangement at more than one point: the treatise is the result of his assessment or opinion (*existimavi*, 61.17), and his system is one of several possibilities. The particular order imposed on money may well be a convention, the result of a choice, a human decision.

Analogous issues were being debated in the legal literature of the second century CE. The epistemic status of jurisprudence itself was questioned: was

73 See Kula (1986); Hocquet (1992); Porter (1995); Pedroni (1996); Grimaudo (1998); Ercolani Cocchi (2001).

74 See Kula (1986).

75 Hyginus 2, *Constitutio limitum* 134.5–6 (Campbell). This kind of position is very common in modern (post-1800) times: see, e.g., Mirowski (1992); Alder (1995); Schaffer (1995).

76 Frontin. *Aq.* 1.26–37. See Alice König in this volume.

it *ars* or *scientia*? Consequently, could it aim at certainty, or was it bound to approximation; were its practitioners technical experts or did they have to derive their authority from their political clout? Crucially, what did the law rest on?[77] Various possibilities were mooted. Tradition was one ground for justification, and one that seems to have been quite powerful in various areas of Roman culture, although it was far from being unquestioned, especially in the period we are talking about. The notion of 'use value' (*utilitas*), often invoked in extant decisions, was far from self-explanatory: the common good was not pellucid, but had to be determined by someone with some sort of authority. The existence of a rationality internal to the law, ultimately congruent with human rationality, and reflecting, if imperfectly, the orderliness of the universe, was also a possibility. We have looked at Maecianus' own mentions of *ratio* (reason, 'rationale'). In fact, attempts to define a *ratio* for law (a *ratio iuris*), and to use it as an underlying, unifying principle have been traced in Roman jurists from at least Pomponius, a contemporary of Maecianus, to the third century CE. Especially in Gaius' work, there is often a juxtaposition of two *rationes*, a natural one and a civic, political one, which ideally should work together.[78] When that is not possible, it is suggested that nature should prevail.[79] Celsus clearly states: '[Testaments] which are forbidden by nature are not endorsed by any law'.[80] Underlying this distinction is the notion of a 'law common to all peoples' (*ius gentium*). Its identification with a sort of 'natural law' (*ius naturale*) is debatable, but, even if the 'law common to all peoples' is the product of convention, then it is a more natural and universal convention than that at the basis of the 'civil law' (*ius civile*), which only binds a specific community.

A good example of the debate is the case of the entitlements of the head of the household (*pater familias*). Jurists of the second and third centuries CE were very aware that the power exerted by the father in a Roman household was a peculiarity of Roman law, i.e., part of their 'civil law', but it was not

[77] Casavola (1980) 54–7; Bretone (1982) 42–3, 268–70; Scarano Ussani (1987) 21–5 and (1997) parts 1 and 3; Ducos (1994).

[78] As they do in *Dig.* 3.5.38, by Gaius (mid- to late second century CE). Cf. also Gai. *Inst.* 1.1; 1.89; 2.66; *Dig.* 8.2.8; 9.4; 13.6.18.2; 41.1.3; 41.1.7.7; 44.7.1.9 (all mentioning *naturalis ratio*, all by Gaius).

[79] See, e.g., Gai. *Inst.* 1.158, *ratio civilis* and *civilia iura* v. *naturalia iura*; *Dig.* 4.5.8, *civilis ratio* v. *naturalia iura*; *Dig.* 7.5.2, *naturalis ratio* v. the authority of the senate; *Dig.* 41.1.1, where the *ius gentium*, based on *naturalis ratio*, is declared older than the *ius civile*, 'being the product of human nature itself'. All the *Dig.* texts mentioned are by Gaius.

[80] *Dig.* 50.17.188.1. Celsus also lived in the second century CE. See also Nocera (1962); Levy (1963); Stein (1974); Archi (1981); Scarano Ussani (1979) 198–9, 200–5 and (1987) 17–20; Bretone (1982) 32–3 and (1989), esp. 323–51; Ducos (1994) 5160–6. For contemporary discussions on whether words are the product of nature or convention, see, e.g., Gell. *NA* 10.4.

found among other peoples, i.e., not in the 'law common to all peoples', and thus arguably it was not based on nature. Its main strength was tradition, but in the course of the second century emperors like Hadrian seemed increasingly willing to put tradition on the side in the name of a different conception of what was legally the right thing to do. On the imperial scene, the sphere of application of any civil law to peoples other than the one that created it required some sort of justification: in metrological terms, in a situation where different units of measure exist, in order to establish a standard, appeal has to be made to something, be it practicality or the claim that the chosen standard is more rational or more natural than the others.

In sum, I would argue that Maecianus' approach to the subdivisions of money reflects contemporary legal debates. Jurists were concerned with the ambiguous nature of money; they, and Maecianus as one of them, reflect a situation where at least to some extent the link between thing and symbol has been problematised, weakened or even severed. Again, jurists were trying to put order in the law, and ground it firmly on a basis of nature, rationality or convention, creating standards, mapping out relations, cases and subcases; Maecianus was trying to do the same in the domain of money. In both cases, history and individual circumstances often got in the way; in both cases, the presence of a supreme authority loomed large in the background: the emperor.

Where did the emperor stand in relation to the law: was he himself subject to it? The question had been discussed throughout the first century and seemed to be more or less settled in the second century CE, with the emperor emerging as the ultimate legal expert.[81] Complications remained, however, as shown by a deliberation process about the inheritance rights of patrons towards freedmen reported by Ulpian and involving Marcus Aurelius and Lucius Verus, Maecianus himself and other jurist friends: 'We . . . followed this opinion (*sententiam*) when we dispatched a rescript in answer to the petition of Caesidia Longina; but likewise, our friend Volusius Maecianus, careful custodian of the civil law, apart from his long and well-grounded expertise in it, was induced by respect for our rescript to declare in our presence that he did not think he ought to say otherwise. But when we discussed the matter more fully with Maecianus himself and other legal experts also friends of ours, who had been summoned, it seemed rather that neither the words nor the meaning (*sententia*) of the law nor the praetor's edict excluded the grandson from the property of his grandfather's

[81] Bretone (1989) 234–7.

freedman; and that such was the view of several legal authorities too, but that it had also been the opinion (*sententia*) of our friend, the most honourable Salvius Julianus.'[82]

Several factors are in play here: legal expertise on the part of various individuals, all reassuringly denoted as 'ours' (*noster* or *nostri*); the literal and not strictly literal interpretation of the law; the edict of a praetor who would have been a member of the Senate and possibly the representative of a political authority other than that of the emperor; the emperors' own opinion. There has been some debate about Maecianus' demeanour in this case: for some, he was being too subservient to the decision of the emperors, for others, he was just being professional, the perfect lawyer-bureaucrat with no political identity, since the imperial will was in fact legally binding.[83] In any case, it is clear that behind the amicable appearances, ever since Augustus the emperor was the gatekeeper on legal expertise: without his sanction, no expert had the authority to express binding legal opinions.[84] In the passage above, the emperors mediate the various sources of authority. *Their* expertise consists in eventually choosing whose expertise ought to be applied to the case in hand.

Rather than having the debate about the origin and justification of legal or metrological order, nature (*physis*) vs. culture (*nomos*), explicitly transferred onto himself, then, the emperor emerges as a figure who stands above others. Take the case of Maecianus' fragment on the Rhodian law: *because* Antoninus Pius is the acknowledged master of the universe, he can sanction the application of a legal order, the law of the sea, other than the normal one. Again, some legislation introduced by Hadrian and Antoninus Pius seems to point in the direction of greater humanity towards women, children oppressed by paternal right (*patria potestas*) and slaves. This has been seen as a reflection of the greater attention they paid to non-Roman laws and customs, which in its turn would be the reflection of a lesser role for Rome as a city in the empire and a greater awareness of the multiculturalism of the empire. The flip side is, in advocating power of interpretation over the law common to all peoples (*ius gentium*) rather than just over civil law (*ius civile*), the emperor was reaffirming his power over the extended domain of the entire world.[85] To put it in metrological terms, while acknowledging

[82] *Dig.* 37.14.17. Cf. Bretone (1989) 219.

[83] Cf. Amarelli (1983) 88–9; Scarano Ussani, (1987) 75–6 and n. 86, with further references.

[84] Bretone (1989) 198, 200, 211–13. Bauman (1989) 236–7, 301–2 thinks that part of the story behind Hadrian's emphasis on juridical administration, reform and greater role for the *consilium principis* is the fact that he wanted to weaken the role of the praetor, and through that indirectly of the senate and of the *senatus consulta*.

[85] Echoes of some of these issues in Dio Chrys. *Or.* 15.20 (mid- to late first century CE); Plut. *De sera* 550b (late first century to early second century CE); Aristid., *Ad Romam* 102–3 (155 CE). Casavola

the existence and utility of other standards, the emperor posited himself as the supreme measure, to which the others were required to refer in case of conflict or when mediation was needed. The process itself by which the emperor became a super-standard can be seen as a kind of inscription device that began with Augustus himself. Even at the level of ritual – through his visage on coins, and the presence of his name and the events of his individual life within the official calendar – the emperor was originally officially a figurehead for the senate and the people of Rome. This link between imperial power and 'real' sources of authority was gradually erased, until the emperor could stand outside debates on rationality, nature or convention because he was not standing in for any further source of authority. From a sort of stand-in he became the ultimate reality of authority.

From this perspective, the fact that the coinage for the Western part of the Empire was in this period and until 192 CE issued from a single centre, the mint of Rome, acquires some significance.[86] Indeed, in mere economic terms Maecianus implies a situation (and this will become more and more the case in late antiquity) where the coin is valuable not so much because of its 'real' value (gold or silver or bronze), but because it is inscribed in a complex trust system, ultimately guaranteed by the state, i.e. the emperor.[87] Or at least it should be. The grounding of order in economics as in law was ultimately contested, subject to recalcitrant moneylenders,[88] the vicissitudes of history and the contingencies of geography. Conflicts ensued which had to be solved: in fact, in land-surveying as in law, most of the administration from the late first century CE seems to be negotiating disputes on the interpretation of previous land-divisions or previous legal decisions.[89] It was in order to measure up to alternative sources of expertise or authority that Marcus Aurelius had to know about the law, and he had to know about money: so that he could afford, like Columella and like the architect described by Columella, not to be an expert, and thus supersede jurists and accountants (*ratiocinatores*) alike.

CONCLUSION

Different peoples will have different measures: some Italic populations used a ten-unit based system for the *as*, which was abandoned by the Romans and

(1980) 215, 222, 226; Marotta (1988) 73–9; van der Waerdt (1994); Amarelli (1996); de Giovanni (1996); Scarano Ussani (1997). Scarano Ussani (1979) 134, 154–5, 200, describes 'a critical attitude' towards traditional Roman legal institutions on the part of members of the ruling class.

86 Hobley (1998) 1. Cf. also Dio Cass. 52.30.9. 87 Cf. Savio (2001) 21. 88 *POxy.* 1411 (260 CE).

89 Salvius Julianus manifests awareness of a conflict between some imperial decisions and the *ratio iuris* according to Scarano Ussani (1987) 150–2.

is mentioned by Maecianus as a possible subdivision which is not in use.[90] Analogously, land-surveyors report that different people will measure, count and divide up land differently.[91] The Imperial administration through its officers had to come to terms with this diversity by finding either a unified system or a way of managing the diversity while partially retaining it. If Frontinus represents an empire where order in the form of measures and standards is being formulated, perhaps Maecianus speaks for a situation where one can, at best, acquire the knowledge to understand an order which is already in place, the result of an ultimately unresolved dialectic between systematising efforts, convention, regional variations, different sedimentations of history and the manifestation of disparate interest groups.

The role of the reader of the *Distributio* (and the emperor is one of the intended readers)[92] is not so much to express an order of one's own, as to grasp and maintain – administer – what is already in place. It is an active role, reinforced by the imperatives and the 'constructive' verbs through which the subdivisions of money are in turn made, the names called out, the signs written down, the account given. Yet, it is not a creative role. The author is almost resigned to the fact that the world is in a certain way, that fringes of deregulation will always be present, that we have the stand-in; in fact, more than one system of stand-ins, but we cannot retrieve with certainty the 'things' behind them and with that, the real cause of the present order(s) of things. The *Distributio*, like many of the legal texts it seems germane to, does not aspire to retrieve the absolute foundations; it does not aim to go back to level zero, as it were, but to create a meta-level from which the others can be adjudicated and regulated. Sheep, if ever they were the 'real' *pecunia*, are not important any more: all that counts, and all that effectively exists, are stand-ins, *pecunia numerata*, and it is this reality that one must try to grasp. Marcus Aurelius may have craved the well-rounded speech that Greek *paideia* could provide, but Maecianus reminds him of the necessity to know what is appropriate for an emperor. The Roman children who learn to divide the *as* into a hundred parts in Horace's vignette may indeed have been training for higher and more momentous imperial tasks.

[90] See Pedroni (1996), esp. 25, 67–8.

[91] Hyginus 1, *De condicionibus agrorum* 80.9, 92.21–22; Hyginus 1, *De generibus controversiarum* 98.11–12; Hyginus 2, *Constitutio limitum* 138.1–28 (Campbell).

[92] Addressing technical books to emperors is not uncommon (see, e.g., Vitruvius, Balbus, Pliny), but that to me does not exclude the possibility to take the dedication at face value *as well*, especially in cases, like Maecianus', where the author was well acquainted with the dedicatee.

CHAPTER 10

Probing the entrails of the universe: astrology as bodily knowledge in Manilius' Astronomica

Thomas Habinek

The upsurge of interest in astrology during the early principate is well attested. Emperors have their horoscopes published. Zodiacal signs take their place on a range of objects, from public monuments to legionary standards. Elite authors advertise their familiarity with astrological lore. Belief in stellar influence thus coincides with – and to a certain extent parallels – the triumph of rationalisation in the realms of law, history and governance. While we cannot rule out an indigenous interest in stellar influence, astrology at Rome is best understood as one more intellectual discipline transported from the Greek East and systematised and legitimised as part of Rome's cultural revolution of the first centuries BCE and CE.[1]

The paradox that a discipline as seemingly irrational as astrology flourishes amidst the rationalising enterprises of its time invites investigation, all the more so since astrology seems to have had a special appeal for the cosmopolitan elite.[2] The traditional explanation for astrology's ascent, which links it to the rise of dominant individuals, is vague and at best partial.[3] Is it concern to identify potential victors in civil struggle that leads Romans to search the stars? If so, then we would expect widespread belief in stellar influence to precede interest in the horoscopes of a Sulla or Pompey – a proposition for which there is little evidence. Moreover, in its insistence on the predestined nature of human affairs, astrology is as likely to

[1] For discussion of astrology's popularity during the period see Bouché-Leclercq (1899) 546–70; Cramer (1954) 44–145; Stierlin (1986) 25–122; Barton (1994a) 27–70; (1994b) 40–60; Beard, North and Price (1998) 231–3. On rationalising disciplines during the late Republic and early principate see Wallace-Hadrill (1997); Moatti (1997).

[2] In addition to the self-advertisement of astrological knowledge by elite poets (e.g., Virg. *G.* 1.3–4; Hor. *Carm.* 2.13, 2.18; Luc. 1.45–59), we might note emperors' interest in astrology (Cramer (1954) 81–231); the law's protection of astrological research (Cramer (1954) 102; further discussion by Beard, North and Price (1998) 1.231–2); and its exemption of the well-connected Thrasyllus and his descendants from periodic expulsions of run-of-the mill astrologers and soothsayers (Cramer (1954): 92–5). In the words of Tacitus, astrologers constitute a class that 'in our state will always be prohibited and always retained' (*Hist.* 1.22) – a good indicator of their indispensability to elite dominance.

[3] E.g., Stierlin (1986) 15; Barton (1994b) 35.

undermine elite claims to authority based on achievement as it is to encourage allegiance to one warlord or another;[4] and its popularity persists unabated even after an imperial decree of 11 CE bans inquiry about individual fates.[5] Is it then that disruption caused by great men and their client armies, or by the transition from republic to principate, prompts lesser mortals to seek consolation in contemplation of cosmic order? On this view, astrology would be a way of accommodating fast-paced change to a prior system of understanding the world. Such an explanation gains credence from surviving horoscopes, which are overwhelmingly retrospective in nature, telling the inquirer why person *x* turned out to live life *y*, rather than offering advice on a specific venture.[6] But it, too, founders on chronology, since elite acceptance of astrology at Rome seems not to predate the problems it allegedly explains, nor does astrology diminish in influence as the turmoil of the late republic subsides. Barton's admirable attempt to understand astrology as a Foucauldian discourse of power improves upon the ascription of astrology's rise to an interest in the fates of great men, but still leaves us to wonder: why a discourse of this sort at this point in time?[7]

Investigation of astrology's hold on the imagination of elite Romans in the early principate can benefit from closer examination of the ways in which it was presented to them. In particular, the five-book hexameter treatise of Manilius, written during the reigns of Augustus and Tiberius, explicitly seeks to justify the ways of heaven to men, who are represented as neither inclined nor disinclined to take them seriously.[8] Comparison of the *Astronomica* (*Astr.*) with earlier works in the Latin didactic tradition brings to light issues and concerns that, at least in Manilius' view, had not been addressed by even the most compelling thinkers and writers of preceding generations. And analysis of his poem in the context of other astrological treatises makes it possible to identify those aspects of astrology that Manilius expected would most engage his audience. Manilius'

[4] This is the inference of Cramer (1954) 60 concerning the revolt of Aristonicus of Pergamum in 133–30 BCE The community of Heliopolis described by Diod. Sic. 2.57.3 combines universal freedom with commitment to astrology.

[5] Dio Cass. 56.25.5; Cramer (1954) 99; Barton (1994a) 54–5.

[6] For extant horoscopes see Neugebauer and van Hoesen (1959). Good examples of retroactive interpretations at 90–1 and 94. Astrology seems to follow the pattern of other ancient modes of explanation, such as medicine, which draw inferences from accumulation of past examples. Failure to make predictions is a key point in the critique of astrology by Favorinus of Arles at Gell. *NA* 14.1.24. Barton (1994b) 93 suggests that catarchic astrology, concerned with the likely outcome of new ventures, may have appealed more to non-elite clientele.

[7] Barton (1994b).

[8] Goold (ed.) (1977) xi–xv; Wilson (1985) 285. On Manilius' participation in the elite tradition of didactic poetry, see Volk (2002), and running comments in Scarcia (ed.) (1996) and (2001).

rhetorical relationship to his subject matter and his audience is itself a kind of testimony to the status of astrology in his day. Unlike Virgil, who in the *Georgics* assumes a shared understanding of the importance of agriculture, or Lucretius, who assumes resistance to the revolutionary nature of the Epicurean enterprise, Manilius claims to merit his readers' attention chiefly on the basis of his virtuosity in presenting the esoteric subject matter of astrology in sophisticated and aesthetically pleasing Latin poetry. He doesn't so much defend astrology as demonstrate its compatibility with his audience's generic and ideological expectations. His emphasis on astrology as both an intellectual and a corporeal practice revitalises a traditional unity of knowledge threatened by the very processes of rationalisation that characterise his era. His poetry transforms astrology into an instrument of domination both in the banal sense of making the status quo seem inevitable and in the more challenging sense of making available to his elite audience 'a strategy for living with the world' that encompasses both mind and body.[9] From the professional astrologers he acquires a defence of astrology as a rational mode of explanation; from the Roman poetic and religious tradition an understanding of the ways in which knowledge is carried by and through bodies. Various scholars have noted the complementarity between decline in importance of augury and haruspication, on the one hand, and rise in interest in astrology, on the other.[10] Manilius helps us to understand not just what was gained by the transition to astrology, but also what was lost by the marginalisation of divination and therefore in need of recovery by some other means.

Manilius' interest in and particular take on the double nature of astrological knowledge is apparent already in his proem. Having just introduced his subject matter and thanked Caesar for the world peace that makes poetic endeavour possible, he proceeds to differentiate between two types of knowledge:

> Nor is it enough to know (*novisse*): there's greater delight in understanding deeply the entrails of the great universe (*scire . . . magni penitus praecordia mundi*) (*Astr.* 1.16–17)

The verbs 'know' (*novisse*) and 'understand' (*scire*) may at first seem to describe modes of intellection. But the addition of the adverb 'deeply' (*penitus*) raises the possibility of a different kind of knowledge altogether, a bodily practice of handling and inspecting, the sort of activity carried out by priests who examine the entrails of sacrificial victims. This suggestion

[9] Taussig (1993) 47 describing mimetic practices more generally.

[10] E.g. Potter (1994); Barton (1994b) 33–40; Beard, North and Price (1998) 230–2, 372–4.

is only strengthened by succeeding verses, which speak of altars and fires and identify Manilius himself as a *vates* – a priest-prophet who, at least in Augustan literature, is also understood to have specialised knowledge of sacrificial victims and procedures.[11] The word *praecordia* (translated here as 'entrails'), describing the object of the second kind of knowing, itself does double duty in the passage, turning knowledge into a physical activity and assimilating the universe to the human body.[12] As Pliny tells us, *praecordia* is the proper term for all of the *exta*, or 'innards', found in humans.[13] While *exta* would be the more precise term in the context of animal sacrifice, *praecordia* is not impossible, and in the passage from the *Astronomica* serves to complicate the image of bodily exploration, transforming the universe into a human figure.[14]

The distinction between the types of knowledge represented by 'know' (*nosco*) and 'understand' (*scio*) recalls the standard ancient differentiation between astronomy, which entails knowledge of the movements of the stars, and astrology, which considers their effect on terrestrial life.[15] Indeed, in an equally charged passage from book 4 of the *Astronomica*, Manilius invokes the same contrast, this time explicitly associating astrology with haruspication, that is the visual and tactile inspection of entrails. In a veritable ode to human ingenuity, Manilius proclaims that the god-like astrologer

> Does not remain content with the external appearance of the gods but searches within the bowels of heaven (*caelum scrutatur in alvo*), and pursuing (*sequens*) a body (*corpus*) kindred to his own he seeks himself (*se quaerit*) among the stars.
>
> I ask for confidence in this process as great as that assigned to birds and to organs (*fibrae*) quivering in the victim's chest. (*Astr.* 4.908–12)

Again, the universe is both sacrificial victim and human analogue. And again the verbs describing the astrologer's knowledge of it vacillate between the intellectual and the physical, with *quaero* (seek, search out) gravitating

11 On *vates*' interest in entrails see Virg. *G.* 3.490–91, *Aen.* 4.60–66, Livy 2.42.10; on their knowledge of sacrificial procedure, Virg. *Aen.* 3.433–40, 6.149–53, Livy 1.45.6. On these and other passages, see Habinek (2005) 226–30, 255. On sacrificial imagery in the proem of *Astronomica* see also Schrijvers (1983); Wilson (1985) 293.

12 Schwarz (1972). 13 Plin. *HN* 30.42.

14 For the Latin vocabulary of sacrifice see Santini (1988). On the animation of the Manilian universe see also Hübner (1984). Several times elsewhere in imperial literature *praecordia* does double duty for human and animal innards: Apul. *Met.* 4.21.1, where a man disguised as a bear is stabbed in the gut; Apul. *Met.* 6.31.5, where a bandit proposes gutting Lucius, who has been transformed into an ass; and Tert. *Apol.* 30.6, where the Christian apologist asks why 'the entrails of the victims rather than of the priests' are to be examined.

15 Wilson (1985) 288.

toward the former, *scrutari* (rummage around in) toward the latter, and *sequor* (pursue, follow, trace) somewhere in between.

Manilius' assimilation of astrology to haruspication is a distinctive and motivated intervention in astrological lore. The most influential of Greek works on astrology, Ptolemy's *Four books* (*Tetrabiblos*: second century CE), also commences with a contrast between astronomy and astrology built around two different verbs of knowing. To Ptolemy, astronomy entails grasping (*katalambanomai*) the movements of the heavens, while astrology relies upon observation (*episkeptomai*) of their influence on human life. The former process, because it pertains to the ethereal realm, yields sureness (*bebaitoēs*) of conclusions, while the latter, concerned as it is with materiality (*hylē*), is less reliable. Ptolemy raises the issue of bodiliness, but only by way of apology: astrology, he concedes, cannot be as dependable a science as astronomy.[16] Knowledge of it will always be at a distance, removed and visual in nature (*episkeptomai* derives from the root *skept-*, implying seeing). Manilius has no such qualms: the bodiliness of astrology and of its objects (the innards of the sky, the lives of human beings) arouses his passion and elevates his subject. Even when he changes metaphors from inspecting entrails to riding a chariot through the skies, he can't escape bodies – his own, or those of the stars and planets that impinge upon him during his travels.[17]

Other passages of Manilius that correspond closely to Greek astrological writings also emphasise knowledge acquired of and through bodies. For example, while Manilius' account of human progress runs parallel to one found in the late-antique body of mystical writings now known as the *Corpus Hermeticum*, especially in its inclusion of religious rites among the civilising practices of early human beings, Manilius expressly identifies haruspication, augury and magic as instances of such rites at the point where the Greek treatise mentions only the swathing of corpses.[18] Manilius' interest in decans-theory (whereby the zodiacal disc is divided into ten-degree sections) differentiates him from Ptolemy and other more philosophical astrologers, but it closely parallels a hermetic treatise that associates each sign of the zodiac (and subsequently each subdivision, or decans, thereof), with a part of the human body.[19] According to both Manilius and the

[16] For analysis of Ptolemy's argument and its relationship to Aristotle, see Long (1982).

[17] On the physicality of Manilius' journey to the skies see Wilson (1985), Landolfi (2003) 23–8.

[18] Stob. *Ecl.* 23.64–8, in reference to the rites taught by Isis and Osiris as part of their civilising endeavours. Cf. the important discussion of Manilius' relationship to the *Corpus Hermeticum* in Vallauri (1954).

[19] Ruelle (1908).

anonymous hermetic treatise, Aries is assigned the head, Taurus the neck, Gemini the shoulders, and so on. By associating limbs of the human body with signs of the zodiac, as opposed to an alternate system in which planets govern body parts, Manilius can maintain the analogy between the totality of the heavens, i.e. the complete circle of the zodiac, and the human body. As he puts it,

> Consider the parts of the human frame distributed among the constellations (*hominis . . . partes*) . . .
>
> and the individual limbs obedient to particular authorities (*singulaque imperiis propriis parentia membra*) . . .
>
> that exercise control over them out of all the body (*toto de corpore*). (*Astr.* 2.453–56)

It would be easy, if misleading, to ascribe Manilius' assimilation of the universe to a human body, and vice versa, to his Stoicising tendencies. As Anthony Long writes, 'the modern consensus on unqualified Stoic support for astrology has alarmingly frail foundations'.[20] Indeed, there was a strain of Stoicism, presumably familiar to Manilius, that overtly rejected astrology. Stoic 'sympathy' (*sympatheia*) implies a correspondence between microcosm and macrocosm that grounds an ethics 'in accordance with nature'. Manilian astrology, in contrast, assumes a mutual mimesis between heaven and earth: human reason imposes terrestrial shapes on the heavens (i.e. the heavens imitate the earth), and human bodies and destinies follow the course set for them by the stars.

If there is an ancient philosophical school to which Manilius must be assigned, Pythagoreanism (which also lies behind the *Corpus Hermeticum*)[21] is as strong a candidate as Stoicism. From the very outset of his poem, Manilius as much as states that it is impossible to transmit astrological lore without the musicality implicit in the term *carmen* ('song') – an idea made explicit in Pythagorean teachings on the foundational power of music.[22] In the words of Walter Burkert, 'the wondrous potency of music, which moves the world and compels the spirit, captured in the net of number – this was a cardinal element of the secret of the universe revealed to the wise Pythagoras'.[23] And so it is to Manilius as well, who proposes to draw the stars down from the heavens and lay bare the ethereal census 'through song' (*per carmina*, 1.12), to register as a singer (*canentem*, 1.22) the 'surrounding clamour' (*circumstrepit*, 1.23) of the universe. For Manilius, the poet's

[20] Long (1982) 172. Does Manilius adopt Stoic terminology in order to legitimise an otherwise suspect discipline? Cf. Baldini Moscadi (1979) on Stoicism, magic and astrology.

[21] Kingsley (1995). [22] Habinek (2005) 86–94. [23] Burkert (1972) 378.

modulation of words to metre is an important manifestation of the 'fixed law' (*certa cum lege*, 1.22) by which the universe operates.

Manilius' claim to poetic authority, in the proem and related passages, is bolder than that of any of his Roman predecessors – so much so that it has led one commentator to describe it (wrongly) as self-contradictory, even impious.[24] Manilius draws upon prior images of vatic power and magical incantation but asks us to take them literally.[25] As he puts it in the opening verses of his poem:

> Through song (*carmine*) I aim to draw down (*deducere*) from the sphere of the heavens divine skills and fate-aware constellations that influence the diverse fates of men – all the work of heavenly reason.
>
> I will be first (*primus*) to set Helicon in motion to new tunes and the woods on its summit nodding from their green treetops.
>
> I bring alien rites unrecounted by any before me. (*Astr.* 1.1–6)

Manilius combines the didactic tradition's interest in knowledge of the cosmos with Roman Alexandrians' pride in adapting alien poetic forms to new contexts. He is both the lucky man of Virgil's *Georgics* (1.490–2), who abandons less worthy concerns for scientific inquiry, and the *alter ego* of Propertius, who uses metaphors of initiation to convey the intensity of his commitment to a new poetic art (e.g., 'I am the first to enter', *primus ingredior*, Prop. 3.1.3).

In combining these two aspects of Augustan poetry Manilius revives and reanimates the metaphor of poet as *vates* (a title he claims for himself at *Astr.* 1.23, as noted earlier). For the Augustan era, *vates* is not only a priest-prophet and a singer and a handler of entrails: he (or she) is also an importer of religious rites from afar.[26] In Augustan poetry appeal to the figure of the *vates* comes close to claiming ritual authority for the poet-singer as means of access to the world beyond the here and now. But the fragmentation of the figure of the *vates* into its various components betrays a secular and rationalising discomfort with any such unified vision of the power of song. Manilius has no such reservations: he really can 'draw down' the universe (like the Thessalian witches who are differentiated from poets by Propertius, Virgil and other predecessors),[27] because the universe is a willing participant in the process. As he puts it elsewhere

[24] Volk (2001).

[25] Cf. Wilson (1985) 290: 'Manilius therefore is claiming literally to set the mountains and the woods in motion and so to rank as an Orpheus or Arion'.

[26] For the *vates* as importer of new rites see Livy 4.30.9, 39.8.4, 39.16.8.

[27] E.g., Prop. 1.1.19, Virg. *Ecl.* 8.69.

Who could know (*nosse*) heaven if not through the gift of heaven?
And find god, unless he were himself a part of the gods?

. . .

Who could deny the sacrilege in laying hold of the universe – if it were unwilling –
and drawing (*deducere*) it down to earth, captive, as to oneself [or itself]?
(*Astr.* 2.115–16, 2.128–9)

Manilius acknowledges the boldness of his earlier appropriation of the Thessalian trick implicit in *deducere* ('draw down') even as he justifies it. He treats the universe as a body to be manipulated through magic or taken by force, but only because the universe wills this treatment. The ambiguity of the phrase *in semet* in the final line quoted above (does it mean 'drag the universe onto itself' or 'drag the universe onto the person of the dragger'?) precisely articulates the commensurability of the universal body and the human subject.[28] By drawing the universe to himself, the poet would also be drawing it in on itself, since the term *mundus* (translated here as universe) can refer both to the totality of creation and to the heavens as a specific portion of creation. In any event, what matters is that Manilius' 'deduction' of the universe falls short of sacrilege because of the universe's willing participation in the process, a point made clear by the emphatic (and in my reading, contrastive) position of the adjective 'unwilling' (*invitum*, *Astr.* 2.127) immediately after the main caesura in its line. Indeed, this entire section of book 2 celebrates the conjunction of man and sky (*hominem coniungere caelo*, *Astr.* 2.105) and their mutual interpenetration by a single spirit (*spiritus unus*, *Astr.* 2.64).

Manilius' adaptation of the earlier poetic tradition raises issues of embodiment both implicitly, in his claim to possess the full range of magical, religious and musical powers of the *vates*, and explicitly, in his revision of earlier accounts of human beings' relationship to the material universe. If Manilius' description of human progress in *Astronomica* 1 draws upon but revises a comparable narrative preserved in the *Corpus Hermeticum*, it bears a similiar relationship to a long-established literary topos concerning the civilising effect of human intellectual endeavours.[29] Like Cicero, Lucretius and Virgil, who provide overlapping lists of human achievements,[30] Manilius too celebrates the power of skill (*sollertia*), utility (*usus*), effort (*labor*),

[28] Volk (2001) clarifies the possible meanings of the phrase *in semet*, but underestimates the logical and rhetorical importance of *invitum*.
[29] For fuller discussion see Romano (1979).
[30] Cic. *Tusc.* 1.62; *Rep.* 3.3; Lucr. 1.62–79, 5.1364–1456; Virg. *G.* 1.121–46.

and prior experience (*experientia*). His evocation of culture-heroes, both divine (Mercury) and human (unnamed kings), plays off Lucretius' praise of Epicurus[31] as well as Virgil's recognition of Jupiter's role in human development. The Virgilian intertext is particularly noteworthy, since both Virgil and Manilius describe a marked before and after in human development ('before Jupiter' (*ante Iovem*), *G.* 1.125 ~ 'before them' (*ante illos*), *Astr.* 1.66), celebrate the importance of labor (*G.* 1.145, *Astr.* 1.80, and 1.114), and refer to the sharpening of mortal wits. Manilius' expression: But when long time sharpened mortal hearts (*sed cum longa dies acuit mortalia corda* (*Astr.* 1.79) clearly echoes Virgil's description, at the outset of his account of cultural progress, of Jupiter as 'sharpening the minds of mortals with cares' (*curis acuens mortalia corda*, *G.* 1.123).

Manilius' references to his literary predecessors serve to highlight his variation on their themes. Whereas Cicero, Lucretius and Virgil collectively refer to the invention of sailing, agriculture, music and astronomy, only Manilius inserts divination and magic among the civilising developments of human history:

> Lest I sing the commonplace (*vulgata*), they began to understand the language of birds
> to consult entrails and rupture snakes with incantation
> to summon shades and set deepest Acheron in motion
> to turn day into night and night into dawn.
> Teachable skill by effort conquered everything (*omnia conando docilis sollertia vicit*).
>
> (*Astr.* 1.91–95).

The passage is replete with allusions to a wide range of earlier literature. Particularly striking is the way Manlius brackets his introduction of the bodily knowledge of haruspicators and magicians. He opens with a paradoxical reference to Virgil's lament over themes that are commonplace, or *vulgata* (*G.* 3.3), only to introduce one that is not. And he concludes by fashioning a new aphorism ('teachable skill by effort conquered everything') out of Virgilian sententia ('effort conquered everything': *labor omnia vicit*, *G.* 1.145) and syntax (the gerund as ablative of means), and pointedly Ciceronian diction (*sollertia*, 'skill', as the human faculty that drives progress through the ages).[32]

From the perspective of the literary tradition, perhaps the most revealing aspect of Manilius' poetics of corporeality is to be found in his literalisation of Lucretius' metaphor of the universe as a body. As we might expect,

[31] Compare Lucr. 5.1–5 and Man. 1.25–33. [32] Baldini Moscadi (1979).

Manilius disavows the randomness of the universe as imagined by Lucretius and his philosophical hero Epicurus. But this rejection of Lucretian atomism includes a reworking of the earlier poet's own imagistic likening of the universe to a body. As Duncan Kennedy notes, Lucretius' description of atoms assimilates them to human body parts and processes of procreation. By giving atoms names such as 'birth-giving bodies' (*genitalia corpora*), 'the seeds of things' (*semina rerum*), and 'the first bodies' (*corpora prima*), and by describing them as acting 'of their own accord' (*sponte sua*), Lucretius seeks to 'reduce the alienating effect of his system. . . The poem [*De Rerum Natura*] thus offers a reconfiguration of nature, granting to nature, in Lucretius' term, a fresh, and, as he believes, definitive *figura*, "form", "outline", "shape".'[33] Manilius, in turn, in his own most explicit declaration of doctrinal allegiance, reworks Lucretius' imagery to suggest that whatever configuration is going on, it results in the animation of the universe. 'A single spirit', he writes, inhabits every sector of the universe, permeates it, and rushing throughout, configures it as an animate body' (*corpusque animale figuret*, 2.65).

As the passage proceeds, Manilius continues to deploy Lucretian images and arguments to his own ends, explaining the cycles of nature as the outcome of the universe's orderly rationality, a rationality that rules human affairs as well:

> Therefore this god and reason (*ratio*), which governs everything,
> leads from heavenly constellations the animate beings of the earth (*terrena animalia*).
> Although the stars are removed at a great distance,
> reason nonetheless compels recognition of how they distribute lives and fates
> among the nations and assign distinctive characters to individual bodies (*singula corpora*). (*Astr.* 2.82–86).

Reason is no longer a tool for analysing the universe but a power that informs it. And the universe, far from being merely figured as a body, consists of multiple bodies, aethereal and terrestrial, that interact meaningfully with one another. Manilius articulates the astrological principle of stellar influence in language that builds upon and transforms the Lucretian image of the universe as a body. Corporeality is not reason's figuration of the universe, but the very feature of reality that reason is called upon to probe and comprehend.

[33] Kennedy (2002) 92.

At a superficial level it is easy to see how Manilius' poem and its subject matter participate in the self-legitimising ideology of Rome's imperial elite. Manilius deftly celebrates the horoscopes of two successive emperors, giving high praise first to Capricorn, which admires itself due to its association with Augustus (*Astr.* 2.507–9), while Libra, the birth sign of Tiberius, governs Italy, and especially Rome, 'which raises and lowers the nations placed in the scales' (*lancibus et positas gentes tollitque premitque*, 5.775). His 'astrology of nations' – a device that in Ptolemy's parallel account defensively explains how individuals born at the same time can have different destinies – instead becomes the basis of a world tour of Roman dominions and explanation for historical enmities between peoples. Indeed, while Ptolemy expressly advises consideration of the general problem of nations and cities prior to investigation of individual destinies (*Tetr.* 2.2), Manilius postpones the astrology of nations until relatively late in the poem, even then treating it only cursorily (*Astr.* 4.711–806). Unlike the *Corpus Hermeticum*, with its 'total absence of references to contingent reality',[34] or the astrological poem of Dorotheus of Sidon (late first or early second century, CE), which focuses on the anonymous lower ranks of society (e.g., 'How many will own the native if he is a slave?', 1.11),[35] Manilius' *Astronomica* celebrates emperors, incorporates astrological exempla from Roman history and identifies the hierarchy of the universe with the orderly ranking of the Roman commonwealth (*Astr.* 5.734–42). The entire poem concludes with an expression of relief that the masses of the stars in heaven are as powerless as the 'populace' (*populus*) is on earth –

> to which had nature given strength to match their number
> the empyrean itself would be unable to endure their flames
> and the universe would blaze atop the Olympian pyre.
> (*Astr.* 5.743–5)

In relating the universe to contemporary social arrangements, Manilius extends a line of thought familiar from earlier Roman epic, such as Ennius' *Annales* and Virgil's *Aeneid*, and characteristic of the domesticated Stoicism that constitutes the dominant ideology of the imperial elites.[36] But – to adopt his own metaphor – he also probes more deeply into the nature of power and knowledge and in so doing points the way to a new understanding of astrology's hold on elite Romans during the early principate.

[34] Vallauri (1954) 167. [35] Quoted from translation of David Pingree (1976).

[36] On the relationship between cosmology and imperialism in Roman epic see Hardie (1986); on Stoicism as elite ideology, Shaw (1985).

Manilius's emphasis on the corporeality of the universe and of the astrological subject's encounter with it restores the unity of body and mind threatened by the processes of rationalisation in which astrology otherwise participates. The dissemination of new disciplines in the late republic and early principate strengthened the hold of what Paul Connerton calls 'inscribing practices', patterns of thinking and knowing based on or analogous to writing, and therefore consistently open to critique.[37] Astrology respects the claims of such practices, even seeking to become one itself, while also producing a kind of knowledge carried by bodies, and thus less susceptible to analysis, resistance and change. As Connerton puts it, summarising Oakeshott, ideologies can only be 'abbreviations of some manner of concrete behaviour . . . [W]hat has to be learned is not an abstract idea, or a set of tricks, nor even a ritual, but a concrete, coherent manner of living in all its intricateness'.[38] As presented by Manilius, this is what astrology provides the urban elites of the early Roman empire: a manner of living in all its intricateness. Astrology is systematic and abstruse enough to be distinctively their possession. But it is sufficiently bodily in orientation, in its teachings concerning the 'exercise' of heavenly forces on the human subject, in its performance (as Manilian song or as professional consultation), and in its intended impact on the everyday experience of its devotee to absorb the functions of divination, magic and other modes of accommodating contingent human experience to the natural world. Precisely in its focus on the responsiveness of the human subject *from birth*, it serves to ease the transition from one generation, one emperor, one set of social circumstances to the next. It tells its followers that they, and all they control, are 'perfectly adapted to the form of cosmic being'[39] in every dimension of experience, and thereby renders their position impervious to critique. Organising knowledge in texts and through bodies, astrology is well-positioned to assume a critical role in the reproduction of social order over time.[40]

[37] Connerton (1989), esp. 102–3. [38] Connerton (1989) 10. [39] Benjamin (1999) 721.

[40] Critical, but perhaps not unique. In time, the spread of sophistic rhetoric, with its own powerful pairing of inscribing and incorporating practices, will pose a significant challenge to the ascendancy of astrology. See Gellius' suggestion that Favorinus' denunciation of astrology may itself be a rhetorical showpiece (*NA* 14.1.2) and Firmicus Maternus' overt hostility to the rhetoricians (*Mathesis* 1.1).

CHAPTER II

Galen's imperial order of knowledge

Rebecca Flemming

Order (*taxis*) is a vital matter for the great imperial physician Galen of Pergamum. Sound method (in all things) depends on it: on beginning at the beginning and proceeding systematically through all the requisite stages until the goal is attained. It is, moreover, a test that most in the medical field fail. Galen's total commitment to good order provides him with a measure against which his rivals (past and present) can be measured and found wanting: it creates an important space within which his superiority can be asserted once again. Thus, for example, he makes order a key dividing line between Rationalists and Empiricists in *On the therapeutic method*, suggesting that it underlies the epistemological gap between these two medical groupings.[1] The latter, he avers, solve problems and make discoveries in a disorderly fashion – through what they happen to observe, through chance experience – while the former lay claim to an orderly and logical approach to the acquisition and consolidation of knowledge. Their delivery is poor, however, and most Rationalists fail to start at the beginning; they also recapitulate received wisdom rather than actually working through a line of reasoning or argument. Two types of taxic failure are thus demonstrated, and duly criticised, allowing the virtues of the Galenic model to shine through all the more clearly. It is stated more positively, and practised, in many of his tracts and treatises: proper order is always asserted, and essayed, in his various enquiries and disquisitions.

Still, as Galen became increasingly aware over the course of his long and illustrious career, especially as his monumental oeuvre began to take on something approaching its final shape, that shape lacked the kind of order he so repeatedly avowed in his individual projects. The sum of well-ordered parts is not necessarily a similarly structured whole; an accusation that could be levelled not just at the sum of his writings, but also at the totality of the

[1] Gal. *MM* 1.4 (x 30–35 K); and on medical sects and Galen's relation to them more generally see, e.g., Frede (1985). A key to abbreviations for Galen's titles can be found at the beginning of this volume.

medical art (*iatrikē technē*) they claimed to encompass, since the two were so closely connected. The former displayed the latter, variously demonstrated Galen's mastery of all the knowledge, methods and skills requisite to medicine; any problems of order could not, therefore, be confined to the literary realm, but might also call into question the authority of his version of the *technē* in a more global sense. This gap, therefore, had to be closed: overall order had to be imposed, and there are recurrent efforts amongst his later works to do just that. The first attempt was made with the short treatise *On the order of my own books*, addressed to one Eugenianus and probably written around the time of Septimius Severus' accession to the imperial throne in 193 CE.[2] Here Galen proposed programmes of reading, structured paths through his oeuvre. Next, the compact compendium on the *Medical art* made a rough stab at a more general ordering of medical knowledge, and supported its summary outline with a bibliographic endpiece that provided a guide to the works that fill in the detail on each topic covered. An exhaustive listing of his entire literary output was, however, deferred to a later occasion, a promise fulfilled by the arrival of *On my own books*, a text that not only lists but classifies, first biographically and then by subject matter. Lastly, *On my own opinions*, is a summation of key Galenic tenets, completed perhaps at the very end of his long life, in the early third century CE.[3]

These last two texts present themselves primarily as guardians of authenticity, as defences against literary fraud or mutilation, and doctrinal error or distortion, respectively. Nor is this a pre-emptive strike. Galen claims that works falsely attributed to him are already on sale in the Sandalarium at Rome, and that his writings, despite their clarity, are currently being traduced by modern readers, ignorant of grammar and the basic tools of understanding as they are.[4] Of course, he also has his eyes fixed firmly on posterity, on the time when he will be unable to come to the aid of his oeuvre in person, and must rely on these textual boundary markers and signposts to police and direct subsequent interpretations. Issues of order

[2] This work is usually located in the period between the death of Marcus Aurelius and the accession of Septimius Severus but has clear connections with works usually placed in Severus' reign, not least the fact that it shares its addressee with the last eight books of *On the therapeutic method* (see X 456 K). On the standard periodisation/chronology of Galen's oeuvre see Ilberg (1889), (1892), (1896), (1897); and Bardong (1942); though various subsequent textual discoveries, and the lengthening of Galen's life (see Nutton (1995b)), have amended the schedule to some extent.

[3] So, at least, the Arabic tradition would have it. Rhazes states, for example, that this was Galen's last work (Muhaqqiq (ed.) (1993) 4.2–4). See the recent edition by V. Nutton (*CMG* V 3.2; 1999) for more detailed discussion of this text.

[4] Gal. *Lib. prop.* pr. (*SM* II 91.1–13), *Prop. plac.* 1 (*CMG* V 3.2 54.19–56.11).

are, therefore, implicit in these productions, submerged under other ostensible objectives; though the threat of disorder is perhaps more palpable than any order Galen imposes. The threat that the border between the genuine and the fake would dissolve, that the fundamental principles to which Galen had been consistently committed throughout his career and had brought to bear on all his literary compositions, might be betrayed, altered beyond recognition, by future generations, is all too real.

Ordering is more openly pursued in the other pair of texts, in response to a different (though interrelated) set of challenges. Thus, *On the order of my own books* opens with Galen's assent to Eugenianus' suggestion that some explanation of the order of his writings would be helpful:

For they do not all have the same aim, function and subject matter. As you know, some were written at the request of friends, aimed specifically at their situation (*hexis*), others were dictated for youthful beginners.[5]

Nor are these the only causes of heterogeneity and confusion. Further works had to be composed in response to criticism received, founded (of course) on error and misunderstanding; while various notes made for Galen's own personal use found their way into the public domain, contrary to his wishes. Indeed, a whole range of Galenic texts passed, unsupervised, from their intended recipients to much wider and less suitable audiences.[6] The diversity inevitably produced by targeted composition thus threatened to degenerate into promiscuous chaos. The inclusion of 'subject matter' among the problematic variables also signals back to the inherent complexity and multiplicity of medicine itself, a further force for literary proliferation and diversification, which is what the summary *Medical art* essentially strives to counter and control.

So too, in its own way, the treatise *On the parts of the art of medicine*, which attempts to rein in, or at least impose some kind of order and reason on, the divisional profligacy within the art. This over-abundance is demonstrated in terms of both the wider range of different methods of partition applied and the myriad branches of medical knowledge and practice that have variously been brought into existence. While the methodical divergences are the product of wider disputes in the learned medical tradition – such as between the Empiricists and Rationalists – the profusion of subdivisions and specialisms, or at least the actual materialisation of so many of the almost endless theoretical possibilities thus created, is more socially and economically determined:

[5] Gal. *Ord. lib. prop.* 1 (*SM* II 80.3–7); cf. *Lib. prop.* 2 (*SM* II 102.10–19).
[6] Gal. *Lib. prop.* pr. and 2 (*SM* II 92.4–93.16 and 97.6–98.11).

> You [i.e. Justus, the treatise's addressee] should not be surprised if the scope of the art of medicine causes it to be divided in a great city into this large number of sections.[7]

It is only in a huge metropolis such as Rome (and Alexandria) that a career as a dedicated tooth-, ear-, or eye-doctor, or as a cutter of hernias, or as a specialist on the stone, or whatever, is viable.[8]

Galen's Rome was particularly awash with such people, who presented, in various ways, a threat to the integrity of the *iatrikē technē*. Firstly, their logical proliferation threatened to burst the boundaries of the art, to render it incoherent through overpopulation and excessive differentiation. For, if being a tooth-doctor and a hernia-cutter are both legitimate professional identities then it follows that a different physician will be required to deal not only with each part of the body, but also for each ailment of each part. Secondly (and interconnectedly), there is the question of the relationship each sub-set of skills has with the art as a whole: where does this leave the unity of medicine? For this is a crucial, foundational, concept for Galen, and indeed other medical writers in a culture that, more broadly, ranked the generalist above the specialist. Parts of medicine must, therefore, be validly and properly derived from the totality; must make clear reference back to their unitary origins. That is, again, to assert the need for order amidst a confusion that might degenerate further; though it must be admitted that the actual *ordering* Galen proposes and performs in *On the parts of the art of medicine* is not as decisive or successful as the situation would seem to demand.

The failures of orderly correspondence between parts and whole in both art and oeuvre are, therefore, derived mainly from a series of circumstances external to Galen himself. The character of medicine itself has a role to play in the story, as does Galen's natural affinity with it, the sense in which he has had valuable things to say on the subject, things people want (or need) to hear, right from the outset of his career, which in turn leads his own output to be heterogeneous, as explained above for *On the order of my own books*.[9] The more serious problems arise, however, from the ways in which medicine's inherent complexity has been exacerbated, allowed to run riot, in the contemporary world: a world of material growth, of increased content,

[7] Gal. *Part. art. med.* 2.3, translation from the Arabic by M. Lyons (*CMG* Supp. Or. II 28.9–10 and 29.13–14; for the Latin version see 120.29–31). On this text, and further discussion of these points, see von Staden (2002).

[8] Gal. *Part. art. med.* 2.3 and 2.2 (*CMG* Supp. Or. II 28.9–18 and 26.21–3; 120.31–121.3 and 120.17–22).

[9] See, e.g., *Ord. lib. prop.* 4 (*SM* II 88.6–89.4) for some of Galen's claims about his innate suitability for medicine, combined, of course, with good education and total commitment; and *Lib. prop.* 2 (*SM* II 97.6–98.11) for his literary precocity.

but falling intellectual and moral standards, a place of much ignorance and error, from which control and sound judgement are too often missing. All of which puts considerable pressure on a man of Galen's educational and ethical formation. The organic development of his own output, driven by his desire for a totalising understanding of all matters relevant to the medical art, and shaped by his own commitment to good order, has thus been multiply disrupted, by his friends and companions, with their requests for clarification and edification, as much as by his enemies and rivals, with their attacks and glaring mistakes: all require (he feels) a response. Nor do the forces that produce this heterogeneity in his work show much sign of letting up thereafter, indeed, various extra entropic tendencies come into operation following production, threatening to dissolve the coherence of Galen's project further. So he is compelled to attempt to redress the situation, to assert his ownership over his own body of writing, and over the *iatrikē technē* itself.

Several themes emerge in this recuperative discourse of order. Some points are very self-referential, and self-serving (though that does not make them entirely untrue). Galen's figuration of this field enables him to complain, and complain vigorously, about his very success; a tactic that he is very partial to. It is his superiority, his abilities, his authority and reputation, which are at the root of many of his problems. The fact that his is a voice people want, indeed need, to hear on such a wide range of topics and issues, that he is so much in demand, is crucial to the loss of control over his oeuvre. However, Galen has also situated himself in the highly competitive and contentious world of classical medicine more broadly, and demonstrated his participation in its complex networks of power and prestige. He has, furthermore, drawn particular attention to certain key aspects of his wider social and cultural environment in this respect, aspects of its imperial formation. Indeed, he has actively involved himself in that formation.

In particular, Galen's struggle for order is a struggle for control over abundance, as also is the ongoing Roman imperial project: indeed, the tension and interplay between the two might be said to characterise processes of conquest and colonial rule more broadly. Empire is a cornucopia, but that richness, that fecundity, must be properly structured and directed, properly arranged and managed. Otherwise it may slip into luxury and excess, be misappropriated and abused, and thus disrupt established patterns of morality and power. It may even come to undermine the mastery of the rulers itself, both practically and conceptually. The alignment between Galen's empire of knowledge and Rome's political dominion in this respect is not just implicit, abstract or figurative, it is positively articulated and concretely grounded in various ways. The world of plenty, productive and problematic as it is, is clearly centred on Rome, as imperial capital, and that is where Galen situates himself as he strives to organise that plenty as it relates to medicine. This is specifically indicated in *On the parts of the art of medicine* and *On my own books*, but there is a general sense of this placement purveyed in the other works mentioned so far too. Galen is, wants and needs to be at the centre of things, at the centre of power: power over a vast empire. Nowhere but Rome could support his ambition, could foster his totalising vision. There is nowhere else he could stand and have both the reach and leverage to bring order to it all, to bring a much better order to so much more than anyone else.

The problems of that location have also been brought out; accusations that abundance is being mismanaged, has become entropic excess, have been made in these same taxic texts. That, however, is very much part of the imperial package, and drawing attention to metropolitan vices, to failures of mastery and control, threats of disorder and devaluation, is an integral part of much writing of the early empire, in Latin as well as Greek. The question has been raised, however, whether Galen's criticisms do not possess a rather different quality to those of, say, Pliny the Elder, or Seneca the Younger, with which they certainly share much content, in that they are lodged in an essentially, avowedly, Hellenic cultural identity, while Pliny's, for example, are ostensibly grounded in old-fashioned Roman values and traditions, and Seneca's are more hybrid products. Simon Swain particularly stresses this point, reading Galen's Greek allegiances as providing 'insulation' from the Roman world, an insulation not bridged by any real interest in the 'Roman idea', or involvement in the imperial government, in contrast to a number of roughly contemporary Greek writers, from Lucian and Pausanias to Aelius Aristides and Arrian.[10] Galen's disapproval of contemporary Rome, his attacks on her anti-intellectualism and poor educational standards as well as her more materialistic failures, has, for Swain, a greater coherence and cogency than his more positive engagements with the city, its inhabitants and endeavours.[11] These are sporadic and superficial, a matter of expediency, about advancing his career, while Galen's true loyalties lie entirely elsewhere. Swain thus concludes that, 'In a very real sense, in what mattered to him, Galen . . . was not in the Roman Empire'.[12]

This whole volume, however, is about how much harder it is to escape from the Roman Empire than that statement would suggest; a point that has

[10] Swain (1996) 377.

[11] Swain does discuss these positive moments (1996) 363–72; and for differently emphasised coverage of some of the same passages see Nutton (1978) and (1991).

[12] Swain (1996) 378–9.

been repeatedly made in much recent scholarship relating to other empires too.[13] Indeed, Swain's suggestion that the intensely Greek identity asserted by men such as Galen in the 'Second Sophistic' was a reaction to Roman control would also seem to undermine the idea of Galen as an author who stands apart from the Roman Empire.[14] Can Galen really be such a clear product of Rome's empire and not participate in it? As already indicated, the argument in this chapter is a different one, in respect to both Galen and the empire in which he operates. Galen may come from Pergamum and remain committed to his essentially Greek cultural and ethical formation, even use it as a basis for his criticisms of the contemporary Roman world, but none of that prevents him from utilising Rome's empire also, from drawing on its material and ideative resources, its scope and structure, in creating, organising and selling his own medical system. There is, moreover, no contradiction here, though there may be tensions and slippages. These kinds of interactions are, rather, constitutive of the Roman imperial project itself; in all their complexity, their multiplicity of perspective and emphasis.[15]

These, then, are the themes that will be explored further in this essay, explored in particular as they emerge around and through questions of order, both in Galen's individual works and in his oeuvre as a whole. For, to find the Roman Empire in the contents of the Pergamene's writings, in the peoples and territories, medical materials and foodstuffs, diseases and cures, referred to and described therein, is too easy and obvious. The claim is rather that specific patterns of empire, the signs of an imperial order that goes beyond simple geography, can be found in, and across, his works. Those patterns do also possess a particular cultural inflection, for Galen's Greek identity and attitudes are not irrelevant here; it is just that they do not allow him to stay detached from the Roman Empire; rather they provide a particular trajectory of involvement, which needs to be examined as part of the overall package.

THE ORDER *IN* THE BOOKS

The methods of organisation and structure adopted in particular texts and treatises, and the reflections on arrangement they contain, will now

[13] As emphasised in the introduction (esp. pp. 3–6); and see also for more thoroughgoing 'imperial' approaches to the Greek literature of the first few centuries CE, Schmitz (1997) and Whitmarsh (2001).

[14] Swain (1996) 411.

[15] As Pliny also demonstrates, for example, with his reliance on, and his manipulation of, Greek knowledge: see, e.g., Beagon (1992); and also Murphy (2004).

be analysed in detail, before returning to the ordering of the cumulative whole at the end. The focus here will be on the major tracts, those covering expansive and complex topics, and comprising multiple books; thus posing rather more acute organisational and presentational challenges than a single, narrowly focused book or booklet. While most of the works in Galen's vast output come in at three books or under, there are plenty that exceed this, with the most voluminous being the monumental forty-eight books of *The words in Attic prose-works*, now lost.[16] More durable have been the seventeen books *On the usefulness of parts*, the fifteen *On anatomical procedures*, and the fourteen *On the therapeutic method*, to mention just a few.

From the surviving large-scale works, as well as indications about those no longer extant, it is possible to discern four main approaches to their overall ordering, although given both the practical exigencies of ancient literary production and Galen's personal predilections, there are always tendencies to disorder operating within, and against, the overarching plan and structure of any of his output. For example, the use of book rolls and dictation, not to mention the lengthy time intervals between the completion of different portions of some treatises, all militate against total coherence.[17] Similarly, Galen's tendency to digress, to follow a current train of thought through, regardless of its precise contextual fit or relevance, and to pursue polemical points at the expense of positive argumentative clarity or development, take their toll too. Nonetheless, the underlying patterns are reasonably clear.

The first order is corporeal. The classic head-to-toe presentation is not Galen's primary organisational mechanism for anatomical or physiological knowledge itself, though some of the more specific or introductory works, such as *On the dissection of the nerves* and *On the dissection of the muscles*, come close, and there is a certain downwards drift in other texts too. But it is employed to structure pathological and therapeutic material. Diseases may be arranged according to the somatic location they afflict, or are seated in, as *On the affected parts* (in six books) demonstrates. A remedial counterpart to this is the eleven-volume compendium *On the compounding of drugs according to places* (*kata topous*). The second approach to order is more categorical or thematic, adopting a framework from a way of breaking

[16] Mentioned at *Ord. lib. prop.* 5 and *Lib. prop.* 17 (*SM* II 90.6–9 and 124.7–8).

[17] Galen refers to a couple of works he dictated to tacheographers sent by the parties who wanted a record of the discourse in question (e.g., at *Praen.* 5.19–20 (*CMG* V 3,1 98.27–100.1) and *Lib. prop.* 1 (*SM* II 95.21–96.1)); and, though he makes no such comments about his regular working practices, it is impossible to believe that he could have been so prolific without the kind of secretarial support employed by, for example, Pliny the Elder (Plin. *Ep.* 3.5).

up the world (or medicine) that is not based so directly on the human body. Thus, the companion tract to *On the compounding of drugs according to places* is that 'according to kind' (*kata genē*): that is, according to an internal pharmacological typology which collects together, for example, all the *emplastra* (plasters), *malagmata* (emollients), and *akopa* (for pain relief and general refreshment). Diseases also have an internal typology (indeed typologies), and *On the therapeutic method*, for instance, operates with a division between maladies based in the homoeomerous (uniform) and anhomoeomerous (non-uniform) parts.[18]

The two other orders are more literary, or at least textual. One takes its structure from a pre-existing work. This is most obviously the case with Galen's 'phrase-by-phrase' commentaries on Hippocratic texts (of which a good number survive), and some philosophical writings; but he also wrote summaries of, for example, the *Anatomical studies* of Marinus, and Heraclides of Tarentum's seven books *On the empiric sect*.[19] The latter apparently took a polemical approach, and other lost but decidedly hostile tracts may well have followed a pattern of roughly 'phrase-by-phrase' refutation. Indeed, within the extant section of Galen's oeuvre, large portions of *On the doctrines of Hippocrates and Plato* are dedicated to systematic argument against the Stoic scholarch Chrysippus' works *On the soul*, and *On affections*, as well as the promised engagement with the teachings of Galen's twin heroes, Plato and Hippocrates; while *On the natural faculties* pursues a sustained critique of Erasistratus' physiology; and it is widely believed that much of his very extensive writing on the pulse is based on that of his, heavily criticised but also heavily relied on, recent predecessor, Archigenes of Apamea (whose career at Rome peaked in the reign of Trajan).[20] The second order under this heading is alphabetical (*kata stoicheion*), an arrangement adopted in most of the books of *On the mixtures (kraseis) and properties (dynameis) of simple drugs* that actually list the simples themselves, as well as in his *Hippocratic glossary* and (presumably) the lost lexical works, including all forty-eight volumes on words used by Attic prose-writers.[21] As

[18] The homoeomeries are those which divide into like pieces, such as blood, bone and arteries, while the anhomoeomeries are not so divisible and include compound parts and organs such as the hand, eye, heart and liver. See, e.g., *MM* 1.6 for a rough explanation, and also 2.6 for the associated pathological schema (x 48 and 125–6 K).

[19] On Galenic exegesis see Flemming (forthcoming); and these abridgements appear at Gal. *Lib. prop.* 3 and 9 (*SM* II 104.12–13 and 115.14–15) respectively.

[20] On Galen and Chrysippus see Tieleman (1996) and (2003); and on Galen and Archigenes see Wellmann (1895).

[21] As suggested at Gal. *Ord. lib. prop.* 5 (*SM* II 89.13–15).

an order of words, the alphabet has more obvious appeal than as an order of things; but it can, and is, applied to both.

The work on simples also clearly illustrates that more than one mode of organisation may be employed in a single, large-scale, literary enterprise. Its first five books lay the foundations of Galenic pharmacology in a methodical fashion: first demonstrating the fallacies and inadequacies of all current approaches to the subject, then expounding the basic building blocks of the system that is to replace them. This exposition begins by establishing that everything in the world is composed of the same four elements, which then combine to produce the humours (in a certain balance or mixture, that is *krasis*) in the human body on the one hand, and properties (*dynameis*) inherent in their mixture (*krasis*) in other things in the world – such as plants, earths, stones and animals – on the other.[22] These *dynameis* can then be grouped in relation to their effect on the human body, through its own mixture of humours: primarily according to whether they are heating or cooling, drying or moistening; and secondarily according to whether they are purgative or productive, softening or hardening, and so forth. Next the things themselves, the external items that can be brought to bear, medically, on the human body, can be organised. The first partition is basically threefold, more or less into the customary categories of animal, vegetable and mineral. The plants then proceed strictly alphabetically (in books 6 to 8), while the minerals (in book 9) and animals (in books 10 and 11) take a more varied course. So, for example, earths are followed by stones, according to their own internal classification, but then come metals *kata stoicheion*. The animal items also initially follow their own typology (rather messily), but revert to alphabetical listing for the 'things generated from the sea' right at the end.[23]

Similarly, the works on compound *pharmaka*, that is those compounded out of numerous simples, comprise a primary structure, as their respective titles announce, and a secondary one, which is more textual in nature. So, within the overall arrangement by 'place' or 'kind', existing pharmacological works are excerpted and reorganised, with some Galenic comment, in the way Galen sees fit. Thus, in the books on *akopa* in *On the compounding of drugs according to kinds*, for example, chapters will be introduced along the lines of '*akopa* and *myrakopa* (that is with myrrh as an ingredient) recorded

[22] For a summary of the fundamentals of Galenic pharmacology see, e.g., Scarborough (1984); and for a more detailed analysis see Harig (1974).

[23] Gal. *SMT* 10.1 and 11.2 (XII 247 and 369–77 K); as Barnes (1997) notes, however, this last alphabetisation is only by first letter, and is more error-prone than the others, which are pretty systematically up to the third letter (10 n. 15).

by Asclepiades Pharmakion in his fourth book *On external (drugs)*', and contain a whole sequence of recipes taken from that source, some of which may themselves have been borrowed from elsewhere.[24] In much the same way, the much briefer treatise *On my own books*, as mentioned, begins with a chronological or biographical listing of his literary products, and then turns to a more thematic mode of organisation.

It is also worth returning to the compositional complexities of *On the doctrines of Hippocrates and Plato* already alluded to, for these indicate the way in which more practical exigencies, and social and political considerations, operate to shape Galen's work, at least as he tells it. The original opening of the treatise is lost, which adds to the difficulties in trying to follow its structure; but, in *On my own books*, Galen explains that he commenced writing it at the urging of the consular Flavius Boethus, a man who combined high political office with philosophical commitments (in his case Peripatetic).[25] Boethus was an important supporter of Galen in his first stay at Rome (between 162 and 166 CE), forming a crucial part of the audience first for his oral performances and anatomical demonstrations, and then, following on from that, for his textual disquisitions and displays, initially (it appears) just as an addressee and subsequently as commissioner.[26] His household also benefited from Galen's prowess as a medical practitioner on more than one occasion, as he proudly recounts in *On prognosis*.[27] With his wealth and class combined with culture and learning, Boethus is exactly the type of man Galen wanted to attract the attention and favour of, particularly in the early stages of his career in the imperial capital: the type of man who would (allegedly) request a work demonstrating the congruence and correctness of the views of Plato and Hippocrates on the powers that govern the human being, their number, nature and location. Boethus, however, took only the first six books of this heavyweight literary project with him when he left Rome to govern his native Syria Palestina (as well as the first book of *On the usefulness of parts*), where he died. Galen too left Rome, for his own reasons, and it was only some time after his return to the city where he was now, basically, going to spend the rest of his long life, that he added the final three books that were to complete the work.

[24] Gal. *Comp. med. gen.* 7.12 (XIII 1009–32 K). This compilatory process is analysed in detail by Fabricius (1972), who also provides biographies and bibliographies for all the major authorities Galen uses, such as this Asclepiades (another reasonably recent – late-first century CE – predecessor, and not to be confused with Asclepiades of Bithynia).

[25] Gal. *Lib. prop.* 1 (*SM* II 96.19–24; and see also 94.16–26 on Boethus).

[26] For more details see Nutton (1973).

[27] Gal. *Praen.* 7–8 (*CMG* V 3,1 104.24–116.23).

Given that about ten years must have elapsed between starting and finishing the project it is not surprising that these last volumes are on somewhat different, though certainly related topics to the earlier portion. Galen also seems to have made some later revisions to the previous parts.[28] It is, however, not just time that serves as a dis-organising force in all this, nor is Boethus the only individual whose influence over the composition of the work is acknowledged. The main problem is the balance between positive presentation and polemic, a polemic that always threatens to take its own course, and often does, leading Galen away from the basic path set down for this literary enterprise. This imbalance, this tendency to slide into a systematic refutation of others, and so lose track of his own argument, is most evident in Books Three and Four of *On the doctrines of Plato and Hippocrates*, and is explained in the preamble to the former. He reports that he was deflected from his original scheme by an 'eminent sophist' who claimed that it was not possible to refute Chrysippus' extensive arguments that only the heart is the source of the ruling power (*hegemonikon*) of the soul, and so the human being.[29] Galen had considered that he had dealt with the matter already, as part of his general survey of previous errors on the subject – mistakes either of fact or demonstrative method – in which Chrysippus had featured, though not exclusively. But he feels forced to rise to the challenge nonetheless, to complete a more comprehensive demolition, which takes up book 3, and spills into book 4. It is not clear whether copies of books 1 and 2 were already circulating, for the anonymous sophist to react to them in this way, or perhaps more likely, whether Galen was presenting their arguments orally and was confronted in person, and in public, so that a response could not be avoided. Either way, Galen again draws attention to the external forces acting on his output. Friends and enemies, supporters and detractors, have all contributed, all have their role to play in the way he constructs his own literary career.

Since previous works play such an important part in the organisation of Galen's own, and that might be considered a challenge to the argument for the operation of a particularly Roman imperial order in them, it is necessary to examine the precedents that Galen is variously following or departing from, adapting or rejecting, rather more closely. Such a discussion also

[28] This, at least, is the explanation offered by Ilberg for the fact that the first six books cross-refer to works only composed later (see P. De Lacy's introduction to his edition of the *PHP* (de Lacy (ed.) (1984)): *CMG* v 4.1,2 47–8). It is also worth bearing in mind that the fate of the actual books Boethus took east with him is unclear, so Galen may have been working with something like a 'draft' version when he came to complete the text anyway.

[29] Gal. *PHP* 3.1.7 (*CMG* v 4.1,2 168.27); and see Rocca (2003) 17–47 for further discussion of the concept of the *hegemonikon* and its development.

enables some further reflection on the manner in which Galen establishes his own patterns, which are then repeated across his oeuvre, reiterated in different works; and which do enact, both through that repetition and through their own positive character and content, his fundamental commitment to right method and good order in all things. The organisational styles already picked out illustrate his orderliness on one level, but there are deeper patterns too.

THE ORDER *BEHIND* THE BOOKS

As already mentioned, organisation *capite ad calcem* was common in a range of classical medical genres. The results of Herophilus' systematic anatomical investigations in early Hellenistic Alexandria, the literary results of all his dissections and vivisections of human beings, seem to have been arranged in this manner; and the surviving anatomical summaries from the early Imperial period also tend to follow this pattern (sometimes taking a double journey from head to toe, first on the outside and then the inside).[30] This corporeal system is also employed in the first part of Scribonius Largus' Latin pharmacological work, *Compounds*, written between 44 and 48 CE; and further informs the prevalent ordering of pathological works in the first two centuries CE.[31] These start from the division between acute and chronic diseases found (along with the external/internal split) in the Hippocratic Corpus, then work roughly downwards in each category (as was the Hippocratic practice also).[32] Thus, chapters on acute diseases proceed from *phrenitis* (by now an illness originating in the head/brain despite its etymology) to *satyriasis* or *diarrhoea* (both ailments involving the lower parts), and coverage of chronic diseases move from *skotōma* (a head-based dizziness) and severe headache to *podagra* (gout, and other similar conditions), affections of the womb, and *elephantiasis* (a skin disease affecting the whole body, these total conditions were added on to the end of the list).[33]

[30] On Herophilus see von Staden (1989) 138–241; and I would count Rufus of Ephesus, *On the naming of the parts of the human being* (133–167) (Daremberg-Ruelle (eds.) (1879)), as well as the relevant sections of the pseudo-Galenic *Introduction* and *Medical definitions* (10–11 and 36–60: XIV 699–720 and XIX 358–62 K respectively) among these summaries.

[31] Scrib. *Comp.* 1–162; and see the preface of the edition by Sconocchia (ed.) (1983) for discussion of the dating (vi–vii).

[32] The Hippocratic writers focused on the acute, as in the *Regimen in acute diseases*, and the internal, as in *On internal affections*, but this clearly implies the other half of the pairing also. Rough head-to-toe orders can be seen in, e.g., *On affections*, and *Diseases* II.

[33] See Aretaeus, *On the signs of acute and chronic diseases* (*CMG* II), the anonymous treatise *On acute and chronic diseases* (Anonymi medici *De morbis acutis et chroniis*) (Garofalo (ed.) (1997)), and Caelius Aurelianus' latinisation of Soranus' *On acute and chronic diseases* (*CML* V, 1).

Therapeutic works might follow these same principles (indeed the same work might cover diagnosis, aetiology and cure), or be structured around their own internal typology; which is also true of their pharmacological sub-set. The initially somatic organisation of Scribonius' *Compounds* then becomes generic, for instance, and Galen clearly draws on both the 'by place' and 'by kind' modes of organisation to be found amongst his other predecessors in the field of complex drugs.[34] Indeed, Archigenes composed a treatise entitled, *On drugs according to kind*, while the first systematic compounder of drugs, Mantias himself, perhaps produced a topological correlate in the Hellenistic period.[35] In relation to simples, the animal, vegetable and mineral division is very widespread, but Galen explicitly states that in taking an alphabetical approach to ordering his plant-based materials he is imitating Pamphilus' *On plants* (*Peri botanōn*), though dramatically improving the quality of the contents.[36] Pamphilus was a grammarian based in first-century CE Alexandria, who was familiar with alphabetisation from his other lexical and philological activities (as also was Galen of course); but, despite Galen's implication to the contrary, it is unlikely that he was the first to apply the *kata stoicheion* arrangement to medical materials. Hippocratic lexicography had long co-existed with pharmacological writing among the Herophileans in Hellenistic Alexandria, so the possibilities of cross-over were certainly present earlier, and the *Suda* reports that Bolus of Mendes' late third- or early second-century BCE work on the sympathies and antipathies of stones was ordered *kata stoicheion*.[37] Moreover, the author of one of the most important ancient collections of medical materials, Dioscorides of Anazarbus, suggests that alphabetisation was reasonably common among his more immediate predecessors, those who worked in the earlier part of the first century CE; a view supported by the structure of parts of the *Natural history* of Pliny the Elder.

In outlining how his work will surpass its predecessors in terms of coverage, accuracy, reliability, precision and order, Dioscorides alleges:

> Mistakes were also made in the organisation of their material [i.e., that of Sextius Niger and the rest], some throwing together incompatible properties, others using a *kata stoicheion* arrangement which splits off genera and properties from what most resembles them. The result is almost impossible to memorise as a whole.[38]

[34] Scrib. *Comp.* 163–271.

[35] On Archigenes see Fabricius (1972) 198–9; and on Mantias see Gal. *SMT* 6 pr. (XI 795 K) and von Staden (1989) 515–18.

[36] Gal. *SMT* 6 pr. (XI 792 K).

[37] Von Staden (1989) 445–62 on the Herophileans; *Suda s.v.* Bôlos Mendêsios; and see for recent discussion of the problems with Bolus' dates and output, Dickie (1999).

[38] Dioscorides, *De materia medica* pr. 3 (I 2.11–15) (Wellmann (ed.) (1906–14)).

Sextius Niger was a Roman citizen who composed medical texts in Greek in the first decades of the first century CE, and 'the rest' are presumably his colleagues among the '*neoi*', the recent writers on the subject who, Dioscorides claims, are prone to different kind of errors than the '*archaioi*', their more distant, Hellenistic, ancestors, such as Heraclides of Tarentum and Crateuas the Rootcutter.[39] Whether Niger himself was among the alphabetisers or not, *kata stoicheion* organisation clearly extends well beyond Pamphilus, even at this juncture. A point also supported by the fact that the final book of botanical medical materials in Pliny's *Natural history* contains an almost alphabetical sequence, some of its deviations indicating a Greek origin.[40] Dioscorides further demonstrates that the organisation of medical knowledge, in particular the organisation of the proliferating knowledge about medically effective things in the widening world, was a topic of debate and dispute, part of the ongoing competition between ancient physicians for prestige and patients, authority and audience.

Galen must have been aware of this, and indeed of Dioscorides' position within the debate, for the Anazarbite was one of the main sources he used in his collection of simples in *On the mixtures and properties of simple drugs*, and he is cited elsewhere also. However, Galen makes surprisingly little reference to the points of organisational dispute themselves. In the preamble to book 6 he contrasts Dioscorides' globalising work, in which all medical materials are included within a single text, with the more specific, thematic, texts of, for example, Mantias; but he says nothing about matters of *internal* structure.[41] His own claim that a *kata stoicheion* order 'is necessary' for this material is never actually substantiated or supported.[42] Moreover, it seems to contradict both some of his general principles and some of the more particular points made in the work itself. Galen has a basic commitment, for example, to ordering according to *physis* rather than *nomos*, that is according to real and meaningful distinctions in the world not conventional categorisations; a commitment that is related to his views on the fallibility of language and problems of terminology.[43] This principle is articulated in the first five books of *On the mixtures and properties of simple drugs*, indeed it is encapsulated in the title itself, and various linguistic challenges are also explicitly recognised. Furthermore, Pamphilus appears as a very unlikely exemplar; one that Galen has nothing good to say about.

[39] Dioscorides, *De materia medica* pr. 1–2 (1 1.4–2.5) (Wellmann (ed.) (1906–14)); Niger's Greek medical writings are included in Pliny's listing of home *auctores* for books 20–34 of the *Natural history*.

[40] See Daly (1967) 35–6. [41] Gal. *SMT* 6 pr. (XI 794–5 K). [42] Gal. *SMT* 6 pr. (XI 792 K).

[43] See, e.g., Hankinson (1994); and also Barnes (1997).

In so far as Galen does set out to justify his catalogue of materials, and does assert its superiority over its precedents and rivals, he does so in terms of content rather than structure. His is best on account of having the greatest coverage without compromising the entry criteria; while Pamphilus (and also Xenocrates of Aphrodisias) have been much less discerning, demonstrating a woeful lack of judgement as they include, 'old wives' tales', flashy but useless 'Egyptian sorcery' (*goētia*), and foolish incantations to mutter while collecting the herbs.[44]

Much, therefore, of Galen's organisational style could actually be subsumed under a broader 'textual' heading. Most of his works mentioned so far have literary precedents and are structured along established lines; though Galen has amended and combined, altered and reworked those models in various ways and to varying degrees. He has also consistently expanded the material encompassed within any given medical domain or genre. His treatises tend to surpass their predecessors in size, and, if not, that may be because he treats the same topic in more than one text. The monumental works on compound pharmacology demonstrate these points particularly clearly. The early Imperial period witnessed a growth in this area, both in terms of the number of collections of compound recipes put into circulation and the number of books comprising each collection.[45] None, however, can match Galen's eighteen-book total in this area, in which everything useful from these previous efforts has been included, within a clearer, more comprehensive and systematic, structure: also borrowed, but also improved.[46] That empire lies behind this growth as it leads up to, and peaks with, Galen is obvious. The physicians who composed these collections all worked in Rome (some – Galen among them – attended on the imperial court), and they all drew on the vast resources of the empire in their compositions. Ingredients from right across the Roman world, and from Rome's trade with places beyond her borders, appear in many rich and complex remedies. So, for example, a *malagma* Galen takes from the writings of Andromachus the Younger (another medical figure of late-first century CE Rome), brings together Tyrrhenian wax, Illyrian iris, Cilician saffron and

44 Gal. *SMT* 6 pr. (XI 792 and 797–8 K).

45 The names attached to such collections between Augustus and Galen include not only those of Archigenes, Asclepiades and Scribonius Largus, already mentioned, but also Heras of Cappadocia, the two Andromachi (Elder and Younger) and Crito, to list just the most important (see Fabricius (1972) for fuller listings); and while it is hard to prove that they were more prolific than their Hellenistic predecessors, few multivolume pharmacological works are definitely attached to the latter, in contrast to some of their sectarian writings.

46 The closest contender seems to be Asclepiades Pharmakion who probably authored ten books, five on external and five on internal remedies: see Fabricius (1972) 192–8.

Indian nard, not to mention more common (but still exotic) items such as cassia, myrrh and terebinth.[47] It is not just diverse materials, but also a very wide range of people, who are thus gathered together and absorbed, along with their recipes, into successive compilations. Precise geographical origins are harder to discern here, but a few more unusual monikers and ethnics, such as those of the (presumably) Persian Rootcutter, Pharnaces, and Fabylla the Libyan, appear amongst crowds of mostly Greek, but also many Roman, names in Galen's collections.[48]

The ways in which Galen's literary compositions reproduce processes and patterns of empire also, crucially, go beyond their magnitude and contents into matters of structure. For the Roman Empire, like so many of the texts mentioned, was an essentially cumulative, compilatory, enterprise. Roughly contiguous territories were accumulated through a series of military victories and more peaceful power-plays, and in attaching these new acquisitions to the centre, constructing a political unity from this diversity, Rome relied heavily on existing patterns of power and governance. The old orders were not destroyed and created anew, but rather amended and adapted, refigured to fit into the overarching structure of Roman rule. This, moreover, was the traditional approach to ancient empire building, in which one of the main effects of conquest on local administrations was that they became integrated into a larger whole, rather than being radically transformed in themselves. Of course, that should not imply that nothing changed: this process of integration and reordering through compilation can be transformative in its own way, so long as it proceeds with a reasonably clear and coherent overall structure.

Now Rome's empire was a larger, and in various ways a more considered, compilation than any other; incorporating more diverse material as it stretched west as well as east, not to mention north and south, and structuring it according to its own unifying system, and in its own style. Part of what was distinctive about that style and system was its inclusiveness, the relative openness of both its politcical and cultural formations. There were limits to this inclusiveness and openness, of course, most strongly on a social level – imperial inclusion was a much more horizontal phenomenon, operating across local elites, than a vertical one – but also on a historical level, as the basic structures were determined, the fundamental principles of order established, prior to their opening up, at least on an imperial scale. Still this was a notable feature of Roman imperial rule, as the career of

[47] Gal. *Comp. med. gen.* 7.7 (XIII 985–6 K).

[48] Pharnaces: XIII 204 K; Fabylla: XIII 250–1 and 341 K.

Galen's patron Boethus illustrates, and Galen too in his own way. For it is not just that his compositional procedures reproduce processes of empire in various rather abstract ways; it is not just that Galen's empire of knowledge and Rome's political empire are constructed along the same lines methodologically, and so come to resemble each other in terms of size and shape; but that there is a more positive ideological overlap too, in the rhetoric and practices of order both employ. This emerges most clearly in some of Galen's departures from previous patterns.

BACK TO THE BOOKS (AND THE BODY)

Galen, then, owes manifold debts to his predecessors, both distant and more proximate, but some of his claims to structural innovation are also justified, particularly in respect to writing about disease and cure in their generality and totality. Here Galen uses much more actively analytical classifications than was traditional. This is most obvious in *On the therapeutic method*, where he employs his own, distinctive, conceptual categorisation of disease as the organisational framework; eschewing the customary division between acute and chronic conditions, and also, to a considerable extent, traditional disease entities like '*phrenitis*' or '*podagra*'. Not that these classes and items have no validity, or utility, but they have no real analytical purchase; they do not go to the heart of the matter, of what being diseased means, and what therapeutics are about. So, they float about on the surface of things, and of his text, rather than contributing to its fundamental structure. *On the affected places* shares some of these features too, though the claim to originality in this case rests with Archigenes, who, according to Galen, was the first to treat localised disease 'systematically' in his own three books by the same name; and these diseased localities are ordered roughly head-to-toe.[49] Galen, of course, has twice as many volumes in his text *On the affected places*, partly in order to give him space to correct Archigenes' many errors.

This leaves, however, the matter of Galen's anatomy and physiology. Little has been said so far about the massively proportioned, and vitally important, works *On the usefulness of parts* and *On anatomical procedures*; except that they do not proceed *capite ad calcem*, and that the former was also requested by the consular Boethus. Or, at least, that is the claim made in *On my own books*, where it is stated that only the first book was ready to accompany Boethus to Syria, while the rest were finished (like *On the*

[49] Gal. *Loc. aff.* 3.1 (VIII 136 K); and see also *Cris.* 2.8 (145.1–146.6) (Alexanderson (ed.) (1967)).

doctrines of Hippocrates and Plato) after Galen's subsequent return to Rome in 169 CE.[50] A rather different account appears in the opening chapter of *On anatomical procedures*, a text which is presented as an expanded and improved version of two books by the same name also given to the consular as he travelled East.[51] These were mere notebooks, however, containing records of Galen's anatomical observations and demonstrations (in which Boethus and other men of his social and intellectual rank had shared) so far. This programme of somatic investigation continued despite the departure of such a keen supporter, so when Galen revisited the subject of anatomy in literary form some years later, a much more detailed and accurate treatise, of greater length and clarity, resulted. The completion of *On the usefulness of parts* in the meantime – indeed its completion in time to send, as a whole, to a still alive and well Boethus, in this version of events – also contributed to the shape and structure of *On anatomical procedures*.

Whichever account is to be believed, the connection between the two major works is a clear and crucial one, and the most immediate impact of *On the usefulness of parts* on *On anatomical procedures*, as Galen himself emphasises, is precisely on its order. The original anatomical pairing had taken their arrangement (*taxis*) from the books of Marinus, which Galen had already epitomised (in four books); but the new improved version will instead follow that of *On the usefulness of parts*, and so begin with the hand.[52] Before turning to the various reasons Galen gives for this point of departure, it is worth saying a bit more about Marinus and the early Imperial intellectual and bibliographical trends he represents. For Marinus, active around the turn of the first into the second century CE, and perhaps based in Alexandria, is a key figure in the medical world of the Roman Empire, certainly for Galen, but also more widely. Galen credits him with reviving, or recovering, the study of anatomy, which had been in a state of neglect since the early Hellenistic era, meaning that Marinus revived the actual practice of dissection and vivisection (albeit on animal rather than human subjects), and pursued a systematic project of investigation into the body through such methods.[53] There is little independent evidence to corroborate Galen's claims about Marinus, but all that survives of his anatomical studies – that is the book-by-book outline provided by Galen as he describes his own abridgment of the text, and his scattered references to more concrete matters of content – indicates that here is expansion and elaboration, not summary and consolidation, of the canonical doctrines

[50] Gal. *Lib. prop.* 1 (*SM* II 96.19–24). [51] Gal. *AA* 1.1 (II 216–18 K). [52] Gal. *AA* 1.3 (II 234 K).
[53] Gal. *PHP* 8.1 (*CMG* V 4.1,2 480.28–30); and see Rocca (2003) 42–6 for further discussion.

of Herophilus and Erasistratus.[54] The magnitude of Marinus' undertaking (his *Anatomical books* were twenty in number), along with its innovative organisation (definitely not *capite ad calcem* – it begins, somatically, with the skin), and the points of positive contribution to anatomical knowledge Galen picks out, all suggest an ambition to outstrip, both quantitatively and qualitatively, what had gone before.

Marinus' influence is also demonstrated by his pupils, most prominent among whom were Quintus and Numisianus. They took up the anatomical baton, and passed it on to their own students in turn: men who were in some cases Galen's teachers, in others his antagonists, those whose dominant position in the field Galen wished to seize for himself.[55] The physician whom he most wanted to depose, and replace, in this respect was Lycus of Macedon, who seems to have died just before Galen arrived in Rome, but who left behind a set of anatomical texts that were widely considered to embody the current state of the art.[56] Part of Lycus' appeal was his direct pedagogical descent from Marinus – via Quintus – but Galen accuses him of squandering that inheritance, indulging in a kind of negligent and degenerative plagiarism.[57] He is reliant on the words of the master, but managed to introduce numerous errors and omissions none the less. Still, Galen deemed it worthwhile to epitomise Lycus, *Anatomical books* (nineteen in number), and to adumbrate their contents in *On my own Books*, before going on to list his works *On what Lycus did not know about anatomy*, and *On differences from Lycus on anatomy*.[58] This outline serves to show that, while Lycus returned to the head-to-toe principle, he added descriptions of 'the dissection of the uterus of a dead woman in which there is a foetus', as well as books on the anatomy of the newborn.[59]

In finding his own physiological order, therefore, Galen is reacting against Lycus as well as absorbing and surpassing Marinus. Neither *capite ad calcem*, nor Marinian, structure was permissible, though he certainly includes accounts of the dissection of pregnant goats in *On anatomical procedures*, and also utilises Marinus' more thematic approach to organisation

[54] Gal. *Lib. prop.* 3 (*SM* II 105.22–108.14, with the lacuna in the Greek filled in the Arabic, see Boudon (2002) which includes an English translation; and, e.g., *AA* 9.3 (II 716 K) and *Nerv. diss.* 5 (II 837 K).

[55] See, e.g., Gal. *AA* 1.1 and 8.3 (2.217–18 and 660 K).

[56] On Lycus' reputation in Rome see e.g. *Lib. prop.* 2 (*SM* II 101.26–102.10).

[57] Gal. *AA* 14.1 (I 232.14–233.5) (Simon (ed.) (1996)).

[58] This section of *Lib. prop.* is preserved only in Arabic, see Boudon (2002) 16–17 for an English translation.

[59] The translation is Boudon's. Despite the phrasing of the headings, which could be taken to imply not only adult human dissection, but also dissection and vivisection of human children, Galen's subsequent discussion refers only to animal dissection and vivisection, mostly of goats (see *AA* 12.3–6; I 144.15–154.7) (Simon (ed.) (1906)).

(Marinus' text, for example, treats the skin and flesh, and veins and arteries, separately, as global rather than local entities). There are some borrowings and reworkings then, but within a distinct overall architecture: an expanded architecture that took the Imperial revival in anatomy further into physiology (the two were always entwined in antiquity), by producing this interlocked pair of heavyweight texts – thirty-two books in total – and so really dominating this territory; and an architecture that is essentially ideological in its approach to the ordering of knowledge about the human body and its functioning, an approach that has much in common with the ordering of empire.

So Galen explains, and emphasises, in the opening sequence of *On the usefulness of parts*. Just as each living thing is a unity in the sense that it has clear borders, is not joined to any other living thing, so also are the parts (*moria*) of which it is composed. Except that these parts – such as the eye, nose and tongue – though having their own boundaries, having their own integrity, are also joined up, joined together to make the whole living thing of which they comprise the parts. These parts are varied in type and size, but the usefulness (*chreia*) of each is related to, depends on, the soul (*psychē*): for 'the body is the instrument (*organon*) of the soul'.[60] Living things with different souls will, accordingly, diverge with respect to their parts. So, the horse has strong hooves and a handsome mane to fit the swift and proud character of its soul, and the fierce lion has teeth and claws while the timid hare is quick but defenceless in its bodily form; but what about man? Man is clever (*sophos*), and even more decisively, shares in the divine (*theion*), so Nature (*physis*) provided him with hands, the best instrument in peace and war. He has no need for teeth or claws, for wielding a sword or spear is much more effective. Nor does he require speed, since, with his skilful hands, he has tamed the horse, which provides not only a means of escape but also a strong position for attack. Indeed, additional protection is offered by the fashioning of clothes and armour, the building of houses and fortifications; while the construction of hunting nets and fish traps demonstrates his lordship of all the creatures of land, air and water. The hands of peaceful (*eirēnikoi*) and social (*politikoi*) human beings, moreover, write laws, raise altars and statues to the gods, build ships, make flutes, fire-tongs and all other instruments of the arts. They even (and perhaps most importantly) compose works about the arts (*technai*), record their reflections on, and theories of, various crucial areas of human activity in writing.

[60] Gal. *UP* 1.2 (I 1.13–14) (=Helmreich (ed.) (1907–9)).

It is not, therefore, that man is the most intelligent of the animals because of his hands (as Anaxagoras had argued); but that, because of his superior intellect he has hands. Indeed, it is the combination of hands and reason that is vital, it is their conjunction that has produced the *technai* and all other human accomplishments. Rationality, as Galen puts it, 'is an art for the arts in the soul', while the hand, 'is an instrument (*organon*) for the instruments in the body'.[61] Furthermore, the hand is ideally constructed for this purpose, with its opposable thumb, its flexible fingers, its delicacy and strength and so on. The detailed elaboration of the hand's excellence takes up the rest of the first book, except for the closing paragraph, in which Galen outlines how the work will now proceed.[62] There will, he says, be movement from hand to arm in the next book, then he will 'explain the skill of Nature (*physis*) displayed in the legs', before advancing to the organs of nutrition, then of the *pneuma* (warm air that has become integrated into somatic functioning), reaching the head in Books eight and nine. More detailed discussion of the eyes and vision, then the rest of the face, will follow, with a journey down and then up the spine to the shoulders in books 12 and 13. The next pair of books cover the generative parts and pelvis; while the sixteenth broadens out to encompass the instruments common to the whole body – the arteries, veins and nerves – and the final book is labelled 'an epode', where all the parts, of body and text, are brought together, the overall utility of both is expounded. For the work itself is useful not just to physicians and philosophers, but also to all men, who will be brought into a better understanding of themselves and their universe by reading it. In particular, they will be brought into an appropriately pious attitude towards 'the power responsible for usefulness itself'.[63]

The journey around the human body that *On the usefulness of parts* describes does, therefore, possess a certain geographical logic: arms, legs, up the torso to the head then down the spine to the pelvis, with two general, totalising books to round things off after the focused start with the hand. However, the real architecture of the text, what gives it shape and structure, is clearly more conceptual and more ideological. It begins with a definition of the parts in relation to their determining whole, in relation to the specifically – rationally, socially, peacefully, intelligently – ensouled human being, and with an assumption about the existence of a beneficent creative force in the universe – Nature (or the Demiurge) – who

[61] Gal. *UP* 1.4 (Helmreich I 6.15–17); cf. Arist. *Part. an.* 687a7–18.

[62] Gal. *UP* 1.25 (Helmreich I 63.9–64.7).

[63] Gal. *UP* 17.2 (Helmreich II 449.17–18); and see Frede (2002) for further exploration of the theme of piety in the *UP*.

has fashioned all living things in accordance with the character and faculties of their souls; indeed, has made each part not only useful, and appropriate, but also for the best, absolutely optimally, in terms of the whole.[64] Optimal in more general terms too, for there is a clear hierarchy of beings at work here also, with man at the top, distinguished sharply from some of his closest rivals (such as the ape) on occasion.[65] The order of the work thus follows on from these points of cosmic order: that is why it opens with the hand, why the organs of nutrition, or generation, are grouped together, that is what makes sense of the sequence, just as the sequence itself makes sense of man.

'In *On the usefulness of parts* my aim was to explain the structure of all the human organs, as far as concerns the art', Galen asserts in one of the many introductory sequences in *On anatomical procedures*:

> In my present work, my aim is twofold; first that each bodily part, the actions of which I explained in the former work, may be accurately observed; and second to promote the proper end of the art.[66]

The objective of providing the means to see, to observe through dissection, the explanation of each part's function and excellence as already described, clearly involves following the same structure (*taxis*), as Galen repeatedly stresses; but this is not just a literary pattern, it reflects the cosmic order too, as is also frequently reiterated.[67] The hand, as 'most characteristic' of man, is the place to start, and the legs 'naturally' come next, as the instrument of man's distinctive upright posture.[68] Then there is a slight deviation from the established order, as Galen covers the whole anatomy of the muscles of the head and torso, and then returns to the pattern of *On the usefulness of parts*, with a final, foetal, addition. This signals the impact of previous works, not his own, on the text, and indeed, there is a running critique of contemporary anatomical inadequacies throughout. The reason the muscles receive such treatment, for example, is that, despite their importance for both understanding the general workings of the body, and ensuring successful surgical intervention, they are woefully neglected by current practitioners who deem them unworthy of serious attention.[69]

Vigorous polemic and self-promotion are permanent features of the Galenic project, but so too is the Roman empire, and this comes very clearly

[64] On this optimising notion (and its problems) see, e.g., Hankinson (1989).
[65] Gal. *UP* 1.22 (Helmreich I 58.13–59.20) [66] Gal. *AA* 4.1 (II 415–16 K).
[67] See, e.g., Gal. *AA* 1.3, 2.3, 4.1 (II 234, 291, 417 K). [68] Gal. *AA* 2.3 and 4.1 (II 291 and 415–16 K).
[69] Gal. *AA* 4.1 (II 416–19 K).

to the fore in *On the usefulness of parts*, ably supported by *On anatomical procedures*. Indeed it is possible to figure the former as a discourse *about* Empire. Its opening definition of a body part, *morion*, as something distinct but joined up with others, something that has its own identity, but within a wider framework, as it is the whole that determines its function and makes it useful, works well also for an imperial part, a province. The similarities are reinforced by the role of the soul in this picture: either in the general, unified, form in which it appears in the introductory sections of *On the usefulness of parts*, or its more specific, ruling aspect – *hegemonikon* – which also makes an occasional appearance in the same work. The basic point, however, is that there is something in charge of all the parts, which has a somatic location, in the brain in Galen's view, and provides a kind of centralised government for the body, as the emperor does for the Empire.[70] All forms of sensation and perception are communicated to the brain through the sensory (*aisthētika*) nerves, while out along the motor (*kinētika*) or deliberative (*prohairētika*) nerves goes the signal for voluntary movement, either in response, or just in general. The central site, or source (*archē*), of this network, the *hegemonikon* itself, has to be engaged in this process, everything has to go through the centre; and it was against this assumption that the key concept of 'the reflex', the idea that action could start and finish at the somatic periphery was developed in the seventeenth and eighteenth centuries.[71]

Patterns of imperial governance seem to be replicated here too, then. The model of provincial report, or petition, then imperial response, of everything going through the centre, is somatically re-enacted. And though there are some decentralising, or, more accurately, multifocal tendencies in the Galenic body, as the brain is not the only bodily *archē* but is accompanied by at least two others – the heart which is the source of the arterial system and the liver which is the source of the venous network – these can also be integrated into the imperial vision.[72] For it is lower level administrative activities that are located at these sites: the management of the basic processes of nutrition and respiration, for example, the ongoing vitalisation, and integration of the body, just as the more mundane business of maintaining the Empire went on outside Rome. Then there is the figure of the

[70] The analogy is made explicit by, e.g., Florus (2.14.5–6), who speaks of Augustus establishing his monarchic rule like that of the soul (*anima*) over the imperial body (*imperii corpus*), and on this notion more widely see McEwen (2003).

[71] See Canguilhem (1955).

[72] The *Ars med.* has four *archai*, with the testicles joining the more standard three, which led to questions being raised about its authenticity: see Kollesch (1988). Her doubts are answered by Boudon (1996).

beneficent and powerful creator – Nature or the Craftsman – who stands above all this, who underwrites the coherence and explicability of the entire system, who gives it meaning, makes sense of it all; that is, who shares some of the same ideological space as the emperor, and also the gods; as indeed both creator and ruler are divine. If it is objected that this is to produce two emperors: the practical rule of the soul has now been displaced by the ideative domination of the Demiurge, then Galen would agree that this is a problem. He wanted and tried to bring the two together, to merge or at least clearly articulate them, in *On the formation of the foetus*, but found it difficult, particularly in terms of giving his conceptual understanding concrete form.[73] Moreover, it could also be said that the divisibility of the emperor as man and god, functional and figurative autocrat, was an issue in the Roman world more broadly.

Still, these reiterations, echoes, of empire in medical form, should not be overplayed. The match is not perfect, there is no exact homology, and many of the key themes and concepts on the medical side, go back not only to Ptolemaic Alexandria (an imperial capital after all), but as far as democratic Athens also. The Demiurge is borrowed from Plato, as also the tripartition of the soul, though many Aristotelian and Stoic ideas and interpretations have also become involved in Galen's system. The centralised conceptualisation of somatic function and control, the *archai* and their networks, belong originally to Herophilus and Erasistratus, though not entirely identically. This too has been added to, amended and reshaped, since: perhaps most importantly through an ongoing engagement with the pneumatology (though *not* the cardio-centrism) of the Stoics. Galen's version probably owes a particular debt to the Stoicising medical lineage founded by Athenaeus of Attaleia, and continued by Archigenes of Apamea (among others), in this respect.[74] In neither case does Galen himself bring much that is new and original to the mix, except in joining them up, in the particularities of the far more encompassing combination in which they participate.

But that is to bring things back to the Roman Empire once again, back to its own processes of formation, organisation and integration. To the Empire as an essentially synthetic political and cultural production itself, and one

[73] Gal. *Foet. form.*4–5 (*CMG* v 3,3 78.12–90.26). The question is: how is the generic, cosmic design and creativity of *Physis* enacted, realised, individually in the construction of the foetus? The direct involvement of *Physis*, on the one hand, and control and guidance by the rational soul, on the other, are the two initially most attractive possibilities, but neither is satisfactory, and Galen is left admitting uncertainty somewhere between the two, unsure how to link the Demiurge and the controlling, causative powers in each human being.

[74] On Athenaeus and 'the *pneumatikoi*' see, e.g., Nutton (2004) 202–5.

that fostered further intellectual synthesis within its borders. The development of 'syncretism', or 'eclecticism' – the pooling of theoretical and conceptual resources, as sectarian boundaries softened (but did not disappear) – in medicine, philosophy and other fields of knowledge and understanding, from the first-century BCE onwards, has been much remarked on.[75] And while the earlier, derogatory, interpretations of both the phenomenon itself and the role of Rome in its appearance have been largely discarded, a sense of connection between the two persists: Rome, the expansion and consolidation of Roman power in the Mediterranean, had some role to play in making a wider range of options available, concurrently and inclusively, to those engaged in a whole host of intellectual endeavours, with divergent approaches. It was, of course, the conflict between Rome and Mithridates that broke the line of authoritative descent in the Athenian philosophical schools, and so disrupted their claims to exclusive ownership of the ideas, and writings, of their founders and successive lineages. Nor was it just philosophical authority that was dispersed and re-located at that time; Actium marked a shift in the centre of medical (and other scholarly) gravity from Alexandria to Rome. More broadly and abstractly, the Roman Empire (following on from its Hellenistic forerunners) encouraged a kind of universalism that is clearly reflected in a variety of intersecting discourses which flourished in the Imperial period. As Rome forged a rough political unity from its conquests, it helped to engender a single community of truth. The diverse sources of information and interpretation it held within itself, historically, geographically and ideatively, all shared a certain status, and so could be drawn on, mobilised, in the service of a range of different systems and projects.[76]

This was not, of course, a world of equality. Some contributions might be adjudged to be more successful or useful than others, and the point was to prioritise, to select, combine and organise, according to individual allegiances, principles and objectives; but in a more flexible and inclusive environment than before. Which is to return the discussion to matters of order, matters which become more pressing given the scale of this imperial community of truth; the sense in which the Empire made more resources available to those involved in generating and mapping knowledge and

[75] The now standard study of 'eclecticism' is the collection edited by Dillon and Long (eds.) (1988); see also Sedley (1989); and, e.g., Gill (2003) for discussion of how these terms are now understood in the context of Roman Stoicism. Galen is an established participant in these 'eclectic' evolutions.

[76] So, at least, many active in a range of intellectual spheres clearly felt; but there were also dissenters, continuing partisans of more particular paths to truth, such as Alexander of Aphrodisias, though he was certainly not entirely unaffected by contemporary trends.

understanding concretely, practically, as well as abstractly and ideologically. Amongst this wealth of imperial resources are, however, organisational forms, patterns of thought and practice, structures of meaning and existence, which provide the means to meet these challenges. Galen draws on, adopts and adapts many of these approaches to order – old and new, medically established, or more externally derived – in his works, and the contours, the texture, of the Roman Empire can be seen both in some specific cases, and in this plurality itself. So, there are some distinctly imperial forms of organisation manifest, textual orders which are original to, or more positively derived from, Rome's empire, and there is a general revelling in its encompassing power, its gathering up, mixing and maintenance of multiple traditions.

Moreover, as countless critics of 'colonial discourse' in other times and places have emphasised, this kind of textual participation in the patterns of empire serves to strengthen imperial rule regardless of actual commitment.[77] Even if Galen is just taking his cue, his models and metaphors, from the way the world is and works, is simply utilising the available means of persuasion, and modes of understanding, his re-inscription of the surrounding structures of domination, his particular retelling of imperial stories, reinforces them through repetition, through the display of their efficacy, through the exclusion of other possibilities. There are some indications of commitment to be found too. Not in terms of explicit political allegiance, though Galen's association with and praise of emperors like Marcus Aurelius and Septimius Severus, as well as his involvement with such leading men of the empire as Boethus, should not be discounted here;[78] but in terms of cosmic adherence and alignment, as *On the usefulness of parts* illustrates. In a structural sense, and in respect to scale, Galen's worldview has a lot in common with that of Rome's rulers. His position and perspective are Roman imperial creations, and, though his theoretical ambitions may be more traditional, many of his ideas about good order converge with the Roman imperial order. On a fundamental level, moreover, he recognises and accepts that, and that recognition is a mutual, and mutually fruitful, one.

Now, Galen's Hellenism has been rather muted in this discussion of the order *in* the books. The fact that his empire of knowledge is in many ways a Greek cultural construction has been left largely unremarked, not subject to much analytical scrutiny so far. After all, all the formulations of the *iatrikē*

[77] See, e.g., the collection of essays edited by Gates (1986).

[78] See, e.g., Gal. *Praen.* 11.1–10 (*CMG* v.8.1 126.16–130.10) on Marcus and *Ther. pis.* II (xiv 218–19 K) on Severus.

technē and the philosophical debates he engages with, all the organisational precedents and literary materials he draws on, were at least articulated and written in Greek, if not by Greeks, nor indeed otherwise uncontaminated by things Roman and imperial. However, these latter caveats are crucial, for they clearly demonstrate the complications, the problems, which attend on the very category of *Greek* culture, or knowledge, itself in the Roman Empire. Some further exploration of these issues will, therefore, help to illustrate the depth of Galen's inevitably imperial entanglements, as they are also shared with, or follow on from those involved in similar intellectual projects around, or before him.

THE GREEK ORDER IN THE BOOKS?

Galen's world of knowledge was one to which freeborn Roman citizens, of impeccably Italian stock, had long contributed, in Greek – as Sextius Niger had done in the field of medicine, as well as his friend Julius Bassus – or otherwise.[79] Dioscorides, moreover, labels both Niger and Bassus 'Asclepiadeans' (followers of the innovative physician and medical thinker, Asclepiades of Bithynia, who had found fame and influence in late Republican Rome), so their participation in the Greek medical tradition was not merely linguistic, a point that Galen himself reinforces with his own respectful reference to Niger in his discussion of pharmacological predecessors and their organisational tactics in *On the mixtures and properties of simple drugs*.[80] Indeed, Galen groups Niger together with Dioscorides, Heraclides and Crateuas, without making any particular distinction between them. Still, Dioscorides himself hints that Niger may have paid more attention to Italian flora than others had done (though without the requisite accuracy); and his ethnic identity was not irrelevant to Pliny the Elder either, who (implicitly) casts him as a traitor to his Quirital status.[81] Insofar as Galen engages with Niger as a medical authority, one who may indeed have presented his simples *kata stoicheion*, just as the Pergamene did, this is then a rather complexly, and surely not exclusively, Hellenic encounter.

[79] Bassus' Greek medical writings appear alongside Niger's in Pliny's lists of authorities in book 1 of the *HN* (for books 20–7 and 33–4); and he is referred to by Caelius Aurelianus as his friend (*CP* 3.16.134).

[80] Dioscorides, *De materia medica* pr.2 (I 1.20–2.5) (Wellmann (ed.) (1906–14)); Gal. *SMT* 6 pr. (XI 797 K).

[81] Dioscorides, *De materia medica* pr. 3 (I 2.5–8) (Wellmann (ed.) (1906–14)); Plin. *NH* 29.17, where Niger is not actually named amongst those few amongst the Quirites to have practised the medical art and 'immediately fled to the Greeks/*statim ad Graecos transfugae*', but given that Pliny explicitly lists his (and Bassus') Greek medical writings amongst the home authorities in book 1, the direction in which the finger points is pretty clear.

Scribonius Largus poses the question of how the Greek status of Galen's (or, indeed, anyone else's) knowledge is to be judged still more acutely. On the one hand, he has a Roman name, his only surviving work is in Latin and it was addressed to a freedman of the Emperor Claudius.[82] On the other hand, Scribonius locates himself firmly within the Greek medical tradition in the dedicatory epistle which prefaces his Latin collection of recipes, and the form, contents and (as has already been mentioned) organisation of those recipes, broadly fits that Hellenic bill, though several members of the Julio-Claudian dynasty, from Octavia to Messalina, have joined the names which give authority and credence to the remedies provided and there are other signs of a certain Roman ambience too.[83] The general point about the character of Scribonius' prescriptions is, however, again emphasised by their Galenic intersections. A number of the same recipes, explicitly attached to the name, and indeed books, of Scribonius, feature in both Galen's works of compound pharmacology, and there is further, tacit, overlap of material too; though this relationship is most likely an indirect one, the result of Scribonius and Galen sharing sources, or just the common currency of certain items in the pharmaceutical repertoire, such as the Mithridatic antidote or theriac.[84]

Most of the recipes positively attributed to Scribonius arrived in Galen's pharmacological compilations via the earlier treatises of Asclepiades Pharmakion, but that is not entirely the case, and it serves only to defer the question of access.[85] If not Galen, then did Asclepiades, and perhaps other Greek physicians of his generation in late-first century CE Rome, read Latin and use and incorporate Latin medical writings? It should be said, again, that Scribonius' is not the only Roman name to feature in Galen's works by any means, especially in his collections of compound *pharmaka*, though explicit literary, rather than just proprietorial or practical, reference is rarer. The only other solidly Roman author with any real medical presence in Galen is Aelius Gallus, the Augustan prefect of Egypt and invader/explorer of Arabia. Andromachus the Younger, explicitly takes various recipes 'from

82 The dedicatee is C. Iulius Callistus, a powerful freedman who successfully made the transition from Caligula's to Claudius' service before his death in (or before) 51 CE (his demise is rather enigmatically mentioned in the epitome of Dio Cass. 61.33.3).

83 Scrib. *Comp.* 59: Octavia's dentifrice; 60: Messalina's, also used by Augustus (35.5–10 and 11–23 Sconocchia).

84 See Gal. *Comp. med. loc.* and *Comp. med. gen.* (XII 683, 764, 774 K; XIII 51, and 737–8, 828 K) for the (rough) reappearance of Scrib. *Comp.* 51/2; 27; 26; 75; and 223; 247/8 respectively. There are other explicit Galenic citations not found in the *Comp.*

85 All those recipes explicitly taken 'from the (books) of Scribonius' (XII 764 and 774; XIII 314 and 828 K) come via Asclepiades, but there are more attributions than that.

his books', and there are other attributions too.[86] It is also worth mentioning that Galen criticises this Andromachus (in contrast to his father) on one occasion for using a Latin, rather than Greek, plant name in a recipe for theriac.[87] His competence in Latin seems assured, therefore, and a similar capability is likely for Asclepiades too; but an equally strong case can be made for the Greek proficiency of Scribonius and Gallus. Indeed it has been argued that Greek was Scribonius' mother tongue, that he was a Greek freedman who wrote the *Compounds* in Latin to curry imperial favour, and that Galen preserves a more linguistically representative sample of his literary output.[88] Such a view is, however, based on assumption and stereotyping rather than any actual evidence, and more recent scholarship has favoured freeborn Roman citizen status, with perhaps Sicilian origins.[89] Even without such a bilingual background, Scribonius' facility with, and mastery of, his Greek material, is manifest in his surviving work anyway, and he could easily have written works in Greek as a second language. Gallus too: as a well-educated Roman aristocrat whose cultural and intellectual interests are indicated by his association with Strabo as well as his medical forays, Greek composition would certainly have been well within his compass.[90]

The real point to take from all these possibilities, this shifting of language and perspective, is, as Vivian Nutton has said, 'the ease with which Latin and Greek information could now be interchanged', an interchangeability which obviously puts the integrity of both categories into question.[91] This reciprocality, this sharing, goes beyond information. Both Scribonius and Dioscorides, for example, associate themselves (rather loosely) with the Roman army as a vehicle through which knowledge of the medical riches of the Roman Empire can be acquired, and it is also worth noting that Dioscorides had both a Roman patron and Roman citizenship (whether inherited or acquired).[92] Amongst his teachers Scribonius counts

[86] This is assuming that all the Gallus references, except that to 'Marcus Gallus the Asclepiadean' (XIII 179 K) are to Aelius, even when not actually thus specified, which is not completely certain, though reasonably secure. The literary references in Galen are then to be found at XII 625; XIII 28, 77, 202, 556 and 838 K; and see also XII 625, 738 and 784; XIII 29, 138, 310, and 472; XIV 114, 158, 189, 203 K.

[87] Gal. *Ant.* 1.7 (XIV 44 K).

[88] See, e.g., Schonack (1912) and Kind (1921).

[89] Kudlien (1986) esp. 23–5; Langslow (2000) 51–3; and, for a more Sicilian perspective, Nutton (2004) 172.

[90] Strabo accompanied his 'friend and companion' Gallus on tours around Egypt (Strab, e.g., 2.5.12; 11.11.5; 17.1.29–46), and also described his Arabian campaign; but whether this description was based on a spoken or written account, and in what language, is unclear. This point, and the general question of Strabo's competence in Latin, is discussed in Dueck (2000) 87–96.

[91] Nutton (2004) 172.

[92] Scribonius was part of Claudius' British expedition, in some capacity (*Comp.* 163: 79.20–22 Sconoocchia); and Dioscorides (infamously) refers to his 'soldierly life' (*De Materia Medica* pr.4: 1

both Trypho, presumably the (Elder) Trypho, who had come to Rome from Cretan Gortyn to make his name in surgery, and Vettius Valens, a Roman *eques* whose medical attendance on the empress Messalina was to get him into trouble; and perhaps also Apuleius Celsus of Centuripae who was certainly Vettius' *praeceptor*.[93] So, not only was medical education – both teaching and learning – a mixed affair in imperial Rome, but so too was medical service at the imperial court.

Many of the texts Galen engages with, reacts against and draws on, come out of this mix, have been shaped by their contact with the institutions and instruments of Roman power, regardless of their language. There is a sense in which learned medicine had already been imperialised, in its scope and structure, its human and material resources, its social location and associations, long before Galen, and he does not reject those developments, that inheritance, as such. He does, of course, deploy the classical Greek past – most especially his interpretation of Hippocratic doctrine – as a basic measure against which to judge all that has followed; but he is well aware that there have been considerable advances, as well as plenty of wrong turnings, since. It is doctrinal and methodological, not temporal or cultural conformity that is the key. Starting from Hippocratic foundations, Galen seeks to build a system that, for example, integrates not only the crucial anatomical discoveries of Herophilus and Erasistratus, but also the gains of Marinus and his more diligent followers; that incorporates both Hellenistic and Roman expansions of the therapeutic repertoire. And, in many ways, it is the newer arrivals who have had the greatest impact on the organisation of his works, the order *in* the books, even if it is far behind them that Galen claims his most fundamental allegiances lie. Still, the shape of *On anatomical procedures* owes more to Marinus (even Lycus) than to Herophilus. Archigenes appears to be the literary model followed, not only in *On the affected places*, but also in Galen's main sphygmological treatises; and his reliance on the more recent, and more manifestly Romanised, pharmacological texts for both material and order, on a number of levels, has also been repeatedly revealed. Even his style of Hippocratic commentary may be a Roman Imperial phenomenon.[94]

2.18) (Wellmann (ed.) (1906–14)). His patron was Laecanius Bassus; and his citizenship is implied by his name – *Pedanius* (or *Pedacius*) Dioscorides: on all these issues see Scarborough and Nutton (1982).

93 For Trypho see Scrib. *Comp.* 175 (and also, e.g., Celsus 6.5.3 and 7.pr.3); Valens appears in the index; and Apuleius in 94 and 171 (Sconocchia (ed.) (1983): 83.8; 9.18; 49.17 and 81.22 – with apparatus – respectively).

94 Though Hippocratic interpretation began in Hellenistic Alexandria, the interests there seem to have been more lexicographical, and the only surviving representative of this exegetical phase – Apollonius

Certainly Galen's general interpretation and understanding of Hippocratic doctrine, his particular construction of this crucial past authority, owes much to his present: most directly to his own teachers, but also to wider recent trends in learned medicine.[95] It is no accident that the previous exegete he speaks most highly of is the Trajanic physician Rufus of Ephesus, who seems to have shared many of Galen's key commitments in respect to both Hippocrates and the medical art more broadly.[96] Galen's Hippocratism provided a foundational link with the prestigious Greek past, therefore, but not in such a way as to occlude subsequent developments, the continuity and cogency of classical medical history right up to the time of his own *didaskaloi*. Rather the reverse: though much rubbish and error has to be rejected and corrected, Galen wants to mobilise the more valuable and worthwhile aspects of this continuity and mould it into an upward spiral. Post-classical progress, properly assessed, acquired and managed, allows him, with all his skills and talents, to return to the Hippocratic point of departure at a level far above that at which the great man himself was forced, by historical circumstance, to operate.

Galen has, therefore, much invested in the association, the complicity, of past and present; and he experiences little nostalgia for archaic forms of textual organisation, for Hippocratic styles and structures, for the inconcinnities and disorder of the Hippocratic Corpus itself.[97] Medicine has come a long way since then, even if faith should be kept with the Hippocratic founding principles of the art. More generally, moreover, Galen is uninterested in denying the fact that Greek culture is now contained within the Roman Empire, has been shaped and structured by Roman power. What he does attempt to do is create, through repeated acts of evaluation and emphasis right across his oeuvre, a certain moral topography of empire that is distinct from its political patterning, and gives ethical precedence to things Hellenic. These patterns mostly co-exist, rather than confronting each other, indeed, they sometimes intertwine and overlap, as well as occasionally conflicting, and, it has to be said, this is all part of the way the Roman Empire worked. Still, insofar as Galen does essay some kind of disaggregation of things Greek from things Roman, or at least tries

of Citium's commentary on the Hippocratic text *On joints* – is a paraphrase, rather than a 'phrase by phrase' exposition, of the work, such had become fashionable by Galen's time.

95 On Galen's particular debt to his teachers see Manetti and Roselli (1994) esp. 1580–93.

96 Gal. *Ord. lib. prop.* 3 (*SM* II 86.13–87.23).

97 See Sluiter (1995) for discussion of Galen's attitude to Hippocratic language and style. He does generally try and defend it, but his defensive posture is itself indicative of the difficulties, which he certainly acknowledges. See also Langholf (2004) for discussion of the 'chaotic' textual structure of many Hippocratic treatises.

to impose a more Hellenic order of knowledge on the hybrid formations of empire, it is to be found in the taxic endeavours directed at his own work, at his own oeuvre and the art it enacts, the endeavours with which this essay began. These now need further examination, in conclusion.

THE ORDER *OF* THE BOOKS

So far the scale and ambition of Galen's overall knowledge project – his drive to cover and connect all the parts and aspects of medicine and their philosophical foundations or framings, as well as the various linguistic issues implicated in the literary presentation of both – has mainly been mapped on to the same features – the reach, scope and integrity – of the Roman Empire. More implicit have been their more theoretical, conceptual underpinnings, as Galen draws on the systemic projects of the most influential currents of Hellenistic thought, the Stoics and the Epicureans. It is these schools that explicitly articulated the ideal of the fully integrated, holistic, philosophical system, in which all the relevant material, methods, approaches and understandings, are encompassed within well-articulated parts that fit together in a seamless whole; and leading figures within them, most especially Chrysippus within Stoicism, attempted to deliver on that promise in literary form.

The Empire, however, enabled and encouraged Galen to exceed these previous efforts in various ways; as it had already acted on other Greek authors involved in large-scale literary projects of knowledge generation, organisation and management under the Principate. The textual, and conceptual, assemblages of Strabo and Plutarch, for example, or indeed Galen's older contemporary, Ptolemy, all illustrate the ways in which Rome continued to expand Hellenistic horizons, to augment resources and multiply the programmatic possibilities, make available more combinations and conjunctions of ideas, disciplines and genres.[98] Traces of the same trajectory can be seen in medicine, despite the loss of so much material from the generations preceding Galen. The early imperial growth in pharmacological and anatomical writings has already been noted, for instance, and there were also renewed debates about the proper partition of the medical art at this time, about how the more synthetic enlargement of its 'rationalist' traditions should be managed.[99] The unity of the *technē* remained a

[98] On Strabo in this context see, e.g., Clarke (1999); for Plutarch see, e.g., Jones (1971); Duff (1999). Ptolemy is less well served as a cultural, rather than scientific, figure, but his disciplinary and methodological combinations are certainly distinctive.

[99] On these debated divisions in the Imperial period see, e.g., Flemming (2000) 90–1 and 185–196.

fundamental commitment, but its increased scope and content, together with the way in which certain concepts, theories and approaches were increasingly held in common, put more emphasis on its internal divisions, on the formation and fit of the parts of medicine, as essential to the maintenance of both the control and coherence of, and identity and difference within, the art.[100]

The challenge for Galen, as also for these other writers, whether medical or not, was, therefore, to control and even harness this excess. Galen wanted to exploit the Roman Empire in order to surpass his Hellenistic predecessors, while remaining true to the Hellenic principles which enabled him to assert that he had outstripped them, on their own terms, rather than entering a different competition. Success in this endeavour also provided something attractive to sell back to the Roman Empire itself, as the best, most advanced and complete rendition of the field; a knowledge project which has drawn on the resources of Rome and has something to offer in return, in the service of Roman power. Galen (and Ptolemy) make that offer much less explicitly than Strabo (or Plutarch), for example, but it is still there: the tacit presumption that encompassing and ordering the whole medical art together with all the neighbouring areas of expertise and understanding on which it depends, will be of benefit to society more broadly, and could strengthen a similarly constructed political formation.

What Galen shares, more openly, with authors such as Strabo and Plutarch is an insistence that a key element of the service offered is ethical; that the engagement between Greek knowledge and Roman power they are involved in has serious moral content, contains moral messages for Rome, her rulers and elite populations more broadly.[101] This is an integral part both of the way Greek culture functions within the Empire, in general – as, *inter alia*, a kind of ethical pole, a complex discourse of evaluative distinctions – and of the particular projects in question. In this latter respect it is, as well as being a point of undoubted personal conviction, a tool of management, of continuity and control. The literary enactment and advocacy of certain Greek values establishes a link with the classical past which can be brought forward into the present, brought to bear on the material being dealt with in any text, and, through the example and teaching of these texts and the

[100] Dissenting, non-eclectic, and more committed sectarian approaches were still possible, however: the Methodists flourished in the Imperial period, and the Empiricists had something of a second-century CE revival in fortunes too. Still, the surviving works of the great Methodic physician Soranus of Ephesus still demonstrate some of the same concerns with organised expansion of the art: see, e.g., Hanson and Green (1994) for an overview of Soranus' oeuvre.

[101] Roman *and* Greek elite populations, of course, both may be, and are, addressed in this context, severally and jointly.

oeuvre they constitute, brought to bear on wider society too. This is a form of managing – reining in, ordering – imperial abundance Greek-style, as distinct from the Roman style adopted by, for example, Pliny the Elder in his *Natural history*.

Which is, broadly speaking, what Galen attempts, in relation to both the art and his oeuvre, in his specifically taxic endeavours. He attempts to assert a basically Greek order, an order constructed according to Greek principles and associated with Greek values, which meets the demands of both continuity and control, within, and in contradistinction to, the disorderly propensities and practicalities of the contemporary Roman world. Some of the ways in which Galen casts the unfortunate present exigencies against which his struggle for order has to be waged as Roman and imperial have, indeed, already been discussed. Rome is, as has been mentioned, the main (but not sole) location of the problems of ignorance, flawed judgement and excess, which Galen confronts in *On my own books* and *On the parts of the art of medicine*. The same tribulations are less localised in *On the order of my own books*, and *On my own opinions*, but their imperial patterning remains implicit, and this theme serves to connect all these treatises. The Greekness of the ordering that these works, and the *Medical art*, strive to establish, and enact, is also implied rather than explicitly asserted, indeed little actual explanation is offered for the various sequences suggested at all. Still, there are, again, certain shared patterns of identification and evaluation that can be clearly discerned.

What emerges from the text of *On the Order of my own books* (at least as it survives in Greek) and from the more summary listing of works at the end of the *Medical art*, is a progression from fundamentals, from works that establish basic epistemological principles and medical methodologies, through the main parts of the art – through knowledge about the human body in health, about disease and sick bodies, and about cures, the recovery (and maintenance) of health – to various reflections on it, mainly in the form of Hippocratic commentaries, then some extra philosophy and philology.[102] The thematic arrangement in *On my own books* also follows roughly the same course. There is, then, an intention to begin at the beginning, with first principles, with what a physician, or anyone who wishes to understand medicine, needs to grasp right at the outset, before proceeding through various logical stages of knowledge acquisition to a final consolidation, elaboration and even ornamentation of the whole, although, as so often, this intention is not entirely realised, since Galen actually identifies three

[102] The Greek text has a lacuna of several pages, though fuller Arabic translations may survive.

possible starting points in *On the order of my own books*: the fundamental pairing of *On the best sect* and *On demonstration* (also cast in the same role at the end of the bibliography in the *Medical art*); the basic set of introductory works (such as *On the sects for beginners* and *On the pulse for beginners*); and (only admitted in the last line of the text) his treatise on the correct use of words.[103] The *Medical art* also begins its listing, as it does its summary of the *technē* itself, with works which describe the constitution of the art as a whole, and in the context of the other *technai* too, for this is the point from which to commence its break down – *dialysis* – into its part and provinces, for definition and description.[104] *On my own opinions* also selects a distinct set of fundamental issues with which to open.

Whatever the precise details, however things work out exactly, Galen is in each case applying, attempting to apply, a logical, orderly, method. In his mind, moreover, this type of systematic approach to things, working from and through first principles, is Greek. It derives from a set of general Greek intellectual values, and has, more specifically, been forged through his engagement with the greatest figures of Greek thought: Plato, Hippocrates and Aristotle, as well as Chrysippus and Epicurus. It possesses, moreover, a kind of timeless truth, an absolute and abstract validity, that contrasts with the mess, the errors, of the Roman present. If, then, on a basic structural level in his works, Galen's idea of good order closely resembles the Roman imperial order; on the higher, more conceptual, level of the *iatrikē technē* itself, these works are to be ordered according to Greek ideals, however hard that may be.

Once again, however, there is no contradiction between the two. The timeless Hellenic truth is, even on Galen's reckoning, a participant in the present Roman mess, albeit a lamentably neglected and downtrodden one; and many scholars of the 'Second Sophistic' would go further, figuring it, in its very timelessness, as a creation of Roman rule.[105] Galen's ideal iatric order nestles neatly within, as much as it transcends, the overarching architecture of Roman power. Space for the *technai*, the *artes*, had been established quite early on in Rome's imperial endeavours; a contested space in many ways, but productively rather than problematically so.[106] All the way through, then, in all his approaches to organising and presenting knowledge, Galen

[103] Gal. *Ord. lib. prop.* 1–2 and 5 (*SM* II 82.16–84.10 and 90.14–17); *Ars med.* 37.14 (392.9–12 Boudon). For further discussion of this plurality see Mansfeld (1994) 117–26.

[104] Gal. *Ars med.* 37.6 (388.4–8) (Boudon (ed.) (2000)).

[105] So Swain (1996), 65–100; and see also, e.g., Bowie (1974).

[106] Varro's *Disciplinae* and Celsus *Artes* testify to this establishment, not to mention the proliferation of technical treatises in both Greek and Latin under the Empire.

remains in the Roman Empire; but this is a dynamic and diverse domain, a complex cultural formation as well as a particular political structure, and the two cannot be separated. Which is to return to the imperial interplay between abundance and control, both for Galen and Rome. Both end up striking a similar balance between the two, exerting their control through formally similar mechanisms, imposing an order that allows plurality but not chaos. So, Galen's writing, the various systematisations he proposes and enacts within, and of, his huge literary output, works for the Empire as much as the Empire works for him.

Bibliography

Adams, J. N. (2003) *Bilingualism and the Latin Language*. Cambridge.

Adams, J. N., Janse, M. and Swain, S. (eds.) (2002) *Bilingualism in Ancient Society: Language Contact and the Written Text*. Oxford.

Aherne, C. M. (1966) 'Late visigothic bishops, their schools, and the transmission of culture', *Traditio* 22: 435–43.

Ahl, F. (1985) *Metaformations: Soundplay and Wordplay in Ovid and Other Classical Poets*. Ithaca.

Alder, K. (1995) 'A revolution to measure: the political economy of the metric system in France', in Wise, M. N. (ed.) *The Values of Precision*. Princeton: 39–71.

Alexanderson, B. (ed.) (1967) *Peri Kriseōn Galenos*. Stockholm.

Amarelli, F. (1983) *Consilia principum*. Naples.

(1996) 'Giuristi e principe da Augusto agli Antonini. Conflitti compromessi collaborazioni', in Mantovani, D. (ed.) *Per la storia del pensiero giuridico romano: da Augusto agli Antonini*. Torino: 183–97.

Amsler, M. E. (1989) *Etymology and Grammatical Discourse in Late Antiquity and the Early Middle Ages*. Amsterdam.

Ando, C. (2000) *Imperial Ideology and Provincial Loyalty in the Roman Empire*. Berkeley.

André, J. (ed.) (1981) *Isidorus Hispalensis Etymologiae XVII*. Paris.

(ed.) (1986) *Isidorus Hispalensis Etymologiae XII*. Paris.

Arafat, K. (1996) *Pausanias' Greece: Ancient Artists and Roman Rulers*. Cambridge.

Archi, G. G. (1981) '*Lex* and *natura* nelle istituzioni di Gaio', in Archi, G. G. (ed.) *Scritti di diritto romano I*. Milan: 139–71 (first published in 1978).

Arévallo, F. (ed.) (1797–1803) *Sancti Isidori Hispalensis Episcopi Opera Omnia* (J. P. Migne, *Patrologia Latina LXXXI–IV*). Rome.

Arnar, A. (1990) *Encyclopedism from Pliny to Borges*. Chicago.

Arnott, W. G. (2000) 'Athenaeus and the epitome: texts, manuscripts and early editions', in Braund and Wilkins (eds.): 41–52.

Aronadio, F. (1990) 'Due fonti laerziane: Sozione e Demetrio di Magnesia', *Elenchos* 11: 203–55.

Arrighetti, G. (2003) 'Filodemo biografo dei filosofi e le forme dell'erudizione', *CErc* 33: 13–30.

Ashby, T. (1935) *The Aqueducts of Ancient Rome*. Oxford.

Asmis, E. (1984) *Epicurus' Scientific Method*. Ithaca.
Astarita, M. (1993) *La cultura nelle* Noctes Atticae. Catania.
Aubenque, P. (1992) 'L'histoire de la philosophie, est–elle ou non philosophique? Oui et non', in Cassin (ed.): 17–36.
Auerbach, E. (1953) *Mimesis: The Representation of Reality in Western Literature* (trans. W. R. Trask; first published in German in 1946). Princeton.
Bakhtin, M. (1981) *The Dialogic Imagination: Four Essays* (ed. M. Holquist, trans. C. Emerson and M. Holquist). Austin.
Baldini Moscadi, L. (1979) 'Magia e progresso in Manilio', *A&R* 25: 8–14.
Baldry, H. C. (1965) *The Unity of Mankind in Greek Thought*. London.
Banniard, M. (1992) *Viva Voce: communication écrite et communication orale du IVe au IXe siècle en occident latin*. Paris.
Barchiesi, A. (1996) '*Extra legem*: consumo di letteratura in Petronio Arbitro', in Pecere, O. and Stramaglia, A. (eds.) *La letteratura di consumo nel mondo greco-latino*. Cassino: 189–208.
(1997) *The Poet and the Prince: Ovid and Augustan Discourse*. Berkeley.
Bardong, K. (1942) 'Beiträge zur Hippokrates- und Galensforschung', *NGG* 7: 577–640.
Barnes J. (1997) 'Logique et pharmacologie: à propos de quelques remarques d'ordre linguistique dans le *de simplicium medicamentorum temperantis ac facultatibus de Galien*', in Debru (ed.): 3–33.
(2002) 'La philosophie entre guillemets', in Canto, M. and Pellegrin, P. (eds.) *Le style de la pensée: recueil de textes en hommage à J. Brunschwig*. Paris: 522–47.
Barrow, R. H. (1967) *Plutarch and his Times*. London.
Barthes, R. (1986) 'From work to text', in *The Rustle of Language* (trans. R. Howard; first published in French in 1984). Oxford: 56–64.
Barton, T. (1994a) *Ancient Astrology*. London.
(1994b) *Power and Knowledge: Astrology, Physiognomics, and Medicine under the Roman Empire*. Ann Arbor.
Bauman, R. A. (1989) *Lawyers and Politics in the Early Roman Empire: A Study of Relations between the Roman Jurists and the Emperors from Augustus to Hadrian*. Munich.
Bayly, C. (1996) *Empire and Information: Intelligence Gathering and Social Communication in India, 1780–1870*. Cambridge.
Beagon, M. (1992) *Roman Nature: The Thought of Pliny the Elder*. Oxford.
Beard, M., North, J. and Price, S. (1998) *Religions of Rome* (2 vols.). Cambridge.
Beck, M. (2002) 'Plutarch to Trajan: the dedicatory letter and the apophthegmata collection', in Stadter and Van der Stockt (eds.): 163–73
Beck, R. (1973) 'Some observations on the narrative technique of Petronius', *Phoenix* 27: 42–61.
(1975) 'Encolpius at the *Cena*', *Phoenix* 29: 271–83.
(1982) 'The *Satyricon*: satire, narrator and antecedents', *MH* 39: 206–14.
Benjamin, W. (1999) 'On the mimetic faculty', in Benjamin, W., *Selected Writings. Volume 2: 1927–1934* (ed. Jennings, M., Eiland, H. and Smith, G.; trans. R. Livingstone *et al.*). Cambridge, MA: 720–2 (this essay first published in 1933).

Bennett, J. (1997) *Trajan: Optimus Princeps*. London.

Berra, A. (2005) 'Le Plutarque d' Athénée: masque, modèle et tradition', *Pallas* 67: 139–52.

Berriman, A. and Todd, M. (2001) 'A very Roman coup: the hidden war of imperial succession, AD 96–8', *Historia* 50: 312–31.

Birley, A. (1981) *The Fasti of Roman Britain*. Oxford.

Bischoff, B. (1966) 'Die europäische Verbreitung der Werke Isidors von Sevilla', in *Mittelalterliche Studien, Vol. 1*. Stuttgart: 171–94.

Blackman, D. R. and Hodge, A. T. (2001) *Frontinus' Legacy: Essays on Frontinus'* De aquis urbis Romae. Ann Arbor.

Bloomer, W. (1997) 'A preface to the history of declamation. Whose speech? Whose history?', in Habinek and Schiesaro (eds.): 199–215.

Bodel, J. (1995) 'Chronology and succession 2: notes on some consular lists on stone,' *ZPE* 105: 279–96.

Bouché-Leclercq, A. (1899) *L'astrologie grecque*. Paris.

Boudon, V. (1996) 'L'*Ars Medica* de Galien est-il un traité authentique', *REG* 109: 111–56.

  (ed.) (2000) *Galien II: Exhortation à la Médecine. Art Médical*. Paris.

  (2002) 'Galen's *On my own books*: new material from Meshed, Rida, Tibb. 5223', in Nutton (ed.): 9–18.

Boulogne, J. (1987) 'Le sens des "Questions romaines" de Plutarque', *REG* 100: 471–6.

  (1992) 'Les "Questions romaines" de Plutarque', *ANRW* 2.33.6: 4682–708.

Bourdieu, P. (1977) *Outline of a Theory of Practice* (trans. by R. Nice; first published in French in 1972). Cambridge.

Bowie, E. L. (1974) 'Greeks and their past in the Second Sophistic', in Finley, M. I. (ed.) *Studies in Ancient Society*. London: 166–209 (revised version; first published in 1970).

Bowler, P. (2002) *The Superior Person's Book of Words*. London.

Boyaval, B. (1971) 'Le P. Ifao 88: problemes de conversion monétaire', *ZPE* 7: 165–8.

Boys-Stones, G. R. (2001) *Post-Hellenistic Philosophy: A Study of its Development from the Stoics to Origen*. Oxford.

Branham, B. (1989) *Unruly Eloquence: Lucian and the Comedy of Traditions*. Cambridge, MA.

Braund, D. (2000a) 'Learning, luxury and empire: Athenaeus' Roman patron', in Braund and Wilkins (eds.): 1–22.

  (2000b) 'Athenaeus, *On the Kings of Syria*', in Braund and Wilkins (eds.): 514–22.

Braund D. and Wilkins J. (eds.) (2000) *Athenaeus and his World: Reading Greek Culture in the Roman Empire*. Exeter.

Bremmer, J. N. (1998) 'Aëtius, Arius Didymus and the transmission of doxography', *Mnemosyne* 51: 154–60.

Bretone, M. (1982) *Tecniche e ideologie dei giuristi romani*. Naples.

  (1989) *Storia del diritto romano* (third edition). Rome.

Brock, R. (2000) 'Athenaeus on Greek wine', in Braund and Wilkins (eds.): 455–65.

Brodersen, K. (1995) *Terra cognita: Studien zur römischen Raumerfassung*. Hildesheim.
Brouwer, R. (2002) 'Sagehood and the Stoics', *OSAPh* 23: 181–224.
Bruun, C. (1991) *The Water Supply of Ancient Rome: A Study of Roman Imperial Administration* (Societas Scientiarum Fennica, Commentationes Humanarum Litterarum 93). Helsinki.
Brunschwig, J. (1992) 'L'histoire de la philosophie, est–elle ou non philosophique? Non et oui', in Cassin (ed.): 37–96.
Bryson, B. (2003) *A Short History of Nearly Everything*. London.
Burkert, W. (1960) 'Platon oder Pythagoras? Zum Ursprung des Wortes "Philosophie"', *Hermes* 88: 159–77.
(1972) *Lore and Science in Ancient Pythagoreanism* (trans. E. L. Minar, Jr; first published in German in 1962). Cambridge, MA.
Burnett, A. (1987) *Coinage in the Roman World*. London.
Cameron, A. (1995) *Callimachus and his Critics*. Princeton.
Campbell, B. (ed.) (2000) *The Writings of the Roman Land Surveyors: Introduction, Text, Translation and Commentary*. London.
Canfora, L. (1994) 'Clemente di Alessandria e Diogene Laerzio', in *Storia, poesia e pensiero nel mondo antico: studi in onore di M. Gigante*. Naples: 79–81.
(ed.) (2001) *Ateneo*: I Deipnosofisti. Rome.
Canguilhem, G. (1955) *La formation du concept de réflexe aux XVIIe et XVIIIe siècles*. Paris.
Carey, S. (2003) *Pliny's Catalogue of Culture: Art and Empire in the* Natural History. Oxford.
Carradice, I. (ed.) (1987) *Coinage and Administration in the Athenian and Persian Empires*. Oxford.
Casavola, F. (1980) *Giuristi adrianei*. Naples.
Cassin, B. (ed.) (1992) *Nos grecs et leurs modernes*. Paris.
Cherniss, H. (1935) *Aristotle's Criticism of Presocratic Philosophy*. New York.
(1944) *Aristotle's Criticism of Plato and the Academy*. New York.
Clark, G. (2000) 'Philosophical lives and the philosophic life', in Hägg and Rousseau (eds.) (2000): 29–51.
Clarke, K. (1999) *Between Geography and History: Hellenistic Constructions of the Roman World*. Oxford.
Clement, P. and Hoffleit, H. (eds.) (1969) *Plutarch's* Moralia (vol. VIII) (Loeb Classical Library). Cambridge, MA.
Clemente, G. (1972) 'Il patronato nei collegia dell'impero romano', *SCO* 21 (1972): 142–229.
Codoñer, C. (1994) '"Origines" o "Etymologiae"', *Helmantica* 45: 511–27.
Cohen, A. (2001) 'Art, myth and travel in the Hellenistic world', in Alcock, S. E., Cherry, J. and Elsner, J. (eds.) *Pausanias: Travel and Memory in Roman Greece*. Oxford: 93–126.
Collison, R. (1964) *Encyclopedias: Their History throughout the Ages*. New York.
Conant, J. (2001) 'Philosophy and biography', in Klagge (ed.): 16–50.
Connerton, P. (1989) *How Societies Remember*. Cambridge.

Connors, C. (1998) *Petronius the Poet*. Cambridge.

Conte, G. B. (1994) *Latin Literature: A History*. Baltimore.

(1996) *The Hidden Author: An Interpretation of Petronius' Satyricon*. Berkeley.

Cook, M. (2004) *A Brief History of the Human Race*. London.

Corti, C. (2001) 'Le misure di capacità', in Corti and Giordani (eds.): 219–25.

Corti, C. and Giordani, N. (eds.) (2001) *Pondera: pesi e misure nell'antichità*. Modena.

Courcelle, P. (1959) 'Isidore de Séville, auteur antique et médiéval' [Review of Fontaine (1959)], *REA* 61: 419–23.

Cox, P. (1983) *Biography in Late Antiquity: A Quest for the Holy Man*. Berkeley.

Cox Miller, P. (2000) 'Strategies of representation in collective biography: constructing the subject as holy', in Hägg and Rousseau (eds.): 209–54.

Cramer, F. H. (1954) *Astrology in Roman Law and Politics (Memoirs of the American Philosophical Society*, vol. 37). Philadelphia.

Crawford, M. H. (1985) *Coinage and Money under the Roman Republic*. London.

Crick, M. (1982) 'Anthropology of knowledge', *Annual Review of Anthropology* 11: 287–313.

Cuomo, S. (2000) 'Divide and rule: Frontinus and Roman land surveying', *Studies in the History and Philosophy of Science* 31: 189–202.

Curtius, E. R. (1953) *European Literature and the Latin Middle Ages*. London.

Dahm, M. (1997) 'The career and writings of Sex. Jul. Frontinus'. Master of Literature Thesis, University of Auckland.

Daly, L. (1967) *Contributions to a History of Alphabetization in Antiquity and the Middle Ages*. Brussels.

Daremberg, C. and Ruelle, E. (eds.) (1879) *Oeuvres de Rufus d'Ephèse*. Paris.

Dawson, D. (1992) *Cities of the Gods: Communist Utopias in Greek Thought*. New York.

de Giovanni, L. (1996) 'Giuristi e principe da Augusto agli Antonini: aspetti e problemi', in Mantovani, D. (ed.) *Per la storia del pensiero giuridico romano: da Augusto agli Antonini*. Torino: 199–214.

de Lacy, P. (ed.) (1984) *Galen*, On the Doctrines of Hippocrates and Plato. Berlin.

Debru, A. (ed.) (1997) *Galen on Pharmacology: Philosophy, History and Medicine*. Leiden.

DeLaine, J. (1996) '"*De Aquis Suis*"? The "commentaries" of Frontinus', in Nicolet (ed.): 117–45.

Della Corte, F. (1991) 'Areio Didimo, Orazio e la dossografia d'età augustea', *Maia* 43: 67–81.

Desbordes, B. A. (1990) *Introduction à Diogène Laërce: exposition de l'* Altertumswissenschaft *servant de préliminaires critiques à une lecture de l'œuvre* (2 vols.). Utrecht.

Díaz y Díaz, M. C. (ed.) (1970) *Los capitales sobre les metales de las* Etimologias *de Isidoro de Sevilla: ensayo de edición critica*. León.

(2000) 'Introduccion general', in Oroz Reta, J. and Marcos Casquero, M. (eds.) *San Isidoro de Sevilla, Etymologías: Edicion bilingüe* (2 vols.) (third edition). Madrid: 1–257.

Dickie, M. (1999) 'The learned magician and the collection and transmission of magical lore', in Jordan, D., Montgomery, H. and Thomasson, E. (eds.) *The World of Ancient Magic*. Bergen: 163–93.

Diels, H. (ed.) (1879) *Doxographi graeci*. Berlin.

Diesner, H.-J. (1978) *Isidor von Sevilla und das westgothische Spanien*. Berlin.

Dilke, O. A. W. (1989) *Mathematics and Measurement* (2nd edn). London.

Dillon, J. M. and Long, A. A. (eds.) (1988) *The Question of 'Eclecticism': Studies in Late Greek Philosophy*. Berkeley.

Dirks, N., Eley, G. and Ortner, S. (1994) 'Introduction', in Dirks, N., Eley, G. and Ortner, S. (eds.) *Culture/Power/History: A Reader in Contemporary Social Theory*. Princeton: 3–45.

Domínguez del Val, U. (1986) *Estudios sobre litteratura latina hispano-cristiana*, vol. 1: *1955–71*. Madrid.

Doody, A. (2001), 'Finding facts in Pliny's encyclopaedia: the summarium of the *Natural History*', *Ramus* 30: 1–22.

Dorandi, T. (1990) 'Filodemo storico del pensiero antico', *ANRW* 2.36.4: 2407–23.

(1991) *Filodemo: storia dei filosofi, Platone e l'Academia, PHerc. 1021 e 164*. Naples.

(1992) 'Considerazioni sull' *Index locupletior* di Diogene Laerzio', *Prometheus* 18: 121–6.

(1994) *Storia dei filosofi: la Stoa da Zenone a Panezio, PHerc. 1018*. Leiden.

(1999) *Antigone de Caryste: Fragments*. Paris.

Drayton, R. (2000) *Nature's Government: Science, Imperial Britain, and the 'Improvement' of the World*. New Haven.

Ducos, M. (1994) 'Philosophie, literature et droit à Rome sous le Principat', *ANRW* 2.36.7: 5134–80.

Dueck, D. (2000) *Strabo of Amasia: A Greek Man of Letters in Augustan Rome*. London.

Duff, T. (1999) *Plutarch's Lives: Exploring Virtue and Vice*. Oxford.

Dumont, J.–P. (1987) 'Les modèles de la conversion à la philosophie chez Diogène Laërce', *Augustinus* 32: 79–97.

Duncan-Jones, R. (1994) *Money and Government in the Roman Empire*. Cambridge.

Dupont, F. (1997) 'Recitatio and the reorganization of public discourse', in Habinek and Schiesaro (eds.): 44–59.

Eagleton, T. (1996) *Literary Theory: An Introduction* (second edition). Oxford.

Eck, W. (1982) 'Die Gestalt Frontins in ihrer politischen und sozialen Umwelt', in Frontinus-Gesellschaft e.V. (ed.) *Wasserversorgung im antiken Rom*. Munich: 47–62.

(2002), 'An emperor is made: senatorial politics and Trajan's adoption by Nerva in 97', in Clark, G. and Rajak, T. (eds.) *Philosophy and Power in the Graeco-Roman World*. Oxford.

Edwards, C. (1993) *The Politics of Immorality in Ancient Rome*. Cambridge.

Edwards, M. J. and Swain, S. (ed.) (1997) *Portraits: Biographical Representation in the Greek and Latin Literature of the Roman Empire*. Oxford.

Elsner, J. (1992) 'Pausanias: a Greek pilgrim in the Roman world', *P&P* 135: 3–29.

(1994) 'From the pyramids to Pausanias and Piglet: monuments, travel and writing', in Goldhill, S. and Osborne, R. (eds.) *Art and Text in Ancient Greek Culture*. Cambridge: 224–54.

(1997) 'Hagiographic geography: travel and allegory in the *Life of Apollonius of Tyana*', *JHS* 117: 22–37.

Elsner, J. and Masters, J. (1994) 'Introduction', in Elsner, J. and Masters, J. (eds.) *Reflections of Nero. Culture, History and Representation*. London: 1–8.

Ercolani Cocchi, E. (2001) 'Misurare e valutare: amministrazione, economia e politica nel mondo romano', in Corti and Giordani (eds.): 113–41.

Erskine, A. (ed.) (2003) *A Companion to the Hellenistic World*. Oxford.

Evans, H. B. (1994) *Water Distribution in Ancient Rome: The Evidence of Frontinus*. Ann Arbor.

Everson, S. (1990) *Epistemology*. Cambridge.

Fabricius, C. (1972) *Galens Exzerpte aus älteren Pharmakologen*. Berlin.

Fairweather, J. (1974) 'Fiction in the biographies of ancient writers', *AncSoc* 5: 231–75.

Fanizza, L. (1982) *Giuristi, crimini, leggi nell'età degli Antonini*. Naples.

Feeney, D. (1998) *Literature and Religion at Rome: Culture, Contexts and Beliefs*. Cambridge.

Festugière, A. (1944–54) *La revelation d'Hermès Trismégiste* (4 vols.). Paris.

Fitzgerald, W. (2000) *Slavery and the Roman Literary Imagination*. Cambridge.

Flacelière, R. and Irigoin, J. (eds.) (1987) *Plutarque: Oeuvres Morales*, vol.1.1 (Budé). Paris.

Flemming, R. (2000) *Medicine and the Making of Roman Women: Gender, Nature, and Authority for Celsus to Galen*. Oxford.

(2003) 'Empires of knowledge: medicine and health in the Hellenistic world', in Erskine (ed.): 449–63.

(forthcoming) 'Galen's Commentaries', in Hankinson, R. J. (ed.) *The Cambridge Companion to Galen*. Cambridge.

Flower, H. (1995) *Ancestor Masks and Aristocratic Power in Roman Culture*. Oxford.

Fögen, T. (ed.) (2005) *Antike Fachtexte: Ancient Technical Texts*. Berlin.

Fontaine, J. (1959; 1983) *Isidore de Séville et la culture classique dans l'Espagne wisigothique, I–II; III*. Paris.

(1988) *Tradition et actualité chez Isidore de Séville*, London: III = 'Isidorus Varro christianus', 89–106; IV = 'Isidore de Séville et la mutation de l'encyclopédisme antique', 519–38; V = 'Cassiodore et Isidore: l'évolution de l'encyclopédisme latin du VIe au VIIIe siècle', 72–91; X = 'Cohérence et originalité de l'étymologie isidorienne', 113–44; XI = 'La situation de la rhétorique dans la culture latine tardive: observations sur la théorie isidorienne de l'étymologie', 197–205; XII = 'Isidore de Séville et l'astrologie', 271–300.

Formisano, M. (2001) *Tecnica e scrittura: le letterature tecnico-scientifiche nello spazio letterario tardolatino*. Rome.

Foucault, M. (1970) *The Order of Things: An Archaeology of the Human Sciences* (trans. anon.; first published in French in 1966). London.

(1972) *The Archaeology of Knowledge* (trans. A. Sheridan-Smith; first published in French in 1969). London.
(1986) 'What is an author?', in Rabinow, P. (ed.) *The Foucault Reader*. Harmondsworth: 101–20.
Frayn, J. M. (1993) *Markets and Fairs in Roman Italy*. Oxford.
Frazier, F. and Sirinelli, J. (eds.) (1996) *Plutarque – Oeuvres Morales* vol.9.3 (Budé). Paris.
Frede, M. (1992) 'Doxographie, historiographie philosophique et historiographie historique de la philosophie', *RMM* 97: 311–25.
(1985) 'Introduction', in Frede, M. and Walzer, R. (eds.) *Galen: Three Treatises on the Nature of Science*. Indianapolis: ix–xxxvi.
(2002) 'Galen's Theology', in Barnes, J. and Jouanna, J. (eds.) *Galien et la Philosophie (Entretiens sur l'Antiquité Classique* 49): 73–126.
French, R. (1994) *Ancient Natural History: Histories of Nature*. London.
Fresina, C. (1991) *Langue de l'etre: essai sur l'étymologie ancienne*. Münster.
Fuhrmann, F. (ed.) (1972) *Plutarque – Oeuvres Morales* vol.9.1 (Budé). Paris.
Gaiser, K. (1988) *Philodems Academica (Supplementum Platonicum* 1). Stuttgart.
Gallardo, M. (1972) 'Estado actual de los estudios sobre los simposios de Platón, Jenofonte y Plutarco', *Cuadernos de Filología Clasica* 3: 127–92.
Garnsey, P. (1999) *Food and Society in Classical Antiquity*. Cambridge.
Garofalo, I. (ed.) (1997) *Anonymi medici* De morbis acutis et chroniis. Leiden.
Gates, H. L. Jr (ed.) (1986) *'Race', Writing, and Difference*. Chicago.
Geertz, C. (1973) *The Interpretation of Cultures*. New York.
George, P. A. (1966) 'Style and Character in the *Satyricon*', *Arion* 5: 336–58.
Georgiadou A. and Larmour, D. H. J. (1998) *Lucian's Science Fiction Novel,* True Histories: *Interpretation and Commentary*. Leiden.
Gerli, E. M. (ed.) (2003) *Medieval Iberia: An Encyclopedia*. New York.
Geus, K. (2002) *Eratosthenes von Kyrene: Studien zur hellenistischen Kultur- und Wissenschaftsgeschichte*. Munich.
Geytenbeek, A. van (1963) *Musonius Rufus and Greek Diatribe* (trans. B. L. Hijmans; first published in Dutch in 1958). Assen.
Giannantoni, G. (1981) 'Pirrone, la scuola scettica e il sistema delle «successioni»', in Giannantoni, G. (ed.) *Lo scetticismo antico*. Naples: 11–34.
Giannattasio Andria, R. (1989) *I frammenti delle «successioni dei filosofi»*. Naples.
Gigante, M. (1986) 'Biografia e dossografia in Diogene Laerzio', *Elenchos* 7: 7–102.
Gigon, O. (1960) 'Das Prooemium des Diogenes Laertios: Struktur und Probleme', in Luck, G. (ed.) *Horizonte der Humanitas: eine Freundesgabe für W. Wili zu seinem 60. Geburtstag*. Stuttgart: 37–64.
Gill, C. (2003), 'The school in the Roman imperial period', in Inwood, B. (ed.) *The Cambridge Companion to the Stoics*. Cambridge: 33–58.
Gillings, R. J. (1972) *Mathematics in the Time of the Pharaohs*. Cambridge, MA.
Gleason, M. W. (1995) *Making Men: Sophists and Self-Presentation in Ancient Rome*. Princeton.

Goldhill, S. (1999) 'Body/politics: is there a history of reading?', in Falkner, T. M., Felson, N. and Konstan, D. (eds.) *Contextualizing Classics: Ideology, Performance, Dialogue*. Lanham, MD: 89–120.

(2002) *Who Needs Greek? Contests in the Cultural History of Hellenism*. Cambridge.

Gómez, P. and Jufresa, M. (1999) 'La risa y el vino en los escritos simposíacos de Plutarco', in Montes Cala, Sanchez Ortiz de Landaluce and Gallé Cejudo (eds.): 255–68.

Goody, J. (1977) *The Domestication of the Savage Mind*. Cambridge.

(1986) *The Logic of Writing and the Organization of Society*. Cambridge.

Goold, G. P. (ed.) (1977) *Manilius,* Astronomica. Cambridge, MA.

Gordon, C. (1980) *Power/Knowledge: Selected Interviews and Other Writings, 1972–77, M. Foucault*. Brighton.

Gould, S. and Vrba, E. (1982) 'Exaptation – a missing term in the science of form,' *Paleobiology* 8: 4–15.

Goulet, R. (1992) 'Des sages parmi les philosophes: le premier livre des "Vies des philosophes" de Diogène Laërce', in Goulet-Cazé, M.-O., Madec, G. and O'Brien, D. (eds.) Σοφίης μαιήτορες: *«Chercheurs de sagesse» Hommage à Jean Pepin*. Paris: 167–78.

(1999) 'Livre I: introduction, traduction et notes', in M.-O. Goulet-Cazé (ed.): 43–158.

Goulet-Cazé, M.-O. (1999) 'Introduction générale', in M.-O. Goulet-Cazé (ed.): 9–31.

(ed.) (1999) *Diogène Laërce: vies et doctrines des philosophes illustres*. Paris.

Goulet-Cazé, M.-O. and Goulet, R. (eds.) (1993) *Le cynisme ancien et ses prolongements*. Paris.

Grainger, J. D. (2003) *Nerva and the Roman Succession Crisis of AD 96–99*. London.

Grant, M. (ed.) (2000) *Galen on Food and Diet*. London.

Grial, J. (ed.) (1599; 2nd edn 1778) *Sancti Isidori Hispalensis Episcopi Etymologiarum Libri XX*. Madrid.

Griffin, M. T. and Barnes, J. (eds.) (1989) *Philosophia Togata I: Essays on Philosophy and Roman Society*. Oxford.

Grimal, P. (1944) *Frontin: les aqueducs de la ville de Rome* (Collection Budé). Paris.

Grimaudo, S. (1998) *Misurare e pesare nella Grecia antica: teorie, storia, ideologie*. Palermo.

Gruen, E. (1993) *Culture and National Identity in Republican Rome*. London.

Gulick, C. (1927–41) *Athenaeus of Naucratis,* The Deipnosophists (5 vols.) (Loeb Classical Library). London.

Gunderson, E. (1996) 'The ideology of the arena', *CA* 15: 113–51.

(2003) *Declamation, Paternity, and Roman Identity: Authority and the Rhetorical Self*. Cambridge.

Guthrie, W. K. C. (1957) 'Aristotle as historian', *JHS* 77: 35–41.

Habinek, T. (1990), 'Towards a history of friendly advice: the politics of candor in Cicero's *De Amicitia*', *Apeiron* 23: 165–85.

(1998) *The Politics of Latin Literature: Writing, Identity and Empire in Ancient Rome*. Princeton.

(2005) *The World of Roman Song: From Ritualized Speech to Social Order*. Baltimore.

Habinek, T. and Schiesaro, A. (eds.) (1997) *The Roman Cultural Revolution*. Cambridge.

Hadot, I. (1997) 'Geschichte der Bildung: *artes liberales*', in Graf, F. (ed.) *Einleitung in die lateinische Philologie*. Stuttgart: 17–34.

Hägg, T. and Rousseau, P. (eds.) (2000) *Greek Biography and Panegyric in Late Antiquity*. Berkeley.

Hahn, J. (1989) *Der Philosoph und die Gesellschaft: Selbstverständnis, öffentliches Auftreten und populäre Erwartungen in der hohen Kaiserzeit*. Stuttgart.

Hall, S. (1992) 'The question of cultural identity', in Hall, S., Held, D. and McGrew, T. (eds.) *Modernity and its Futures*. Cambridge: 273–316.

Hankinson, R. J. G. (1988) 'Galen explains the elephant', in Matthen, M. and Linsky, B. (eds.) *Philosophy and Biology*. Calgary: 135–56.

(1989) 'Galen and the best of all possible worlds', *CQ* 39: 206–22.

(1994) 'Usage and abusage: Galen on language', in Everson, S. (ed.) *Language*. Cambridge: 166–87.

Hanson, A. E. and Green, M. H. (1994) 'Methodicorum princeps: Soranus of Ephesus', *ANRW* 2. 37.2: 968–1075.

Hardie, P. R. (1986) *Vergil's Aeneid: Cosmos and Imperium*. Oxford.

(1992) 'Plutarch and the interpretation of Myth', *ANRW* 2.33.6: 4743–87.

Harig, G. (1974) *Bestimmung der Intensität im medizinischen System Galens*. Berlin.

Harrauer, H. and Sijpesteijn, P. J. (1985) *Neue Texte aus dem antiken Unterricht*. Vienna.

Harries, J. (1999) *Law and Empire in Late Antiquity*. Cambridge.

Harrison, G. (2000) 'Tipping his hand: Plutarch's preferences in the *Quaestiones Naturales*', in van der Stockt, L. (ed.) *Rhetorical Theory and Praxis in Plutarch*. Louvain: 237–49.

Harrison, S. J. (ed.) (1999) *Oxford Readings in the Roman Novel*. Oxford.

Hartog, F. (2001) *Memories of Odysseus: Frontier Tales from Ancient Greece* (trans. J. Lloyd; first published in French in 1996). Edinburgh.

Heilbron, J. L. (1990) 'The measure of enlightenment', in Frängsmyr, T., Heilbron, J. L. and Rider, R. E. (eds.) *The Quantifying Spirit in the Eighteenth Century*. Berkeley.

Helmreich, G. (ed.) (1907–9) *Galeni De Usu Partium libri XVII* (2 vols.). Leipzig.

Hemelrijk, E. (2004) *Matrona Docta: Educated Women in the Roman Elite from Cornelia to Julia Domna*. London.

Henderson, Jeffrey (1991) *The Maculate Muse: Obscene Language in Attic Comedy*. New York.

Henderson, John (1993) 'The pupil as teacher: Persius' didactic satire', *Ramus* 20: 123–48.

(2002a) *Pliny's Statue: The Letters, Self-Portraiture, and Classical Art*. Exeter.

(2002b) 'Columella's living hedge: the Roman gardening book', *JRS* 92: 110–33.

(2004) *The Roman Book of Gardening*. London.
(2007) *The Medieval World of Isidore of Seville: Truth from Words*. Cambridge.
Hillgarth, J. N. (1983) 'The position of Isidorian studies: a critical review of the literature, 1936–75', *Studi Medievali* 24.2: 817–905.
(1990) 'Isidorian studies, 1976–1985', *Studi Medievali* 31: 925–73.
Hobley, A. S. (1998) *An Examination of Roman Bronze Coin Distribution in the Western Empire, AD 81–192*. Oxford.
Hobsbawm, E. and Ranger, T. (eds.) (1983) *The Invention of Tradition*. Cambridge.
Hocquet, J.-C. (1992) *Anciens systèmes de poids et mesures en Occident*. Aldershot.
Hodge, A. T. (1991) *Future Currents in Aqueduct Studies*. Leeds.
(2002) *Roman Aqueducts and Water Supply*. London (2nd edn).
Holford-Strevens, L. (2003) *Aulus Gellius: An Antonine Scholar and his Achievement* (revised edition; first published in 1988). Oxford.
(1997) 'Favorinus: the man of paradoxes', in Barnes, J. and Griffin, M. (eds.) *Philosophia Togata II: Plato and Aristotle at Rome*. Oxford: 188–217.
Holford-Strevens, L. and Vardi, A. (eds.) (2004) *The Worlds of Aulus Gellius*. Oxford.
Holzberg, N. (1995) *The Ancient Novel: An Introduction* (trans. C. Jackson-Holzberg; first published in German in 1993). London.
Honderich, T. (2001) *Philosopher: A Kind of Life*. London.
Horster, M. and Reitz, C. (eds.) (2003) *Antike Fachshriftsteller: Literarischer Diskurs und sozialer Kontext* (*Palingenesia* 80). Stuttgart.
Howgego, C. (1992) 'The supply and use of money in the Roman world, 200 BC to AD 300', *JRS* 82: 1–31.
Hübner, W. (1984) 'Manilius als Astrologe und Dichter', *ANRW* 2.32.1: 126–320.
Hüllen, W. (2005) *A History of Roget's Thesaurus: Origins, Development, and Design*. Oxford.
Hultsch, F. (ed.) (1866) *Metrologicorum scriptorum reliquiae*. Leipzig.
Hunter, R. (2003) 'Literature and its contexts', in Erskine (ed.): 477–93.
Ilberg, J. (1889, 1892, 1896, 1897) 'Ueber die Schriftstellerei des Klaudios Galenos I–IV', *RhM* 44: 207–39; 47: 489–514; 51: 165–96; 52: 591–623.
Jacob, C. (2000) 'Athenaeus the librarian', in Braund and Wilkins (eds.): 85–110.
(2001) 'Ateneo o il dedalo delle parole', in Canfora (ed.): xi–cxv.
Jaggar, A. M. and Bordo, S. (eds.) (1989) *Gender/Body/Knowledge: Feminist Reconstructions of Being and Knowing*. New Brunswick, NJ.
Jeanneret, M. (1991) *A Feast of Words: Banquets and Table Talk in the Renaissance* (trans. J. Whiteley and E. Hughes; first published in French in 1987). Cambridge.
Jones, C. P. (1966) 'The teacher of Plutarch', *HSCP* 71: 205.
(1971) *Plutarch and Rome*. Oxford.
Jones, F. (1987) 'The narrator and narrative in the *Satyricon*', *Latomus* 46: 810–19.
Jong, A. de (2003) 'Dions Magierhymnen: zoroastrischer Mythos oder griechische Phantasie?', in Nesselrath, H.-G., Bäbler, B., Forschner, M. and de Jong, A. (eds.) *Dion von Prusa: menschliche Gemeinschaft und göttliche Ordnung: Die Borysthenes-Rede*. Darmstadt: 157–78.

Jouanna, J. and Boudon, V. (1997) 'Remarques sur la place d'Hippocrate dans la pharmacologie de Galien', in Debru (ed.): 213–34.

Kaibel, G. (ed.) (1887–90) *Athenaei Naucratitae Dipnosophistarum libri XV*. Leipzig.

Kaster, R. (ed.) (1995) *C. Suetonius Tranquillus*: De Grammaticis et Rhetoribus. Oxford.

Kaster, R. (1988) *Guardians of Language: The Grammarian and Society in Late Antiquity*. Berkeley.

Keil, H. (ed.) (1857–80) *Corpus Grammaticorum Latinorum*, I–VII. Leipzig.

Kennedy, D. (2002) *Rethinking Reality: Lucretius and the Textualization of Nature*. Ann Arbor.

Kenney, E. J. and Clausen, W. V. (eds.) (1982) *The Cambridge History of Classical Literature, vol. 2: Latin Literature*. Cambridge.

Ker, J. (2004) 'Nocturnal writers in imperial Rome: the culture of *lucubratio*,' *CPh* 99: 209–42.

Kienle, W. von (1961) *Die Berichte über die Sukzessionen der Philosophen in der hellenistichen und spätantiken Literatur*. Berlin.

Kilgour, M. (1990) *From Communion to Cannibalism: An Anatomy of Metaphors of Incorporation*. Princeton.

Kind, F. (1921) 'Scribonius Largus', *RE* II, 2. Stuttgart: 876–80.

Kingsley, P. (1995) *Ancient Philosophy, Mystery, and Magic: Empedocles and Pythagorean Tradition*. Oxford.

Kiple, K. F. and Ornelas, K. C. (eds.) (2000) *The Cambridge World History of Food*. Cambridge.

Klagge, J. C. (ed.) (2001) *Wittgenstein: Biography and Philosophy*. Cambridge.

Koerner, L. (1999) *Linnaeus: Nature and Nation*. Cambridge, MA.

Kollesch, J. (1988) 'Anschauungen von der ἀρχαί in der *Ars medica* und die Seelenlehre Galens', in Manuli, P. and Vegetti, M. (eds.) *Le opera psicologiche di Galeno*. Naples: 215–29.

Kollesch, J. and Nickel, D. (eds.) (1993) *Galen und das hellenistiche Erbe*. Stuttgart.

König, J. (2005) *Athletics and Literature in the Roman Empire*. Cambridge.

Kudlien, F. (1986) *Die Stellung des Arztes in der römischer Gesellschaft: freigeborene Römer, Eingebürgerte, Peregrine, Sklaven, Freigelassene als Ärzte*. Stuttgart.

Kula, W. (1986) *Measures and Men* (trans. R. Szreter; first published in Polish in 1970). Princeton.

Kunderewicz, C. (ed.) (1973) *Sex. Iulii Frontini* De aquaeductu urbis Romae (Teubner). Leipzig.

Kunkel, W. (2001) *Die römischen Juristen: Herkunft und soziale Stellung*. Köln (first published in 1967).

Kurke, L. (1999) *Bodies, Games, Coins, and Gold: The Politics of Meaning in Archaic Greece*. Princeton.

Laird, A. (1999) *Powers of Expression, Expressions of Power: Speech Presentation in Latin Literature*. Oxford.

Laks, A. (1993) 'Annicéris et les plaisirs psychiques: quelques préalables doxographiques', in Brunschwig, J. and Nussbaum, M. (eds.) *Passions and Perceptions: Studies in Hellenistic Philosophy of Mind*. Cambridge: 18–49.

Landolfi, L. (2003) *Integra prata: Manilio, i proemi*, Bologna.
Lang, M. and Crosby, M. (1964) *The Athenian Agora, vol. 10: Weights, Measures and Tokens*. Princeton.
Langholf, V. (2004) 'Structure and genesis of some Hippocratic treatises', in Horstmanshoff, H. and Stol, M. (eds.) *Magic and Rationality in Ancient Near Eastern and Graeco-Roman Medicine*. Leiden: 219–75.
Langslow, D. R. (2000) *Medical Latin in the Roman Empire*. Oxford.
Latour, B. (1987) *Science in Action: How to Follow Scientists and Engineers Through Society*. Milton Keynes.
Levy, E. (1963) 'Natural law in Roman thought', in Levy, E. *Gesammelte Schriften*, vol. 1. Köln: 3–19 (first published in 1948).
Lindsay, W. M. (ed.) (1911a) *Isidori Episcopi Etymologiarum sive Originum Libri XX*. Oxford.
(1911b) 'The editing of Isidore's *Etymologiae*', *CQ* 5: 42–5.
Lloyd, G. (1979) *Magic, Reason and Experience: Studies in the Origins and Development of Greek Science*. Cambridge.
(1996) *Adversaries and Authorities: Investigations into Ancient Greek and Chinese Science*. Cambridge.
Long, A. A. (1982) 'Astrology: arguments pro and contra', in Barnes, J. *et al.* (eds.) *Science and Speculation: Studies in Hellenistic Theory and Practice*. Cambridge: 165–92.
(1986) 'Diogenes Laërtius, *Life of Arcesilaus*', *Elenchos* 7: 429–49.
Long, A. A. and Sedley, D. (1987) *The Hellenistic Philosophers*. Cambridge.
Long, P. (2001) *Openness, Secrecy, Authorship: Technical Arts and the Culture of Knowledge from Antiquity to the Renaissance*. Baltimore.
Lynch, C. H. (1938) *Saint Braulio, Bishop of Saragossa (1631–51): His Life and Writings*. Washington.
McArthur, T. (1986) *Worlds of Reference: Lexicography, Learning and Language from the Clay Tablet to the Computer*. Cambridge.
McCabe, M. M. (2000) *Plato and his Predecessors*. Cambridge.
MacCormack, S. (1998) *The Shadows of Poetry: Vergil in the Mind of Augustine*. Berkeley.
McElwain, M. B. (ed.) (1925) *Frontinus*, The Stratagems. The Aqueducts of Rome. (With an English translation by C. E. Bennett) (Loeb Classical Library). Cambridge, MA.
McEwen, I. K. (2003) *Vitruvius: Writing the Body of Architecture*. Cambridge, MA.
MacFarlane, K. N. (1980) *Isidore of Seville on the Pagan Gods (Origines VIII, 11)* (*TAPhS* 70.3). Philadelphia.
McGinn, C. (2003) *The Making of a Philosopher*. London.
McInerney, J. (2003) 'Plutarch's manly women', in Rosen, R. and Sluiter, I. (eds.) *Andreia: Studies in Manliness and Courage in Classical Antiquity*. Leiden: 319–44.
Macleod, R. (1993) 'Passages in imperial science: from empire to commonwealth', *Journal of World History* 4: 117–50.

MacMullen, R. (1992) *Enemies of the Roman Order: Treason, Alienation and Unrest in the Roman Empire* (2nd edn; first published in 1966). London.
McNay, L. (1994) *Foucault: An Introduction*. Cambridge.
Magallón García, A.-I. (1996) *La tradición gramatical de* differentia *y* etymologia *hasta Isidoro de Sevilla*. Saragossa.
Maltby, R. (1991) *A Lexikon of Ancient Latin Etymologies*. Leeds.
(1999) 'Late Latin and etymology in Isidore of Seville', in Petersmann, H. and Kettemann, R. (eds.) *Latin vulgaire – Latin Tardif, VI: actes de 5[e] Colloque international sur le Latin vulgaire et Tardif (Heidelberg, 5–8 Septembre 1997)*. Heidelberg: 441–50.
Manetti, D. and Roselli, A. (1994) 'Galeno commentatore di Ippocrate', *ANRW* 2.37.2: 1527–634.
Manfredini, A. D. (1983) 'Il naufragio di Eudemone (*D.* 14.2.9)', *Studia et Documenta Historiae et Iuris* 49: 375–94.
Mansfeld, J. (1994) *Prolegomena: Questions to be Settled before Study of an Author, or a Text*. Leiden.
Mansfeld, J. and Runia, D. T. (1997) *Aetiana: The Method and Intellectual Context of a Doxographer*. Leiden.
Marotta, V. (1988) *Multa de iure sanxit: aspetti della politica del diritto di Antonino Pio*. Milan.
Marshall, P. K. (ed.) (1983) *Isidorus Hispalensis Etymologiae II*. Paris.
Martin, R. (1998) 'The seven sages as performers of wisdom', in Dougherty, C. and Kurke, L. (eds.) *Cultural Poetics in Archaic Greece*. New York: 108–28.
Matthews, J. F. (1984) 'The tax law of Palmyra: evidence for economic history in a city of the Roman East', *JRS* 74: 157–80.
Mattingly, H. (1928) *Roman Coins*. London.
Meiggs, R. & Lewis, D. (eds.) (1969) *A Selection of Greek Historical Inscriptions to the End of the Fifth Century* BC. Oxford.
Meissner, B. (1999) *Die Technologische Fachliteratur der Antike: Struktur, Überlieferung und Wirkung technischen Wissens in der Antike (ca. 400 v. Chr.–ca. 500 n. Chr.)*. Berlin.
Mejer, J. (1978) *Diogenes Laertius and his Hellenistic Background (Hermes Einzelschriften* 40). Wiesbaden.
(1992) 'Diogenes Laërtius and the transmission of Greek philosophy', *ANRW* 2.36.5: 3556–602.
Merkle, S. (1994) 'Telling the story of the Trojan war: the eyewitness account of Dictys of Crete', in Tatum, J. (ed.) (1994) *The Search for the Ancient Novel*. Baltimore: 183–96.
Merleau-Ponty, M. (1962) *The Phenomenology of Perception* (trans. by C. Smith; first published in French in 1945). London.
Millar, F. (1999) 'The Greek East and Roman law: the dossier of M. Cn. Licinius Rufinus', *JRS* 89: 90–108.
Miller, D. and Reill, P. (eds.) (1996) *Visions of Empire: Voyages, Botany, and Representations of Nature*. Cambridge.

Mirowski, P. (1992) 'Looking for those natural numbers: dimensionless constants and the idea of natural measurement', *Science in Context* 5: 165–88.

Miyoshi, M. (1994) *Off Center: Power and Culture Relations between Japan and the United States*. Cambridge, MA.

Moatti, C. (1988) 'Tradition et raison chez Cicéron', *MEFRA* 100: 385–430.

(1991) 'La crise de la tradition à la fin de la république romaine à travers la littérature juridique et la science des antiquaires', in Pani, M. (ed.) *Continuità e trasformazione fra repubblica e principato*. Bari: 30–45.

(1997) *La raison de Rome: naissance de l'esprit critique à la fin de la république*. Paris.

Moles, J. L. (1993) 'Le cosmopolitisme cynique', in Goulet-Cazé and Goulet (eds.): 259–80 (translated and reprinted in Branham, R. B. and Goulet-Cazé, M.-O. (eds.) (1996) *The Cynics: The Cynic Movement in Antiquity and its Legacy*. Berkeley: 105–20).

Möllendorff, P. von (2000) *Auf der Suche nach der verlogenen Wahrheit: Lukians* Wahre Geschichten. Tübingen.

Momigliano, A. (1974) *Alien Wisdom: the Limits of Hellenization*. Cambridge.

(1993) *The Development of Greek Biography* (extended edition; first published in 1971). Harvard.

Mommsen, T. (1857) *Abhandlungen der sächsische Gesellschaft der Wissenschaften* III.

Monk, R. (1990) *Ludwig Wittgenstein: The Duty of Genius*. London.

(1996) *Bertrand Russell: The Spirit of Solitude*. London.

(2000) *Bertrand Russell, 1921–1970 (vol. 2): The Ghost of Madness*. London.

(2001) 'Philosophical biography: the very idea', in Klagge (ed.): 3–15.

(2002) 'Philosophical biography: in conversation with Ray Monk', in Baggini, J. and Stangroom, J. (eds.) *New British Philosophy: The Interviews*. London: 9–24.

Montes Cala, J. (1999) 'Εὐφροσύνη convival en Plutarco', in Montes Cala, Sanchez Ortiz de Landaluce and Gallé Cejudo (eds.): 3–28.

Montes Cala, J., Sanchez Ortiz de Landaluce, M. and Gallé Cejudo, R. (eds.) (1999) *Plutarco, Dioniso y el vino*. Madrid.

Moore, H. (1996) 'The changing nature of anthropological knowledge', in Moore, H. (ed.) *The Future of Anthropological Knowledge*. London: 1–15.

Morgan, T. (1998) *Literate Education in the Hellenistic and Roman Worlds*. Cambridge.

(2004) 'Educational values', in Holford-Strevens and Vardi (eds.): 187–205.

Mrozek, S. (2001) *Faenus: Studien zu Zinsproblemen zur Zeit des Prinzipats*. Stuttgart.

Muckensturm, C. (1993) 'Les Gymnosophistes étaient-ils des cyniques modèles?', in Goulet-Cazé and Goulet (eds.): 225–39.

Muhaqqiq, M. (ed.) (1993) *Kitab al-shukuk 'ala Jalinus li-Muhammad ibn Zakariyya al-Razi*. Tehran.

Murphy, T. (2003) 'Pliny's *Naturalis Historia*. The prodigal text', in Boyle A. and Dominik, W. (eds.) *Flavian Rome: Culture, Image, Text*. Leiden: 301–22.

(2004) *Pliny the Elder's Natural History: The Empire in the Encyclopaedia*. Oxford.

Naas, V. (2002) *Le project encyclopédique de Pline l'Anciene* (*Collection de l'école française de Rome* 303). Rome.
Nachmanson, E. (1941) *Der griechische Buchtitel: einige Beobachtungen*. Göteborg.
Natali, C. (2000) 'Schools and sites of learning', in Brunschwig, J. and Lloyd, G. (eds.) *Greek Thought: A Guide to Classical Knowledge*. Cambridge, MA: 191–217.
Navarra, L. (1987) *Leandro di Siviglia: profilo storico-letterario*. Rome.
Neugebauer, O. and van Hoesen, H. B. (1959) *Greek Horosocopes (Memoirs of the American Philosophical Society* 48). Philadelphia.
Newlands, C. E. (1995) *Playing with Time: Ovid and the Fasti*. Cornell.
Nicolet, C. (1991) *Space, Geography, and Politics in the Early Roman Empire* (trans. H. Leclerc; first published in French in 1988). Ann Arbor.
(ed.) (1995) *Les littératures techniques dans l'antiquité romaine: statut, public et destination, tradition (Entretiens sur l'antiquité classique* 42). Geneva.
Nightingale, A. W. (1995) *Genres in Dialogue: Plato and the Construct of Philosophy*. Cambridge.
Nijf, O. van (1997) *The Civic World of Professional Associations in the Roman East*. Amsterdam.
Nutton, V. (1972) 'Galen and medical autobiography', *PCPhS* 198: 50–62.
(1973) 'The chronology of Galen's early career', *CQ* 23: 138–71.
(1978) 'The beneficial ideology', in Garnsey, P. and Whittaker, C. R. (eds.) *Imperialism in the Ancient World*. Cambridge: 209–21.
(1988) *From Democedes to Harvey*. London.
(1991) 'Style and content in the *Method of Healing*', in Kudlien, F. and Durling, R. (eds.) *Galen's Method of Healing*. Leiden: 1–25.
(1995a) 'Galen and the Traveller's Fare', in Wilkins, J., Harvey, D. and Dobson, M. (eds.) *Food in Antiquity*. Exeter: 359–70.
(1995b) 'Galen *ad multos annos*', *Dynamis* 15: 25–39.
(ed.) (2002) *The Unknown Galen (BICS Supplement* 77). London.
(2004) *Ancient Medicine*. London.
Nocera, G. (1962) *Jus naturale nella esperienza giuridica romana*. Milan.
Nock, A. D. (1933) *Conversion: The Old and New in Religion from Alexander the Great to Augustine of Hippo*. Oxford.
Obbink, D. (1996) *Philodemus*, On Piety. Oxford.
O'Donnell, J. J. (1979) *Cassiodorus*. Berkeley.
O'Hara, J. J. (1996) *True Names: Vergil and the Alexandrian Tradition of Etymological Wordplay*. Michigan.
Oliver, J. (1971) 'Epaminondas of Acraephia', *GRBS* 12: 221–37.
Opelt, I. (1965) 'Etymologie', *Reallexicon für Antike und Christentum* 1: 797–844.
Pailler, J.-M. (1998) 'Les *Questions* dans les plus anciennes *Vies* romaines: art du récit et rhétorique de la fondation', in Payen, P. (ed.) *Plutarque: Grecs et Romains en 'Questions'*. Saint-Bertrand-de-Comminges: 77–94.
Panayotakis, C. (1995) *Theatrum Arbitri: Theatrical Elements in the Satyrica of Petronius (Mnemosyne Supplement 146)*. Leiden.
Paschalis, M. (1997) *Virgil's Aeneid: Semantic Relations and Proper Names*. Oxford.

Peachin, M. (2004) *Frontinus and the Curae of the* Curator aquarum. Stuttgart.
Pedroni, L. (1996) *Asse romano e asse italico: momenti di un'integrazione difficile.* Naples.
Petrochilos, N. K. (1974) *Roman Attitudes to the Greeks.* Athens.
Pfeiffer, R. (1968) *A History of Classical Scholarship from the Beginnings to the End of the Hellenistic Age.* Oxford.
Pingree, D. (ed.) (1976) *Dorothei Sidonii Carmen astrologicum.* Leipzig.
Pintaudi, R. and Sijpestijn, J. P. (1989) *Tavolette lignee e cerate da varie collezioni.* Florence.
Porter, T. (1995) *Trust in Numbers: The Pursuit of Objectivity in Science and Public Life.* Princeton.
Potter, D. (1994) *Prophets and Emperors: Human and Divine Authority from Augustus to Theodosius.* Cambridge, MA.
Powell, O. (ed.) (2003) *Galen,* On the Properties of Foodstuffs. Cambridge.
Puech, B. (2002) *Orateurs et sophists grecs dans les inscriptions d'époque impériale.* Paris.
Preston, R. (2001) 'Roman questions, Greek answers: Plutarch and the construction of identity', in Goldhill, S. (ed.) (2001) *Being Greek under Rome: Cultural Identity, the Second Sophistic and Development of Empire.* Cambridge: 86–119.
Pyke, S. (1993) *Philosophers.* London.
Rathbone, D. (1991) *Economic Rationalism and Rural Society in Third-Century* AD *Egypt.* Cambridge.
Rawson, E. (1985) *Intellectual Life in the Late Roman Republic.* London.
(1989) 'Roman rulers and the philosophic adviser', in Griffin and Barnes (eds.): 233–57.
von Reden, S. (1997) 'Money, law and exchange: coinage in the Greek polis', *JHS* 117: 154–76.
Rée, J. (1978) 'Philosophy and the history of philosophy', in Rée, J., Ayers, M. and Westoky, A. (eds.) *Philosophy and its Past.* Hassocks: 3–39.
Reeve, M. (1983) 'Petronius', in Reynolds, L. D. (ed.) *Texts and Transmission: A Survey of the Latin Classics.* Oxford: 295–300.
Relihan, J. (1992) 'Rethinking the history of the literary symposium', *ICS* 17: 213–44.
Reydellet, M. (ed.) (1984) *Isidorus Hispalensis Etymologiae IX.* Paris.
Rice, E. E. (1983) *The Grand Procession of Ptolemy Philadelphus.* Oxford.
Ridings, D. (1995) *The Attic Moses: The Dependency Theme in Some Early Christian Writers.* Göteborg.
Riggsby, A. (2003) 'Pliny in space (and time),' *Arethusa* 36: 167–86.
Rimell, V. (2002) *Petronius and the Anatomy of Fiction.* Cambridge.
Roberts, C. and Skeat, T. (1983) *The Birth of the Codex.* London.
Rocca, J. (2003) *Galen on the Brain: Anatomical Knowledge and Physiological Speculation in the Second Century* AD. Leiden.
Rodríguez Noriega-Guillén, L. (2000) 'Are the fifteen books of the *Deipnosophistae* an excerpt?', in Braund and Wilkins (eds.): 244–55.
Rodgers, R. H. (1986) '*Copia Aquarum*: Frontinus' measurements and the perspective of capacity', *TAPA* 116: 353–60.

(ed.) (2004) *Frontinus*, De Aquaeductu Urbis Romae. Cambridge.
Roget (2002) = Davidson, G. (ed.) *Roget's Thesaurus of English Words & Phrases* (150th anniversary edition) (first published in 1852 (=Roget (1852); 2nd edn in 1879 (=Roget (1879)). London.
Rohden, P. and Dessau, H. (1898) *Prosopographia imperii romani*. Berlin.
Romano, E. (1979) 'Teoria del progresso ed età dell'oro in Manilio (1.66–112)', *RFIC* 107: 394–408.
Romeri, L. (2002) *Philosophes entre mots et mets: Plutarque, Lucien et Athénée autour de la table de Platon*. Grenoble.
Romm, J. (1991) *The Edges of the Earth in Ancient Thought: Geography, Exploration, and Fiction*. Princeton.
Rorty, R. (1984) 'The historiography of philosophy: four genres', in Rorty, Schneewind and Skinner (eds.): 49–75.
Rorty, R., Schneewind, J. and Skinner, Q. (eds.) (1984) *Philosophy in History*. Cambridge.
Rudich, V. (1993) *Political Dissidence under Nero: The Price of Dissimulation*. London.
(1997) *Dissidence and Literature Under Nero: The Price of Rhetoricisation*. London.
Ruelle, C.-E. (1908) 'Hermes Trismégiste: le livre sacré sur le décans (texte, variantes et traduction française)', *RPh* 32: 247–77.
Russell, B. (1961) *History of Western Philosophy* (2nd edn). London.
(1967–9) *The Autobiography of Bertrand Russell* (3 vols.). London.
Russell, D. (1973) *Plutarch*. London.
Rütten, U. (1997) *Phantasie und Lachkultur: Lukians* Wahre Geschichten. Tübingen.
Saastamoinen, A. (2003) 'The Literary Character of Frontinus' *De aquaeductu*', in Bruun, C. and Saastamoinen, A. (eds.) *Technology, Ideology, Water: From Frontinus to the Renaissance and Beyond (Acta Instituti Romani Finlandiae 31)*. Rome: 15–39.
Said, E. (1978) *Orientalism*. New York.
(1993) *Culture and Imperialism*. London.
Sallares, R. (1991) *The Ecology of the Ancient Greek World*. London.
Santini, C. (1988) 'Il lessico della spartizione nel sacrificio romano', in Grottanelli, C. and Parise, N. F. (eds.) *Sacrificio e società nel mondo antico*. Rome: 293–302.
Santini, C., Mastrorosa, I. and Zumbo, A. (eds.) (2002) *Letteratura scientifica e tecnica di Grecia e Roma*. Rome.
Savio, A. (2001) *Monete romane*. Rome.
Scarano Ussani, V. (1979) *Valori e storia nella cultura giuridica fra Nerva e Adriano: studi su Nerazio e Celso*. Naples.
(1987) *L'utilità e la certezza: compiti e modelli del sapere giuridico in Salvio Giuliano*. Milan.
(1997) *L'ars dei giuristi: considerazioni sullo statuto epistemologico della giurisprudenza romana*. Torino.
Scarborough, J. (1984) 'Early Byzantine pharmacology', in Scarborough, J. (ed.) *Symposium on Byzantine Medicine*. Washington, DC: 213–32.

Scarborough, J. and Nutton, V. (1982) 'The Preface of Dioscorides' *Materia Medica*: introduction, translation, and commentary', in *Transactions and Studies of the College of Physicians of Philadelphia* (5th series) 4: 187–227.
Scarcia, R. (ed.) (1996) (2001) *Manilio: il poema degli astri (Astronomica)* (2 vols.): Milan.
Schaffer, S. (1995) 'Accurate measurement is an English science', in Norton Wise, M. (ed.) *The Values of Precision*. Princeton: 135–72.
Schenkenveld, D. (1996) 'Plutarch's first *Table Talk* (612E-615C)', in Van der Stockt, L. (ed.) *Plutarchea Loveniensia: A Miscellany of Essays on Plutarch*. Louvain: 257–64.
(1997) 'Plutarch's *Table Talks* VIII.4 as a literary work of art', *Mnemosyne* 50: 597–602.
Schiesaro, A. (1997) 'The boundaries of knowledge in Virgil's *Georgics*', in Habinek and Schiesaro (eds.): 63–89.
(2002) 'Ovid and the professional discourses of scholarship, religion and rhetoric', in Hardie, P. (ed.) *The Cambridge Companion to Ovid*. Cambridge: 62–78.
Schmidt, T. (1999) *Plutarque et les barbares: la rhétorique d'une image*. Leuven.
Schmitt-Pantel, P. (1992) *La cité au banquet: histoire des repas publiques dans les cités grecques*. Rome.
Schmitz, T. (1997) *Bildung und Macht: zur sozialen und politischen Funktion zweiten Sophistik in der griechsichen Welt der Kaiserzeit*. Munich.
Schofield, M. (1991) *The Stoic Idea of the City*. Cambridge.
Schofield, M., Burnyeat, M. and Barnes, J. (eds.) (1980) *Doubt and Dogmatism: Studies in Hellenistic Epistemology*. Oxford.
Schonack, W. (1912) *Die Rezeptsammlung des Scribonius Largus*. Jena.
Schott, B. (2002) *Schott's Original Miscellany*. London.
(2003) *Schott's Food and Drink Miscellany*. London.
Schrijvers, P. H. (1983) 'Le chant du monde: remarques sur *Astronomica* I, 1–24 de Manilius', *Mnemosyne* 36: 143–50.
Schwarz, W. (1972) 'Praecordia Mundi: zur Grundlegung der Bedeutung des Zodiak bei Manilius', *Hermes* 100: 601–14.
Sconocchia, S. (ed.) (1983) *Scribonius Largus*: Compositiones. Leipzig.
Seaford, R. (2004) *Money and the Early Greek Mind: Homer, Philosophy, Tragedy*. Cambridge.
Sear, D. S. (2000) *Roman Coins and their Values* (2 vols.). London.
Seckel, E. and Kübler, B. (eds.) (1908) *Iurisprudentiae anteiustinianae reliquiae*. Leipzig.
Sedley, D. (1989) 'Philosophical allegiance in the Graeco-Roman world', in Griffin and Barnes (eds.): 97–119.
(2003) 'Philodemus and the decentralisation of philosophy', *CErc* 33: 31–41.
Shaw, B. (1985) 'Stoicism as ideology', *Latomus* 44: 16–54.
Shelton, J. (1998) *As the Romans Did*. Oxford.
Sherwin-White, A. (1966) *The Letters of Pliny: A Historical and Social Commentary*. Oxford.
Shields, C. (2002) *Unless*. London.

Sijpesteijn, P. J. (1980) 'Wiener mélange', *ZPE* 40: 98–9.
Simon. M. (ed.) (1906) *Sieben Bücher Anatomie des Galen* (2 vols.). Leipzig.
Slater, N. (1987) 'Against interpretation: Petronius and art criticism', *Ramus* 16: 165–76.
(1990) *Reading Petronius*. Baltimore.
Sluiter, I. (1995) 'The embarrassment of imperfection: Galen's assessment of Hippocrates' linguistic merits', in van der Eijk P., Horstmanshoff, H. F. J. and Schrijvers, P. H. (eds.) *Ancient Medicine in its Socio-cultural Context*. Amsterdam: 519–35.
Smith, M. S. (1975) *Petronii Arbitri Cena Trimalchionis*. Oxford.
Sperber, D. (1974) *Roman Palestine: Money and Prices*. Ramat-Gan.
Spoerri, W. (1959) *Späthellenistische Berichte über Welt, Kultur und Götter*. Basel.
Staden, H. von (1989) *Herophilus: The Art of Medicine in early Hellenistic Alexandria*. Cambridge.
(2002) 'Division, dissection and specialization: Galen's "On the parts of the medical techne"', in Nutton (ed.): 19–45.
Stadter, P. A. (1980) *Arrian of Nicomedia*. Chapel Hill.
(1999) 'Drinking, *Table Talk*, and Plutarch's contemporaries', in Montes Cala, Sanchez Ortiz de Landaluce and Gallé Cejudo (eds.): 481–90.
(2002) 'Introduction: setting Plutarch in his context', in Stadter and van der Stockt (eds.): 1–26.
Stadter, P. and van der Stockt, L. (eds.) (2002) *Sage and Emperor: Plutarch, Greek Intellectuals, and Roman Power in the Time of Trajan (98–117 AD)*. Leuven.
Stafford, R. (1989) *Scientist of Empire: Sir Roderick Murchison, Scientific Exploration and Victorian Imperialism*. Cambridge.
Stanton, G. R. (1968) 'The cosmopolitan ideas of Epictetus and Marcus Aurelius', *Phronesis* 13: 183–95.
Stein, P. (1974) 'The development of the notion of *naturalis ratio*', in Watson, A. (ed.) *Daube Noster: Essays in Legal History for David Daube*. Edinburgh: 305–16.
Stierlin, H. (1986) *L'astrologie et le pouvoir: de Platon à Newton*. Paris.
Strack, P. L. (1937) *Untersuchungen zur römischen Reichsprägung des zweiten Jahrhunderts* (3 vols.). Stuttgart.
Striker, G. (1996) *Essays on Hellenistic Epistemology and Ethics*. Cambridge.
Sullivan, J. P. (1968) *The* Satyricon *of Petronius: A Literary Study*. Bloomington, IND.
Swain S. (1990) 'Hellenic culture and the Roman heroes of Plutarch', *JHS* 110: 126–45.
(1996) *Hellenism and Empire: Language, Classicism, and Power in the Greek World AD 50–250*. Oxford.
Talbert, R. J. A. (1984) *The Senate of Imperial Rome*. Princeton.
Taub, L. (2003) *Ancient Meteorology*. London.
Taussig, M. (1993) *Mimesis and Alterity: A Particular History of the Senses*. New York.
Taylor, C. (1984) 'Philosophy and its history', in Rorty, Schneewind and Skinner (eds.): 17–30.

Teixera, E. (1992) 'Remarques sur l'esprit scientifique de Plutarque d'après quelques passages des *Propos de Table*', in Gallo, I. (ed.) *Plutarco e le scienze*. Genova: 211–23.

Teodorsson, S.-T. (1999) 'Dionysus moderated and calmed: Plutarch on the convivial wine', in Montes Cala, Sanchez Ortiz de Landaluce and Gallé Cejudo (eds.): 57–69.

Tieleman, T. (1996) *Galen and Chrysippus on the Soul: Argument and Refutation in De Placitis Books II–III*. Leiden.

(2003) *Chrysippus'* On Affections: *Reconstruction and Interpretation*. Leiden.

Too, Y. L. (2000) 'The walking library: the performance of cultural memories', in Braund and Wilkins (eds.): 111–23.

Trapp, M. B. (1995) 'Sense of place in the orations of Dio Chrysostom', in Innes, D., Hine, H. and Pelling, C. (eds.) *Ethics and Rhetoric: Classical Essays for Donald Russell on his Seventy-Fifth Birthday*. Oxford: 163–75.

Usener, H. (ed.) (1887) *Epicurea*. Leipzig.

Vallauri, G. (1954) 'Gli *Astronomica* di Manilio e le fonti ermetiche', *RFIC* 82: 133–67.

Vardi, A. (2004) 'Genre, conventions, and cultural programme in Gellius' *Noctes Atticae*', in Holford-Strevens and Vardi (eds.): 159–86.

Veyne, P. (1964) 'Le "Je" dans le Satyricon', *REL* 42: 301–24.

Volk, K. (2001) 'Pious and impious approaches to cosmology in Manilius', *MD* 47: 84–117.

(2002) *The Poetics of Latin Didactic: Lucretius, Vergil, Ovid, Manilius*. Oxford.

Waerdt, P. A. van der (1994) 'Philosophical influence on Roman jurisprudence? The case of Stoicism and natural law', in *ANRW* 2.36.7: 4851–900.

Wallace-Hadrill, A. (1983) *Suetonius: The Scholar and his Caesars*. London.

(1988) 'Greek knowledge, Roman power' (review of Rawson (1985)), *CPh* 83: 224–33.

(1997) '*Mutatio morum*: the idea of a cultural revolution', in Habinek, T. and Schiesaro, A. (ed.) *The Roman Cultural Revolution*. Cambridge: 3–22.

Warnock, M. (2000) *M. Warnock: A Memoir*. London.

Warren, J. I. (2002) *Epicurus and Democritean Ethics: An Archaeology of Ataraxia*. Cambridge.

Walsh, P. G. (1970) *The Roman Novel: The Satyricon of Petronius and the Metamorphoses of Apuleius*. Cambridge.

Washbrook, D. (1999) 'Orients and occidents: colonial discourse theory and the historiography of the British Empire', in Winks, R. (ed.) *The Oxford History of the British Empire*. Oxford: 596–611.

Weeks, A. R. (2004) *Ancient Knowledge, Roman Politics and Frontinus' 'Technical' Treatises*. PhD Thesis, University of Cambridge.

Wellmann, M. (ed.) (1895) *Die pneumatische Schule bis auf Archigenes*. Berlin.

(ed.) (1906–14) *Pedanius Dioscorides. De Materia Medica* (3 vols.). Berlin.

West, L. C. and Johnson, A. C. (1967) *Currency in Roman and Byzantine Egypt*. Amsterdam (first published in 1944).

White, S. A. (2001) '*Principes sapientiae*: Dicaearchus' biography of philosophy', in Fotenbaugh, W. W. and Schütrumpf, E. (eds.) *Dicaearchus of Messana: Text, Translation and Discussion*. New Brunswick: 195–236.

Whitmarsh, T. (2000) 'The politics and poetics of parasitism: Athenaeus on parasites and flatterers', in Braund and Wilkins (eds.): 304–15.

(2001) *Greek Literature and the Roman Empire: The Politics of Imitation*, Oxford.

(2004) *Ancient Greek Literature*. Cambridge.

(2005a) *The Second Sophistic*. Oxford.

(2005b) 'Quickening the classics: the politics of prose in Roman Greece', in Porter, J. (ed.) *Classical Pasts: The Classical Traditions of Greco-Roman Antiquity*. Princeton: 378–406, 622–30.

Wilkins, J. (2000) 'Athenaeus and the *Fishes* of Archippus', in Braund and Wilkins (eds.): 523–35.

(2003) 'Foreword', in Powell (ed.) (2003): ix–xxi.

(forthcoming) 'Towards a sympotic history', in Lenfant, D. (ed.) *Athénée et les fragments des historiens grecs*. Strasburg.

Wilson, A. M. (1985) 'The Prologue to Manilius I', *Papers of the Liverpool Latin Seminar* 5: 283–98.

Woolf, G. (1994) 'Becoming Roman, staying Greek: culture, identity and the civilizing process in the Roman East', *PCPS* 40: 116–43.

Wooton, C. (1976) 'Petronius, the mime, and rhetorical education', *Helios* 3: 67–74.

Yeo, R. (2001) *Encyclopedic Visions: Scientific Dictionaries and Enlightenment Culture*. Cambridge.

Zadorojnyi, A. (2002) 'Safe drugs for the good boys: Platonism and pedagogy in Plutarch's *De Audiendis Poetis*', in Stadter and Van der Stockt (eds.): 297–314.

Zanker, P. (1995) *Die Maske des Sokrates: das Bild des Intellektuellen in der antiken Kunst*. Munich.

Zecchini, G. (1989) *La cultura storica di Ateneo*. Milan.

Zeitlin, F. I. (1971) 'Petronius as paradox: anarchy and artistic integrity', *TAPA* 102: 631–84.

Index

For EU product safety concerns, contact us at Calle de José Abascal, 56–1°, 28003 Madrid, Spain or eugpsr@cambridge.org.

www.ingramcontent.com/pod-product-compliance
Lightning Source LLC
LaVergne TN
LVHW052251200926
848481LV00002B/331

* 9 7 8 0 5 2 1 2 9 6 9 3 9 *